The molecular biology of the
mammalian genetic apparatus

Volume 2

The molecular biology of the mammalian genetic apparatus

Volume 2

Edited by:

Paul O.P. Ts'o

Division of Biophysics, Johns Hopkins University
School of Hygiene and Public Health,
Baltimore, Md. 21205, U.S.A.

1977

North-Holland Publishing Company
Amsterdam · New York · Oxford

© Elsevier/North-Holland Biomedical Press, 1977

All rights reserved. No parts of this publication may be reproduced, stored in a retrieval system, or transmitted, in any form or by any means, electronic, mechanical, photocopying, recording or otherwise, without the prior permission of the copyright owner.

ISBN: 0 7204 0626 9

Published by:
Elsevier/North-Holland Biomedical Press
335 Jan van Galenstraat, P.O. Box 211
Amsterdam, The Netherlands

Sole distributors for the U.S.A. and Canada:
Elsevier/North-Holland Inc.
52 Vanderbilt Avenue
New York, N.Y. 10017

Library of Congress Cataloging in Publication Data

International Symposium on the Molecular Biology of the
    Mammalian Genetic Apparatus—Its Relationship to Cancer, Aging and Medical Genetics, California Institute
    of Technology, 1975.
    The molecular biology of the mammalian genetic apparatus.
    Includes bibliographies and indexes.
    1. Molecular genetics—Congresses. 2. Mammals—Genetics—Congresses. 3. Cancer—Genetic aspects—Congresses. 4. Aging—Genetic aspects—Congresses.
I. Ts'o, Paul On Pong, 1928– II. Title.
QH426.16 1975   599'.01.'51   77–22374
ISBN 0–7204–0625–0 (v. 1)
ISBN 0–7204–0626–9 (v. 2)

Printed in The Netherlands

PREFACE

With enthusiasm and optimism, we introduce Volume II of 'The Molecular Biology of the Mammalian Genetic Apparatus'. As indicated in the Preface to Volume I of this series, the chapters in this volume are also from the international symposium bearing the same name as this series which was held in December of 1975 at the California Institute of Technology. This symposium not only provided an opportunity to examine the recent achievements and future developments in the research on the mammalian genetic apparatus, but also recognized the contributions made to this field by the Caltech people (faculty, students, alumni, associates and friends) and, particularly, those of Professor James Bonner's laboratory on the occasion of his 65th birthyear. This volume is divided into two parts: Part A. DNA Organization and Gene Expression, and Part B. Relationship to Somatic Genetics, Cancer and Aging. Thus, this volume extends the contents of Volume I on Structure to the organizational and functional aspects of the genetic apparatus, and describes the relationship and the relevance of the research on the genetic apparatus to the challenging problems in cell biology and biomedical science. This volume closes with a chapter by Professor James Bonner entitled 'My Life as a Chromosomologist'. In this final chapter, James describes the events, the strategy, the human aspects and his personal evaluation and feelings on the histone/chromatin research over the past two decades.

With sadness and remembrance, we pay a special tribute in this volume to Professor Jerome Vinograd, a member of the Organizing Committee of this Symposium, a close colleague and a very dear friend. His passing away on July 3, 1976 was a great loss and sorrow for all of us who knew and loved him. To me in particular, he was a great teacher; from him personally I have learned English, physical chemistry, science, and most of all, how to be a man of understanding, compassion and quality. I vividly remember our first meeting in the Crelin Chemistry Building in 1953, when I had just entered Caltech as a graduate student, and he had just returned to an academic institution. It was a turning point in my life when such a bond of learning and friendship was welded between us which has lasted nearly a quarter of a century. During the past 23 years, I have often asked myself, particularly when writing a challenging scientific paper, 'Would Jerry regard this as a paper of excellent quality, both in terms of science and presentation?' From him, I have learned to understand, appreciate, and strive for excellence.

With these two volumes, the Symposium has now become history. We have honored and benefitted from the pioneers who diligently tilled the soil of this virgin field of research on the genetic apparatus. More than 70% of the participants in this symposium of 400 people are under the age of 30. The best is yet to come, in terms of future symposia, future publications and future researchers. To paraphrase Newton, we can indeed build a greater structure of science, since we build on the foundation laid by scientists such as J. Bonner, J. Vinograd, and N. Davidson, to name just a few.

Again, we wish to thank those agencies and individuals responsible for the support and assistance given to the Symposium and to these two volumes as described in the Preface to Volume I.

Paul O.P. Ts'o

JEROME VINOGRAD (1913–1976)

Jerome Vinograd was born February 9, 1913 in Milwaukee, Wisconsin, and died July 3, 1976. After undergraduate work at the University of Minnesota, he studied colloid chemistry with H. Freundlich from 1931 to 1935, first in Berlin, and then in London. He completed his doctoral work with J.W. McBain at Stanford from 1937 to 1939. He was associated with the Shell Development Co. from 1941–1951 where he worked mainly on problems of surface films and colloids. At that time, he gave up a secure industrial career and came to the California Institute of Technology because of his intense desire to work in the exciting new area of molecular biology. His early work at Caltech dealt mainly with the physical chemistry of proteins.

His first outstanding contribution to molecular biology was the invention, in collaboration with Meselson and Stahl, of the method of equilibrium density gradient centrifugation of nucleic acids in cesium chloride solutions. This method has been elaborated in Vinograd's laboratory and other laboratories to include the effects of base composition, denaturation, alkaline titration, binding of metal ions and of dyes, and hydration on buoyant density. All told, equilibrium buoyant banding has been one of the key techniques in the explosive development of the nucleic acid side of molecular biology since the mid-1950's.

Vinograd's excellent work in the last 12 years has been in the study of closed circular DNA. The distinction between supercoiled closed circles and open circles was recognized in 1965. The ethidium bromide method of buoyant banding for isolation of closed circular DNA was invented in 1967. This method, which makes it possible to analyze for, and to isolate, closed circular DNA in the presence of massive quantities of linear and open circular DNA is widely used and is one of the main reasons why there has been such exciting progress in the study of closed circular DNAs in different kinds of cells.

Later work has dealt with the properties of mitochondrial DNA from malignant cells and with the study of the replication of mitochondrial DNA.

This latter topic was initiated by the discovery of D-loops in Vinograd's laboratory. More recently, his students have made several important contributions to the studies of the properties of the nicking and closing enzyme which relaxes supercoiled DNA while leaving it covalently closed. The Vinograd group, as well as the groups of Keller at Cold Spring Harbor and Wang at Berkeley observed that one can use gel electrophoresis and relaxation by an enzyme to count the number of turns in a superturn in molecules which are closed under an equilibrium condition.

Vinograd was by early training and in his approach to molecular biology a physical chemist. He had a flair for recognizing when an anomalous and unexplained observation, if subjected to fundamental physical chemical analysis, would lead to an unexpected and important result in molecular biology. The 'special Vinograd touch' we called it.

A simple example is the development of the velocity band centrifugation method by Vinograd, Bruner, Kent, and Weigle in 1963. Shortly before that Jean Weigle discovered that he could layer a dilute solution of viruses in aqueous buffer onto a sucrose solution (not a gradient) in a centrifuge tube, centrifuge and get a good sharp virus band. Vinograd was puzzled as to why the band was sharp and reasoned that there must be some effect providing convective stability. He recognized that there had to be a self-generating (by diffusion) density gradient to provide convective stability; a careful study then led to the development of this general and widely used method.

At a Caltech seminar in late 1966 or early 1967 J.B. LePecq described his studies on the binding of ethidium bromide (Etd Br) to DNA. An important point was that, unlike the common acridine dyes, there was a reasonable amount of binding at high salt concentration. Vinograd pointed out at a discussion immediately following that if the dye ion bound in 6 M CsCl it would cause a large shift in the buoyant density of the DNA because it would displace a Cs^+ ion. Experiments by W.R. Bauer (then a graduate student) showed that Etd Br would indeed decrease the buoyant density of DNA, but that the shift was much larger for a relaxed than for a closed circular DNA. On learning of this result, Vinograd immediately perceived the correct explanation in terms of the topological constraint on unwinding of closed circular DNA, thus leading to the development of an extremely useful method for isolating closed circular DNA, as well as for studying the free energy of supercoiling.

One of the many distinguished honors which Vinograd received throughout his scientific career was the election to membership in the National Academy of Sciences in 1968. He was a recipient of the Kendall Award in Colloid and Surface Chemistry in 1970 from the American Chemical Society, as well as the recipient of the Helen Hay Whitney Foundation T. Duckett Jones Award in 1972. He was also a national lecturer at the 1972 Biophysical Society Annual Meeting. Vinograd also served on the editorial boards of BBA, JCCP, BP, ABMP, and

Intervirology. At the time of his death, Vinograd was the Ethel Wilson Bowles and Robert Bowles Professor of Chemical Biology at Caltech.

He suffered a major heart attack in 1954 and a second one in 1969. He accepted the resulting restrictions on his overall activities philosophically. But his condition did not affect the intensity with which he devoted himself to his scientific work. We remember him through all these years being in the laboratory regularly on evenings and weekends discussing and analyzing experimental data in painstaking detail with his students, seeking fundamental explanations for unexpected and unexplained results. His students and colleagues will miss his penetrating analyses of technical problems and his wise counsel on general policy issues.

Norman Davidson
James Bonner

CONTENTS TO VOLUME 1

Preface	V
The life and times of James Bonner	IX
Chapter 1 Histone conformation: predictions and experimental studies by G.D. Fasman, P.Y. Chou and A.J. Adler	1
Introduction	1
H1	3
Prediction of secondary structure	3
Other predictive schemes	8
Physico-chemical studies: correlation with predicted structure: optical rotatory dispersion (ORD), circular dichroism (CD) and nuclear magnetic resonance (NMR) studies	11
H2A	15
Prediction of the secondary structure	15
Other predictive schemes	15
Physico-chemical studies: correlation with predicted structure: ORD, CD, NMR	20
H2B	22
Prediction of secondary structure	22
Other predictive schemes	25
Physico-chemical studies: correlation with predicted structure: ORD, CD, NMR	26
H3	27
Prediction of secondary structure	27
Other predictive schemes	29
Physico-chemical studies: correlation with predicted structure: ORD, CD, NMR	29
H4	32
Prediction of the secondary structure	32
Other predictive schemes	34
Physico-chemical studies: correlation with predicted structure: ORD, CD, NMR	35
Conformational studies on histones in chromatin and whole histone mixtures	38
Chromatin	39
Histone–histone interactions	40
Speculation on histone–histone interactions	40
Summary	42
Acknowledgements	42
Appendix	42
Critique of predictive methods	42
Critique of interpretation of optical rotatory dispersion and circular dichroism data	43
Critique of NMR spectral data	44
Conformational parameters for helical and β-sheet residues based on 29 proteins	45
References	45

Chapter 2 Histones and chromatin structure
by E.M. Bradbury, R.P. Hjelm, B.G. Carpenter, J.P. Baldwin
G.G. Kneale and R. Hancock   53

Histones	53
Structural studies of chromatin	55
Neutron studies of deuterated chromatin	58
Neutron scatter from chromatin subunits	61
Histone H1	66
Concluding remarks	67
Acknowledgements	68
Note added in proof	68
References	69

Chapter 3 The dual structure of concentrated nucleohistone gels from different sources: a quantitative analysis of X-ray diffraction patterns
by J.A. Subirana, F. Azorin, J. Roca, J. Lloveras, R. Llopis
and J. Cortadas   71

Summary	71
Introduction	72
Experimental procedures	72
Calf thymus nucleohistone	74
Sea urchin nucleohistone	77
Nucleohistone depleted of H1: diffraction in the meridional direction	78
An analysis of the diffraction given by other samples	82
The dry state of nucleohistone	83
Nucleohistone depleted of H1: diffraction in the equatorial direction	84
The molecular structure of the globular component	87
Discussion	89
Acknowledgements	91
References	91

Chapter 4 Special features of the structure of H1 histones
by R.D. Cole   93

Acknowledgements	103
References	103

Chapter 5 Histone H5 in developing chick erythroid cells
by P.C. Huang, L.P. Branes, C. Mura, V. Quagliarello and
P. Kropkowski-Bohdan   105

Occurrence	105
Physico-chemical properties	107

Biological properties	111
Synthesis and turnover	114
Cellular studies	114
Specificity of the anti-H5 antibody	115
Quantification of the immunofluorescent assay	118
Histone H5 and chromatin structure	120
Histone H5 and DNA synthesis	120
Summary	122
Acknowledgement	122
Note added in proof	122
References	122

Chapter 6 The in situ distribution of *Drosophila* non-histone chromosomal proteins
by S.C.R. Elgin, L.M. Silver and C.E.C. Wu   **127**

Abstract	127
Acknowledgements	140
References	140

Chapter 7 Evidence for the presence of contractile-like non-histone proteins in the nuclei and chromatin of eukaryotes
by A.S. Douvas, A. Bakke and J. Bonner   **143**

Introduction	143
Methods for non-histone isolation	143
Rat liver	143
Slime mold	144
Results and discussion	145
Properties of contractile-like non-histones	145
Interspecies and intertissue homologies of non-histones	151
Possible contamination of chromatin proteins by extra-chromosomal contractile proteins	154
Quantitation of contractile-like proteins in chromatin	157
Summary	160
References	161

Chapter 8 Structural aspects of low molecular weight RNA and the implications of the 5′ Cap for messenger RNA and protein synthesis
by H. Busch, D. Henning, F.W. Hirsch, M.S. Rao, T.S. Ro-Choi, W.H. Spohn and B.C. Wu   **165**

Relationship of proteins to mRNA	167
The relationship of rRNA to mRNA	168
Methylated nucleotides	168
Isolation of low molecular weight nuclear RNA(s)	170

Unusual features of 'U-rich' RNA species	170
Two 3'-termini in U-rich RNA	171
U2 RNA sequence	171
The U1 RNA 5' terminus	172
The 5' terminus of U2 RNA	173
$m_3^{2,2,7}$Guanylic acid	174
mRNA	174
Viral mRNA	175
Eukaryotic mRNA	175
Mechanism of synthesis of the 5'-Cap of mRNA	176
The role of the 5' terminal structure in translation	177
Rigid structural requirements for translational activity of mRNA	178
RNP particles	179
Comparison of nuclear and cytoplasmic mRNA	180
Profiles of mRNA readouts of cancer cells and other cells	183
Discussion	186
Acknowledgements	188
References	188

Chapter 9 Structure of chromatin
by R.D. Kornberg   **195**

References	197

Chapter 10 Oligomers of the histones: an octamer of the four main types
by J.O. Thomas   **199**

The histone octamer in chromatin	200
The proximity of histone octamers in chromatin	201
The arrangement of histones within the octamer	202
The histone octamer free in solution	203
The properties and stability of the octamer in solution. Is the free octamer an artifact of cross-linking?	203
Is the histone octamer unique and ubiquitous?	205
The possible fate/role of the histone core in replication	207
References	207

Chapter 11 On nu models for chromatin structure
by A.L. Olins, J.P. Breillatt, R.D. Carlson, M.B. Senior, E.B. Wright and D.E. Olins   **211**

Introduction	211
Visualization of the chromatin subunits	212
Isolation and properties of chromatin v bodies	218

Contents to volume 1   XV

Internal structure of v bodies	227
Higher order packing of v bodies	230
Historical perspective	233
Acknowledgements	233
References	234

Chapter 12   The structural subunit of chromatin
by K.E. van Holde, B.R. Shaw, K. Tatchell, J. Pardon, D. Worcester, J. Wooley and B. Richards   **239**

Some recent history: evidence leading to a subunit model for chromatin	239
Arrangement of DNA within the subunits	240
Overall structure and composition of the core particles	246
Internal structure of the core particles: localization of the DNA and protein	249
Summary	252
Node added in proof	252
References	253

Chapter 13   Organization of chromatin proteins
by G. Felsenfeld, B. Sollner-Webb and R.D. Camerini-Otero   **255**

References	262

Chapter 14   On the use of psoralen cross-linkage in the study of chromatin structure
by J.E. Hearst, C.-K.J. Shen and C.V. Hanson   **265**

Results	267
Acknowledgement	271
References	271

Chapter 15   The structure of condensed DNA: similarities between higher cells and bacteria
by J.D. Griffith   **273**

Notes and acknowledgements	279
References	279

Chapter 16 Some problems in dealing with chromatin structure
by V. Jackson, P. Hoffmann, R. Hardison, J. Murphy, M.E. Eichner and R. Chalkley — 281

Levels of chromosome structure	281
Studies with nuclease	282
Studies with cross-linking reagents	288
A new model for chromatin structure	292
References	298

Chapter 17 On the role of histone–histone interactions in controlling chromatin structure and genetic expression
by E.N. Moudrianakis, P.L. Anderson, T.H. Eickbush, D.E. Longfellow, P. Pantazis and R.L. Rubin — 301

Studies with isolated chromatin	302
Causes of compaction of the DNA helix	309
Histone–DNA and histone–histone interactions	314
Structure of the chromatin fiber	318
References	321

Chapter 18 Chromatin structure – a model
by H.J. Li — 323

Introduction	323
Concept of subunits in chromatin	324
Concept of a string of beads in chromatin	326
A model for chromatin structure in histone-bound regions	328
Comparison with other models	331
A model for chromatin structure	336
Step 1. Specific binding of nonhistone proteins or RNA to N regions of a newly synthesized DNA	336
Step. 2. Nonspecific and cooperative binding of histone subunits to H regions not bound by other macromolecules	338
Acknowledgements	339
References	339

Chapter 19 Chromatin organization and the existence of a chromatin-associated DNA-generating system in human sperm
by S.S. Witkin, D.P. Evenson and A. Bendich — 345

Sperm head purification	346
Endogenous DNA polymerase activity	350
Sperm chromatin	352
References	354

Contents to volume 1

Chapter 20 The expression of protamine genes in developing trout sperm cells
G.H. Dixon, P.L. Davies, L.N. Ferrier, L. Gedamu and
K. Iatrou **355**

Introduction	355
Acknowledgement	376
References	377

Chapter 21 Isolation and properties of the expressed portion of the mammalian genome
by J.M. Gottesfeld and J. Bonner **381**

Chromatin fractionation	382
Sequence composition of the active chromatin fraction	384
DNA-RNA hybridization	386
Composition of chromatin fractions	387
Chromatin structure and transcriptional activity	389
Thermal denaturation	390
Subunit structure of active chromatin	390
Conclusions	396
Acknowledgements	397
References	397

Chapter 22 Chromosome proteins and metaphase chromosome structure
by W. Wray, M.L. Mace Jr. and Y. Daskal **401**

Introduction	401
Materials and methods	401
Results	402
Discussion	411
Acknowledgements	414
References	414

Chapter 23 Nuclear and cytoplasmic RNA complexity in *Drosophila*
by B. Levy W., C.B. Johnson and B.J. McCarthy **417**

Introduction	417
Materials and methods	418
Preparation of nuclear RNA	418
Preparation of nuclear and cytoplasmic poly(A)-containing RNA	418
Preparations of *Drosophila* DNA	419
Isolation of unique sequence ^{3}H-DNA	419
Degradation of DNA by shearing and partial depurination	419
Synthesis of cDNA	419
Hybridization reactions	419

Nuclear cDNA fractionation	420
Hydroxyapatite chromatography	420
Results	420
The size of polyadenylated nuclear RNA	420
RNA excess unique sequence hybridization	421
Renaturation of nuclear cDNA with DNA	423
Complexity of Schneider's cell nuclear polyadenylated RNA	423
Hybridization of nuclear cDNA with cytoplasmic RNA	426
Hybridization of cytoplasmic cDNA with various RNAs	426
Isolation of fractionated nuclear cDNA probes	428
Discussion	432
References	434

Subject index   **437**

CONTENTS TO VOLUME 2

Preface   V

Jerome Vinograd (1913–1976)   VII

Contents to volume 1   XI

Part A   DNA ORGANIZATION AND GENE EXPRESSION

Chapter 1   Gene mapping by electron microscopy
by N. Davidson   1

Acknowledgment	12
References	12

Chapter 2   Genetic and physical selection of eukaryotic genes cloned in *E. coli*
by R.W. Davis, M. Thomas, J.R. Cameron, P. Philippsen, R. Kramer, T. St. John, K. Struhl and J. Ferguson   15

Introduction	15
Cloning eukaryotic DNA in *E. coli* with λt	15
*Eco*RI and *Hind*III restriction spectra of *Saccharomyces cerevisae* DNA	16
Genetic selection	17
Lytic selection	17
Lysogenic selection	21
Physical selection	21
Physical screening with complementary RNA	21
Physical selection with complementary RNA	22
References	25

Chapter 3 DNA sequence organization and expression of structural gene sets in higher organisms
by R.C. Angerer, W.R. Crain, B.R. Hough-Evans, G.A. Galau, W.H. Klein, M.M. Davis, B.J. Wold, M.J. Smith, R.J. Britten and E.H. Davidson   29

The two patterns of DNA sequence arrangement in eukaryotes	29
Detailed investigation of sequence organization in the genomes of *Musca* and *Apis*	33
Functions of short interspersed repetitive sequences	36
Specificity of structural gene transcription	37
Acknowledgments	40
References	41

Chapter 4 The highly repeated DNA sequences of *Drosophila melanogaster*
by D.L. Brutlag   43

Acknowledgements	49
References	49

Chapter 5 Analysis of sequence structure of the rat genome
by J.R. Wu, W.R. Pearson, M. Wilkes and J. Bonner   51

The renaturation kinetic classes of rat DNA	52
Foldback DNA	53
Size distribution of repetitive sequences	55
Interspersion of repetitive with single copy sequences	58
Discussion	61
Acknowledgements	62
References	62

Chapter 6 Dispersal of satellite DNA sequences throughout the muskmelon genome and the nature of families of repeated DNA sequences in plants
by A.J. Bendich   63

How do we measure DNA base sequence repetition?	64
Homogeneous families of sequences	65
Muskmelon satellite DNA sequences interspersed with main band DNA	70
The evolutionary history of homogeneous families	73
On the value of sequence repetition for the investigator and for the cell	75
Summary	76
Acknowledgements	77
References	77

Contents to volume 2

Chapter 7	**The ribosomal DNA of** *Tetrahymena* by J.G. Gall, K. Karrer and M.-C. Yao	**79**

Extrachromosomal location 79
Detailed structure of the rDNA 80
Amplification 83
References 84

Chapter 8	**Anatomy of the gene cluster coding for the five histone proteins of the sea urchin, a progress report** by M.L. Birnstiel, K. Gross, W. Schaffner, R. Portmann and E. Probst	**87**

Introduction 87
Regulation of histone mRNA production 87
   Coordination of histone protein production 88
   Production of histone variants 88
Identification of the repeat unit of histone DNA 89
Isolation of the five messengers coding for the five kinds of histone 89
Dissection of the 5.6 kilo base and ordering of the coding sequences 91
Polarity and asymmetry of the coding sequences in *Psammechinus miliaris* 92
Distribution of spacer and genes 93
Acknowledgements 97
References 97

Chapter 9	**Some aspects of the interaction between the mouse (LA9) nicking-closing enzyme and closed circular DNA** by J. Vinograd	**99**

References 109

Chapter 10	**Towards an in vivo assay for the analysis of gene control and function** by J.B. Gurdon and D.D. Brown	**111**

Abstract 111
Introduction 112
The design of a living-cell assay system 112
DNA injection 113
Toxicity of injected DNA 113
Intracellular location of injected DNA 115
Persistence of injected DNA 115
Transcription of injected DNA 118
Discussion and conclusions 121
Acknowledgements 123
References 123

Chapter 11 Transcriptional unit of mouse myeloma
by R.C.C. Huang — 125

Establishment of an in vitro transcriptional system — 126
Synthesis of kappa chain mRNA (mRNA$_k$) in vitro — 128
A comparison between nuclear transcription and chromatin transcription by *E. coli* RNA polymerase — 131
Some unsolved problems in analyzing transcription of the kappa chain gene — 132
References — 133

Chapter 12 Transcriptional control in higher eukaryotes
by J. Paul, N. Affara, D. Conkie, R.S. Gilmour, P.R. Harrison, A.J. MacGillivray, S. Malcolm and J. Windass — 135

Erythroid versus non-erythroid cells — 136
Mature versus immature erythroid cells — 136
Studies with Friend erythroleukaemia cells — 138
Transcription from chromatin — 141
Fractionation of chromatin — 143
Acknowledgments — 145
References — 146

Chapter 13 Structural analysis of native and reconstituted chromatin by nuclease digestion
by M.J. Savage and J. Bonner — 149

Summary — 149
Introduction — 149
Methods — 150
  Chromatin preparation and reconstitution — 150
  Nuclease digestion — 151
Results — 151
  Staphylococcal nuclease — 151
  Deoxyribonuclease II — 155
Discussion — 158
References — 160

PART B RELATIONSHIP TO SOMATIC GENETICS, CANCER AND AGING

Chapter 14 New approaches to cell genetics cotransfer of linked genetic markers by chromosome mediated gene transfer
by F.H. Ruddle and O.W. McBride — 163

References — 168

Chapter 15 Studies in mammalian cell regulation: cell surface antigens and the action of cyclic AMP
by T.T. Puck — 171

Introduction	171
Biochemical genetic studies of tissue-specific cell membrane structures	173
The action of cyclic AMP and various hormones	176
References	180

Chapter 16 Thymidine kinase: some recent developments
by J.W. Littlefield — 181

Purification and properties	181
Electrophoretic forms of the kinase	182
Relation between cytosol and mitochondrial kinases during growth	183
Uptake of thymidine	184
Herpes virus – induced kinase	185
Summary	187
Acknowledgement	187
References	187

Chapter 17 Image analysis, flow fluorometry and flow sorting of mammalian chromosomes
by M.L. Mendelsohn, A.V. Carrano, J.W. Gray, B.H. Mayall and M.A. van Dilla — 191

Introduction	191
Absorption cytophotometry and image analysis	192
Flow fluorometry and sorting	196
Conclusion	203
References	204

Chapter 18 Chromosomal alterations in carcinogen transformed mammalian cells
by J.A. Dipaolo — 205

Immediate effects of carcinogens	207
Conclusions	227
References	224

Chapter 19 The importance of chromosomal changes in the expression of malignancy
by W.F. Benedict — 229

Summary	229
Introduction	229

Chromosomal changes associated with malignant expression in mammalian cells	230
Diploid hamster embryo cells	230
Mouse cells	232
Diploid human cells	232
A general hypothesis on specific chromosomal changes and malignant expression produced by all carcinogens	235
DNA oncogenic viruses	235
RNA oncogenic viruses	235
Chemical or physical agents	236
Role of chromosomes with suppressor genes	236
Conclusion	237
Acknowledgements	238
References	238

Chapter 20 The relationship between neoplastic transformation and the cellular genetic apparatus
by P.O.P. Ts'o, J.C. Barrett and R. Moyzis   241

Introduction	241
Concepts	242
Experimental design	243
System	243
Perturbation to the genetic apparatus	244
Responses	245
Biological responses	245
Cytological and molecular responses	247
Current results	248
Somatic mutation	248
Relationship between neoplastic transformation and somatic mutation	253
Changes in DNA sequence organization and gene expression in neoplastic transformation	254
A direct demonstration of the involvement of DNA damage in neoplastic transformation	259
Concluding remarks	260
References	263

Chapter 21 Cellular modifications of transfecting SV40 DNA
by D. Nathans and C.-J. Lai   269

Cyclization of linear SV40 DNA	269
Repair of partial heteroduplex molecules	270
Evidence for correction of mismatched bases in SV40 DNA	271
Conclusion	274
Acknowledgements	275
References	275

Contents to volume 2

Chapter 22	The transforming gene of avian sarcoma virus by J.M. Bishop, D. Stehelin, J. Tal, D. Fujita, D. Spector, D. Roulland-Dussoix, T. Padgett and H.E. Varmus	**277**

Introduction	277
Preparation of specific cDNAs	278
Specificity of the nucleotide sequences in the transforming gene of ASV	280
The origin of nucleotide sequences in the transforming and glycoprotein genes of ASV	281
Expression of *sarc* nucleotide sequences in normal cells	284
Conclusion	285
Acknowledgements	285
References	286

Chapter 23	Somatic cell genetic investigations of clonal senescence by G.M. Martin, T.H. Norwood and H. Hoehn	**289**

Acknowledgements	300
References	301

Chapter 24	Endocrine and neural factors of reproductive senescence in rodents by C.E. Finch	**303**

Endocrine and neural factors of reproductive senescense	303
Cell functions in the liver	303
Cell functions of the female reproductive system	304
Age-related changes of catecholamines	307
Conclusion	310
Acknowledgements	310
References	310

Chapter 25	Introduction of Professor James Bonner, California Institute of Technology, Pasadena, California – a symposium honoring the occasion of his 65th birthyear by J.L. Liverman	**313**

Chapter 26	My life as a chromosomologist by J. Bonner	**317**

References	324

Subject index   **327**

Chapter 1

GENE MAPPING BY ELECTRON MICROSCOPY

NORMAN DAVIDSON

Division of Chemistry and Chemical Engineering, California Institute of Technology, Pasadena, Calif. 91125, U.S.A.

The present status of several techniques that are under development and/or in use in our laboratory for gene mapping by electron microscopy is presented in this communication.

Fig. 1. Schematic diagram of RNA:DNA hybrids. The reasonably long RNA:DNA duplex on the left (>500 nucleotides) can be recognized and measured in electron micrographs. In order to recognize a very short RNA:DNA hybrid as for tRNA, a ferritin label is attached to the tRNA, as shown on the right.

A statement of the general problem is given in fig. 1. We are given a DNA strand containing a gene coding for a particular RNA molecule. If the RNA can be isolated in a fairly pure state, the gene can be physically mapped by hybridizing the RNA onto the DNA and observing in the electron microscope the position of the RNA : DNA hybrid region relative to the ends of the molecule or to other features. In the standard formamide, cytochrome *c* spreading method, duplex regions are thicker and smoother than single strand regions. If the duplex region is fairly long (say >500 nucleotides) it is usually easy to discriminate between DNA : DNA duplexes and single strands. The contrast between RNA : DNA hybrids and DNA single strands is not as good as for DNA : DNA hybrids but in cases where it is possible to prepare and observe many molecules sufficient data can be obtained to give reliable results [1,2,3].

However if the RNA : DNA hybrid region is quite short – for example the length of a tRNA gene or of a 5S RNA gene (approximately 80 or 120 nucleotides respectively) – the duplex region is much too short to be recognized in a cytochrome *c* spread. Some time ago a ferritin labeling method for mapping tRNA

The Molecular Biology of the Mammalian Genetic Apparatus:
edited by P. Ts'o © 1977, Elsevier North-Holland Biomedical Press

and other short RNA genes was introduced. In this procedure the electron opaque label ferritin was covalently attached to the 3′ end of the tRNA by suitable in vitro chemical reactions [4]. The tRNA-ferritin conjugate was hybridized to the DNA strand. The position of the ferritin along the DNA strand can be readily recognized in the electron microscope and defines the position where the tRNA has hybridized.

This ferritin labeling procedure was not entirely satisfactory because the preparation of the tRNA-ferritin conjugate in high yield required meticulous control of the reaction conditions and because the hybridization of the tRNA-ferritin molecule in vast excess to DNA containing tRNA genes never resulted in more than 50% saturation of the genes.

We have therefore investigated an alternative procedure. The philosophy of this new approach is to attach the small molecule, biotin, to the tRNA. The tRNA-biotin conjugate is then hybridized to the DNA. Because biotin is fairly small it is anticipated that the rate and yield of hybridization will be the same as for unlabeled tRNA. Heitzmann and Richards have independently conceived of and developed a method based on the biotin-avidin reaction for ferritin labeling of cell surface components [5].

The protein, avidin, is covalently attached to the electron opaque label ferritin. It is known that biotin reacts with avidin to form a very strong stable non-covalent complex with an estimated dissociation constant of 10^{-15} M [6]. The DNA : tRNA-biotin hybrid is then incubated with ferritin-avidin, and thus, the ferritin label is attached to the position of the tRNA genes.

The overall procedure is as follows:

1) Preparation of tRNA-biotin: a) periodate oxidize the 3′ terminus of the tRNA to give a dialdehyde. This is a good reaction with a yield of 90 to 100%; b) form a Schiff base by reaction of the tRNA with excess 1,5-pentanediamine; c) stabilize the Schiff base against dissociation and β elimination by BH_4-reduction. Reactions (b) and (c) together are somewhat unsatisfactory in our hands, in that the yield varies uncontrollably from 20 to 80%; d) attach biotin to the newly attached terminal amino group on the tRNA by acylation with the N-hydroxy succinimide ester of biotin; e) purify the tRNA-biotin by affinity chromatography on avidin-agarose. Because of the strength of the biotin-avidin bond, the elution of the tRNA-biotin requires extreme conditions, 6 M guanidium hydrochloride, pH 2.5. It is probable that the combination of step c and the acidic elution in step e results in chain breakage of some tRNA's at 7-methyl guanine and dihydrouridine positions.

2) Preparation of ferritin-avidin: a) acylate ferritin with the N-hydroxy succinimide ester of bromoacetic acid [7] to give about 10 active bromide groups per ferritin. Add SH groups to avidin by acylation with the N-hydroxy succinimide ester of dithiodiglycollic acid, followed by exposure of the SH groups by treatment with DTT. Bromo-ferritin is then treated with SH-avidin to give a ferritin-avidin

conjugate. The ferritin-avidin conjugate is purified from uncoupled avidin by several cycles of sucrose gradient velocity sedimentation. The resulting product can bind ca 0.2 to 3.5 moles of biotin per mole of ferritin.

3) Hybridization and labeling: a) DNA-tRNA hybrids are made by standard methods using tRNA-biotin. The hybrids are purified from free tRNA by gel filtration with 4% agarose in a micro column; b) the hybrid is then incubated with ferritin-avidin. For unknown reasons, a vast excess of ferritin-avidin is needed. The excess can be removed by buoyant banding (not quite to equilibrium) in a suitable sodium iothalamate gradient [8]. Ferritin bands at a much higher density than does the DNA hybrid; c) the DNA fraction is dialyzed into spreading solution, mounted in a cytochrome c film by standard methods, shadowed lightly, and examined in the electron microscope.

As already stated, the positions of the ferritin molecules along the DNA strands map the positions of the tRNA genes. In test reactions, about 50% of the genes on a $\emptyset 80$ psu $_3^+/\emptyset 80$ heteroduplex are labeled.

This new mapping technique, which has been developed by T.R. Broker, L. Angerer, N.D. Hershey, M. Pellegrini, and P. Yen has been applied by Dr. Angerer to reinvestigate the problem of the number and location of 4S RNA genes (presumably mostly tRNA genes, but perhaps one of them is the mitochondrial equivalent of 5S rRNA) on HeLa mitochondrial DNA. She has enjoyed a collaboration with D. Lynch, G. Attardi, W. Murphy, D. Ojala and P.J. Flory, who provided the necessary reagents.

Earlier electron microscope mapping studies [1] supported by early radioactive labeling hybridization studies had indicated that there were nine 4S RNA genes on the H strand and three on the L strand. Lynch and Attardi [9] have measured the hybridization of tRNA's charged with individually labeled amino acids to mitochondrial DNA. They found that 12 different tRNA-^{3}H amino acids hybridized to the H strand, and 5 hybridized to the L strand.

In the previous EM mapping study [1], 12 and 16S rRNA was hybridized to the H strand, as were ferritin labeled 4S RNA's. The long rRNA: DNA duplex regions could be recognized and provided reference regions for mapping the 4S genes. No reference markers were available for the L strand and a relative map was constructed using only interferritin distances.

In the present study we had the advantage of having available as a reference marker for the L strand a choice of either: a) an HpaII duplex fragment of HeLa mitochondrial DNA known to map at a region spanning the two rRNA genes. b) rDNA from the H strand, prepared by hybridizing rRNA to mitochondrial DNA and digesting away all single strand material with S1 nuclease.

A preparation containing purified circular and linear L strand DNA was hybridized to the reference DNA fragments and to biotin labeled mitochondrial 4S RNA. The ferritin-avidin labeling procedure was carried out and electron microscope grids were scanned for molecules which contained some ferritin labels

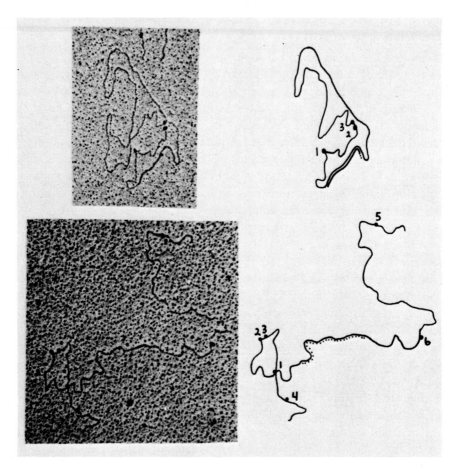

Fig. 2. Electron micrographs of L strand HeLa mitochondrial DNA hybridized to Hpa II fragment DNA (shown as a duplex in the explanatory drawings on the right) and to biotin labeled mitochondrial 4S RNA, and then incubated with ferritin-avidin. In the explanatory drawings, the several ferritins are numbered according to the map of fig. 4. Micrographs by L. Angerer.

and a recognizable duplex region due to hybridization of the reference marker to the L strand. Fig. 2 shows several electron micrographs of the resulting L strand DNA molecules, when the HpaII fragment is the reference marker. As is typical for such experiments, the ferritin positions are easily recognizable; the unlabeled DNA : DNA reference duplex regions are recognizable but the discrimination from single strand regions is not very great. A large number of such micrographs were analyzed. A histogram of the positions of the several ferritin markers relative to the position of the reference DNA : DNA duplex region is shown in fig. 3.

Similar experiments were carried out to map the 4S RNA genes on the H strand. In this case, since a preparation of separated H strand DNA was not

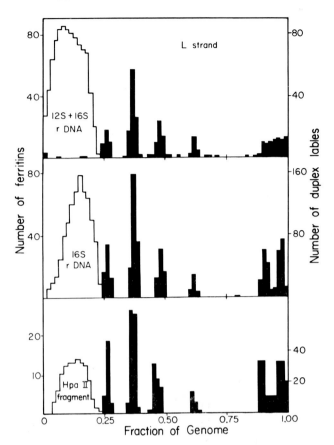

Fig. 3. Histogram of duplex and 4S-ferritin labels on L strand mitochondrial DNA. The filled-in areas are the 4S-ferritin labels; the open areas are the H strand rDNA or the HpaII labels.

available, we hybridized denatured total DNA with rRNA and with biotin-labeled 4S RNA. Only those strands which had rRNA : DNA hybrid regions (and were therefore H strands) and ferritin labels were mapped. A histogram (not shown) comparable in quality to that in fig. 3 was obtained for the positions of the 4S genes on the H strand.

The map giving our final assignments is shown in fig. 4. The results may be summarized as follows:

1) In this study the overall efficiency of labeling per gene was about 30 to 50%, varying in different experiments. There is only a very small background of non-specific binding and map positions which can be assigned with reasonable confidence.

2) There are seven identifiable 4S RNA binding sites on the L strand. They are distributed all around the genome in an irregular manner. There is one

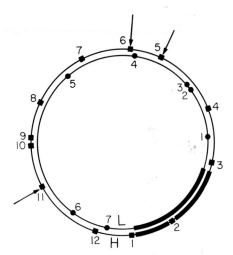

Fig. 4. Map of the positions of the 4S genes on HeLa mitochondrial DNA. The reference duplex regions are the rRNA genes on the H strand and the HpaII fragment on the L strand. Arrows point to 3 probable positions on the H strand, where ferritin labels were found, but a a lower frequency than for the other positions.

closely spaced doublet (which gives rise to the double size peak in the histogram of fig. 3 at the position of about 0.33 fractional lengths).

3) The H strand map shows 9 positive assignments. There are also 3 probable assignments at sites where ferritins were observed at a lower frequency. The 9 sites agree quite accurately with those previously found [1].

4) We wish to note an important technical point. In the previous mapping study without a reference marker for the L strand, three interferritin distances were found, and it was concluded that there were only three 4S RNA genes on the L strand [1]. We now find seven genes. However, a histogram (not shown) of L strand interferritin distances, constructed from the map of fig. 4, shows three strong peaks that agree with those previously recognized [4] and one additional longer interferritin distance. Thus in order to obtain a reliable map, it is highly desirable to have a fixed reference marker on a DNA strand and/or to have a very high efficiency of labeling.

5) The H strand map has the interesting property, noted previously, of having 4S genes on each end of the rRNA genes and in the short spacer between them. It is of interest to note that *E. coli* rRNA genes have a tRNA gene in the spacer between them [10].

6) We cannot be certain that all of the 4S RNA binding sites have been identified. Some of the 4S RNA's may be present in too low a concentration to hybridize efficiently and/or may be degraded by the chemical modification procedure.

Table 1 gives a history of the determination of the number of 4S RNA genes

TABLE 1

History of the determination of the number of 4S RNA genes on the HeLa mitochondrial genome.

	Heavy strand	Light strand	Total
Saturation hybridization, 1971 [16]	8	3	11
E.M. Mapping with tRNA-ferritin conjugates, 1972 [1]	9	3	12
Hybridization with all different aminoacyl tRNA's, 1976 [12]	12	5	17
E.M. Mapping with ferritin-avidin: biotin-tRNA complexes, 1976 (L. Angerer, personal communication)	12	7	19

on HeLa mitochondrial DNA. The number of recognizable genes is now almost equal to the number of different amino acids needed in protein synthesis. Quite conceivably, later work will find additional tRNA genes on HeLa mitochondrial DNA. Thus we suspect that the mitochondrial genome codes for tRNA's for all or almost all of the amino acids so that importation from the cytoplasm of nuclear coded tRNA's is not needed for mitochondrial protein synthesis.

We now turn our attention to the problem of obtaining more positive identification of genes that code for reasonably long RNA molecules. The two methods I wish to discuss are the gene 32 method for identifying RNA : DNA (and DNA : DNA) hybrid regions along a single strand of DNA and the R loop method of mapping genes on DNA duplexes.

A greatly improved technique for discriminating between duplex regions and single-strand regions has recently been described [11]. It is based on the fact that T4 gene 32 protein binds selectively and cooperatively to single strand DNA. The procedure is as follows: RNA corresponding to one or several genes is hybridized to its coding sequence(s) on a single strand of DNA. This preparation is treated with gene 32 protein, which binds to all of the single-stranded DNA regions, but not at all to duplex. The protein is fixed to the DNA with glutaraldehyde. The DNA is then spread in the presence of ethidium bromide, but without cytochrome c. The single strands are now relatively thick and the duplex region is thinner (about 8 and 3.5 nm respectively). The excellent contrast between single strand and double strand DNA achieved by this method is illustrated in the electron micrographs in fig. 5 of single and double stranded ϕX174 DNA. The contour length of either the single or double stranded form is 2.4 μm, corresponding to a spacing of 0.46 nm per base or base pair.

We have applied this technique to study the distribution of histone genes on the sea urchin genome. Kedes et al. [12,13] have derived two chimeric plasmids, pSp2 and pSp17, which contain histone genes of the sea urchin *Strongylocentrotus*

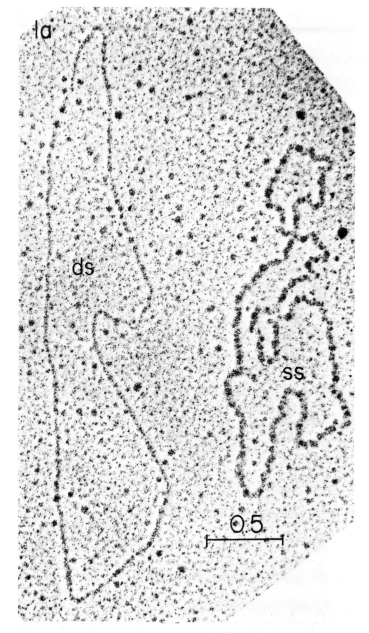

Fig. 5. Electron micrographs of single-stranded (ss) and double-stranded (ds) ϕX-174 DNA mounted by the gene 32 - ethidium bromide method. Micrographs by M. Wu.

purpuratus and which have pSC101 as the procaryotic vector. The eukaryotic fragments of the two plasmids have respective lengths of 4.6 and 2.1 kb and have no detectable sequence homology. Further studies have shown that these two fragments are arranged in tandem on the sea urchin genome and comprise one complete 7.0 kb repeat unit of histone genes [13].

It has been shown that there are several electrophoretic size classes of histone mRNA's [13,14]. In the electrophoresis experiments of Kedes et al. [13] these are listed, in the order of increasing size, as C, B4, B2 plus B3 (unresolved), B1 and A. It has been shown that the messenger RNA species C and A code for histones H4 and H1 respectively. The other three genes code for histones H2a, H2b, and H3; but at the time of this writing, positive assignments have not been made. Hybridization experiments have shown that the mRNA's A, B4, and C hybridize to pSp2, whereas B1 and the mixture B2 plus B3 hybridize to pSp17.

M. Wu, in collaboration with D.S. Holmes, has performed experiments to map the mRNA's on the plasmids by the gene 32 method. A mixture of all the mRNA's was hybridized to single strands of the respective plasmids. In the particular experiments, for which electron micrographs are shown in fig. 6, the

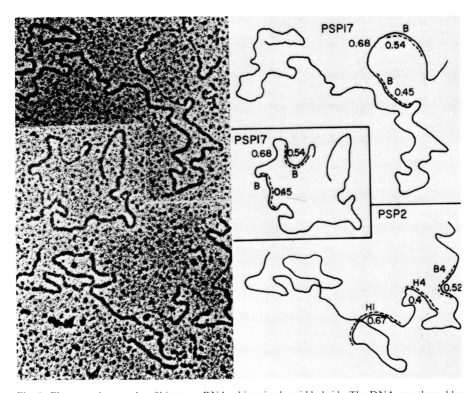

Fig. 6. Electron micrographs of histone mRNA : chimeric plasmid hybrids. The DNA was cleaved by Hind III. Single strands of DNA were hybridized to a preparation of total histone mRNA and mounted by the gene 32 - ethidium bromide method. The explanatory drawings on the right give the assignments relative to the map shown in fig. 9. The two B genes on pSp17 are B1 and B2 plus B3, but we cannot distinguish which is which. Micrographs by M. Wu.

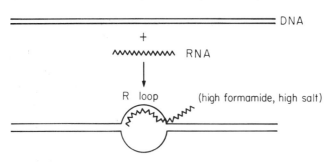

Fig. 7. Schematic illustration of the formation and nature of R loops.

plasmids were first cleaved with the Hind III restriction endonuclease, which makes a double strand cut in the DNA of either plasmid very close to one junction between the prokaryotic pSC101 piece and the eukaryotic part.

The micrographs show three RNA:DNA hybrid regions on pSp2 and two RNA:DNA hybrid regions on pSp17. The messenger length data and hybridization data quoted above permit us to unambiguously assign the three different sizes of RNA:DNA duplexes in pSp2 as the genes for A (histone H1), C (histone H4) and B4, which are linked in the order listed. We see from the figure that the spacing between A and C has a length of 0.69 kb, whereas the spacing between C and B4 has the length of 1.14 kb.

There is only a small difference in the lengths of the two RNA:DNA hybrid regions on pSp17. The region closer to the Hind III site appears to be slightly larger, but this identification is not absolutely certain. The spacing between the two genes is 0.68 kb. One of the two genes is the B1 gene and the other is the B2 plus B3 gene.

There is an alternative method of preparing RNA:DNA hybrids which is useful for gene mapping. This is the method of R loops which was invented by White and Hogness at Stanford (personal communication). It is based on the apparent fact that RNA:DNA hybrids are more stable than DNA:DNA duplexes in solvents containing a high percentage of formamide. If DNA:DNA *duplexes* are incubated in the presence of RNA molecules complementary to a segment on one of the DNA strands in high formamide solvents, at temperatures close to but just below the denaturation temperature of the DNA, the DNA can open up so that the appropriate part of one strand can form an RNA:DNA hybrid. This process is schematically depicted in fig. 7. D.S. Holmes has found that under suitable conditions it is possible to obtain a moderate efficiency of R loop formation to the duplex DNA in the plasmids pSp2 and pSp17 when incubated with total histone mRNA. An illustration of R loops for the pSp2 plasmid is shown in fig. 8. There are three R loop regions corresponding to the

Gene mapping by electron microscopy

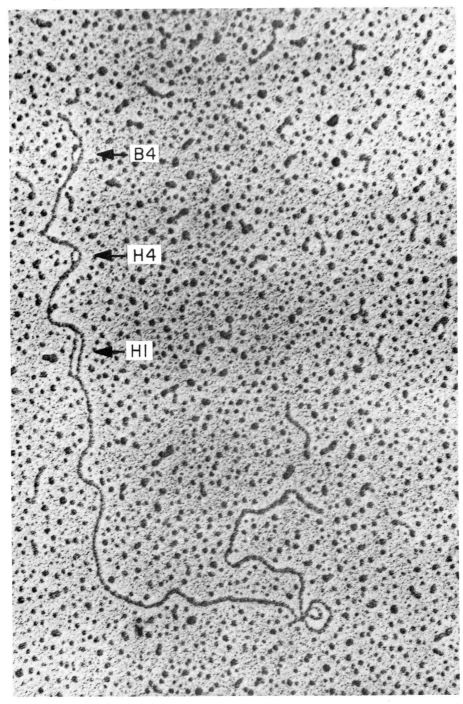

Fig. 8. R loops on Hind III cleaved duplex pSp2. Identifications relative to the map of fig. 9 and the single strand hybrid study of fig. 6 are indicated. Micrographs by D.S. Holmes.

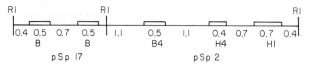

Fig. 9. Approximate map of lengths and spacing of histone genes on one of the tandem repeats in *Strongylocentrotus purpuratus* DNA. The map is based on electron microscope studies by D.S. Holmes and M. Wu and restriction mapping from the laboratory of L. Kedes (13; R. Cohn, J.C. Lowry, L. Kedes, personal communication).

three duplex regions shown in fig. 6 and, therefore, to the genes for the sequences A(H1), C(H$_4$), and B4. The R loop experiments provide an independent method of mapping the histone genes and confirm the results obtained by the gene 32 method. A combination of these electron microscope experiments and restriction endonuclease mapping experiments on total sea urchin DNA (12; Cohn, Lowry, and Kedes, personal communication) leads to the overall map of the organization of the histone genes, as shown in fig. 9. This map is in agreement with that obtained by other authors [14,15]. The electron microscope experiments provide quantitative data on the spacing between genes that is not readily obtainable by restriction endonuclease studies. The combination of the two techniques provides a good quantitative picture of the sequence organization of histone genes.

Acknowledgment

The relevant work in our laboratory has been supported by grants GM 19991 and GM 20927 from the United States Public Health Service. Contribution No. 5269 from the Department of Chemistry, California Institute of Technology, Pasadena, Ca. 91125, U.S.A.

References

[1] M. Wu, N. Davidson, G. Attardi and Y. Aloni, Expression of the mitochondrial genome in HeLa cells. J. Mol. Biol. 71, 81–93 (1972).
[2] A.B. Forsheit, N. Davidson and D.D. Brown, An electron microscope heteroduplex study of the ribosomal DNAs of *Xenopus laevis* and *Xenopus mulleri*. J. Mol. Biol. 90, 301–314 (1974).
[3] R.W. Hyman, Physical mapping of T7 messenger RNA. J. Mol. Biol. 61, 369–376 (1971).
[4] M. Wu and N. Davidson, A technique for mapping transfer RNA genes by electron microscopy of hybrids of ferritin-labeled transfer RNA and DNA: the $\phi 80 \text{hpsu}_{III}^{+-}$ system. J. Mol. Biol. 78, 1–21 (1973).
[5] H. Heitzmann and F.M. Richards, Use of the avidin-biotin complex for specific staining of biological membranes in electron microscopy. Proc. Nat. Acad. Sci. (Wash.) 71, 3537–3541 (1974).

[6] N.M. Green, Avidin, the use of [^{14}C] biotin for kinetic studies and for assay. Biochem. J. 89, 585–591 (1963).

[7] M. Pellegrini, H. Oen and C.R. Cantor, Covalent attachment of a peptidyl-transfer RNA analogue to the 50S sub-unit of *Escherichia coli* ribosomes. Proc. Nat. Acad. Sci. (Wash.) 69, 837–841 (1972).

[8] P. Serwer, Buoyant density sedimentation of macromolecules in sodium iothalamate density gradients. J. Mol. Biol. 92, 433–438 (1975).

[9] D.C. Lynch and G. Attardi, The amino acid specificity of the tRNA species coded by HeLa cell mitochondrial DNA. J. Mol. Biol. 102, 125–141 (1976).

[10] E. Lund, J.E. Dahlberg, L. Lindahl, S.R. Jaskunas, P.P. Dennis and M. Nomura, Transfer RNA genes between 16S and 23S rRNA genes in rRNA transcription units of *Escherichia coli*. Cell, in press (1976).

[11] M. Wu and N. Davidson, Use of gene 32 protein staining of single-strand polynucleotides for gene mapping by electron microscopy: application to the $80d_3il$vsu^{+7} system. Proc. Nat. Acad. Sci. (Wash) 72, No. 11, 4506–4510 (1975).

[12] L.H. Kedes, A.C.Y. Chang, D. Houseman and S.N. Cohen, Isolation of histone genes from unfractioned sea urchin DNA by subculture cloning in *E. coli*. Nature 255, 533–537 (1975).

[13] L.H. Kedes, R.H. Cohn, J.C. Lowry, A.C.Y. Chang and S.N. Cohen, The organization of sea urchin histone genes. Cell 6, 359–369 (1975).

[14] M.L. Birnstiel, W. Schaffner, K. Gross and J. Telford, Anatomy of the gene cluster coding for the 5 histone proteins of the sea urchin. FEBS Proceedings, 10th Meeting, Paris, 38, in press (1975).

[15] E.S. Weinberg, G.C. Overton, R.H. Shutt and R.H. Reeder, Histone gene arrangement in the sea urchin, *Strongylocentrotus purpuratus*. Proc. Nat. Acad. Sci, 72, 4815–4819 (1975).

[16] Y. Aloni and G. Attardi, Expression of the mitochondrial genome in HeLa cells. J. Mol. Biol. 55, 271–276 (1971).

Chapter 2

GENETIC AND PHYSICAL SELECTION OF EUKARYOTIC GENES CLONED IN *E. COLI*

RONALD W. DAVIS, MARJORIE THOMAS, JOHN R. CAMERON, PETER PHILIPPSEN, RICHARD KRAMER, THOMAS ST. JOHN, KEVIN STRUHL and JILL FERGUSON

Department of Biochemistry, Stanford University School of Medicine, Stanford, California 94305, U.S.A.

Introduction

Cloning eukaryotic DNA in E. coli with λgt

Techniques have been recently developed for the construction and cloning of viable molecular hybrids between modified bacteriophage λ DNAs and any foreign DNA [1,2,3,4]. As shown in fig. 1, one of these modifications of λ DNA involves the genetic removal of two *Eco*RI restriction sites (sites 4 and 5), a genetic deletion of non-essential DNA in the right arm, and the biochemical deletion of the λ *Eco*RI C fragment [2,3]. Some of the unique advantages of this modified λ, called λgt, as a vector for foreign DNA are: 1) a segment of DNA must be inserted between the right and left end fragments in order to have sufficient DNA to make a viable virus particle. 2) The DNA can be modified for K restriction so that it can infect any K12 strain of *E. coli* which is sensitive to infection by λ. 3) The phage can grow lytically with active anti-terminated λ promotors which will probably transcribe the foreign DNA. 4) The DNA can be integrated into the bacterial chromosome with the λ promotors repressed. 5) The strands of λ DNA can be readily separated. 6) Deletions that occur in the inserted DNA can be readily selected.

Large pools of hybrids containing many different fragments of DNA from any given genome can be generated. Fig. 2 shows the construction of hybrids with *Saccharomyces cerevisiae* (yeast) DNA. The vector used is λgt-λB, which carries the λ *Eco*RI B fragment for the initial propagation of the vector [3]. This fragment can be removed by preparative agarose gel electrophoresis prior to hybrid construction; however, its removal is not essential, as its reinsertion

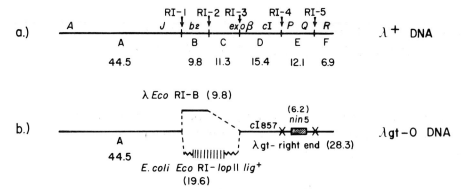

Fig. 1. a) Structure of bacteriophage λ DNA. The five EcoRI sites are labeled RI-1 through RI-5. A number of the λ gene positions are labeled A through R. The λ exonuclease gene is at EcoRI site 3. The EcoRI fragments are lettered sequentially A through F. The size of these fragments is given in fractional units of whole λ DNA which contains about 46,500 base pairs for a mass of 30.8×10^6 daltons. b) Structure of λgt-λB DNA. The two extreme right EcoRI sites have been removed by mutation and genetic selection. A non-essential region, nin5, has been genetically deleted. The EcoRI-C fragment between RI-2 and RI-3 has been biochemically deleted by EcoRI endonuclease cleavage and rejoining. This fragment contains the attachment site and genes necessary for the establishment of stable lysogens. The CI857 mutation renders the λ repressor temperature sensitive. Also shown is the hybrid which contains the E. coli DNA sequence coding for DNA ligase. This gene also carries the lop11 mutation which causes an overproduction of DNA ligase.

after cleavage can be competed with EcoRI cleaved yeast DNA [2]. The biochemically constructed hybrid DNA is used to infect calcium treated E. coli cells [5] that lack a restriction modification system. The plaques that are produced are combined to produce a hybrid pool. The hybrids formed with yeast DNA are designated λgt-Sc, with either an isolation number or a descriptive term following the Sc. We have prepared three independent λgt/Sc pools, each consisting of about 10,000 pooled plaques from the DNA infection.

EcoRI and HindIII restriction spectra of Saccharomyces cerevisae DNA

The variety of DNA sequences that are found in yeast DNA and in the cloned λgt-Sc DNA pools are readily visualized by separating the restriction endonuclease generated DNA fragments by agarose gel electrophoresis [6,7,8,9]. The display of DNA fragments, visualized by fluorescence from bound ethidium bromide, is termed a restriction spectrum (fig. 3). The HindIII and EcoRI restriction spectra of S. cerevisiae DNA are shown in fig. 3a and c. Fig. 3b shows the HindIII, EcoRI double digestion restriction spectrum. The arrows indicate faint bands that probably represent unique single copy sequences. The darker bands result from the coincidence of a number of different single copy sequences or from one sequence found in multiple copies. The EcoRI restriction spectrum

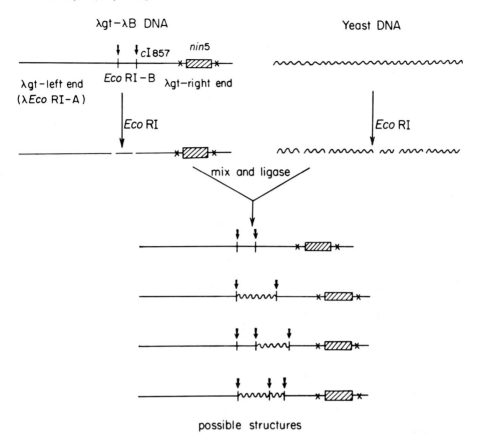

Fig. 2. Construction of hybrids. λgt-λB and yeast (*Saccharomyces cerevisiae*) DNA's were separately cleaved to completion with *Eco*RI endonuclease (partial cleavage products of yeast DNA were used in some cases). They were then mixed at equal concentrations of vector and yeast DNA for covalent joining of the *Eco*RI cohesive ends by *E. coli* DNA ligase. This was carried out at 10°C for 18 h. Four possible resulting structures of viable phage DNA are shown.

of the λgt-Sc DNA pool is very similar to that of whole yeast DNA, with the exception of the λ *Eco*RI B fragment which is found in the pool. This serves as evidence that most of the *Eco*RI DNA fragments from the yeast genome can be cloned in *E. coli*.

Genetic selection

Lytic selection

Direct genetic selections can be applied to the K modified λgt-Sc pools. One of the simplest questions to ask is whether the inserted yeast DNA fragments

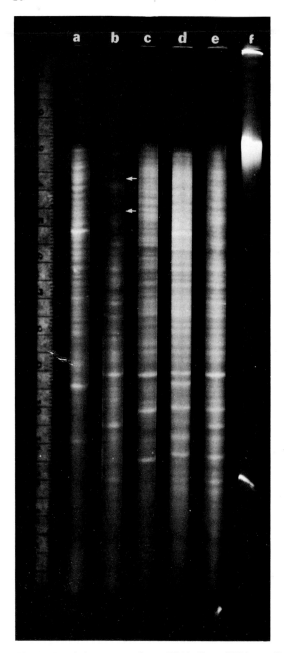

Fig. 3. Restriction spectra of yeast DNA. Yeast DNA was digested to completion with HindIII and/or EcoRI endonuclease in 50 mM Tris–HCl, pH 7.5, 10 mM MgCl$_2$, 50 mM NaCl at 37°C. Aliquots were loaded on 20 cm-long cylindrical 0.7% agarose gels. Electrophoresis was performed at 1.5 v/cm for 30 h. a) 0.8 µg DNA cleaved with HindIII. b) 0.8 µg DNA cleaved with HindIII and EcoRI. c) 0.8 µg DNA cleaved with EcoRI. d) Mixture of samples a and c. 2) Mixture of samples b and c. f) 0.4 µg DNA incubated without enzyme.

affect λ growth or whether they are completely neutral. The simplest way to answer this question is to continuously passage the phage pool and follow the restriction spectra as a function of the number of rounds of infection. It is quite clear that after only a few rounds of infection, a few DNA bands of the restriction spectra become enriched, while most DNA bands become diminished [10]. After 30 sequential infections, about 95% of the λgt-Sc hybrids in the pool contain one yeast DNA fragment of about 8 kb. Heteroduplex analysis shows that this fragment is inserted in λgt in either direction with approximately equal frequency (λgt-Sc1000 and λgt-Sc1000^1). Since there is no polar effect of this insertion, transcription from the λ promotors is probably not germane in its selection. It is not clear if this yeast fragment aids λ growth or if it is the least detrimental of the yeast fragments to λ growth. In any event, it grows better than λgt-λB, but not as well as wild type λ$^+$, the burst of λgt-Sc1000 being about one-half that of λ$^+$. Therefore, it was of interest to determine if a yeast DNA fragment could be found that would aid λ growth in a more restrictive host. This is easily accomplished by plating on a DNA polymerase I deficient cell (*polA amber* mutant), since the λ exonuclease gene has been deleted in all the λgt-Sc phage, and λ*exo*$^-$ phage will not plaque on a DNA polymerase I deficient cell. Therefore, if the yeast hybrid pool contains phage which can complement the mutation in the DNA polymerase gene or the λ exonuclease gene, then these phage should give plaques on a *polA* lawn. The EcoRI restriction spectrum of the pool, examined after three successive infections of a DNA polymerase I deficient cell, shows about 30 different sizes of fragments [10]. A number of these have been cloned and shown to have different DNA sequences by heteroduplex analysis. The yeast fragment that confers the best growth advantage to a λ *exo*$^-$ phage on a *polA amber* cell was selected by 20 successive infections. This phage, called λgt-Sc1001, gives a burst equal to that of a λ *exo*$^+$ phage on this *polA* cell. It contains two fragments of 2.2 and 1.6 kb. Unexpectedly, it gives a lower phage burst on infection of a wild type cell than on a DNA polymerase I deficient cell, indicating that there is an adverse interaction between this phage or phage product and wild type DNA polymerase I levels in a cell. If the inserted DNA fragment codes for a diffusible product, this would be the first evidence of functional genetic expression of eukaryotic DNA in a prokaryotic cell. A *trans* complementation test was conducted to determine whether the presence of the foreign DNA fragment can help a genetically distinguishable λ *exo*$^-$ phage during coinfection. λgt-Sc1001 clearly acts in *trans*, while a number of other hybrids do not show a marked *trans* effect [10]. It would appear that λgt-Sc1001 complements the λ exonuclease defect rather than the polymerase defect since this phage can also grow on a DNA ligase ts cell, which also does not plate *exo*$^-$ phage.

The selections thus far discussed have focused on phage functions. A method for the direct selection of a phage which supplies a function essential to cell growth is also needed. However, even if the infecting phage carries a genetic element which allows the cell to grow, what method will determine this if the cell dies from the infection? A plaque cannot be easily seen since the uninfected cells cannot grow, and therefore, cell growth around the plaque cannot be observed. The infected cell should support phage growth if it carries into the cell a function essential for cell growth, and the infected cells will lyse and release the DNA of the cell. This released DNA is easily localized by spraying the plate with ethidium bromide and then viewing with ultraviolet light. The fluorescent spots indicate the plaques from phage which perform an essential function for the cell. An example is shown in fig. 4 of a λtrp (which carried the *E. coli* tryptophan operon) infecting an *E. coli trp* cell lawn on a minimal plate. One of the advantages of this type of selection is that the inserted DNA is likely to be transcribed from one of the λ antiterminated promotors, thus allowing the genetic selection of DNA sequences which do not contain their own promotors.

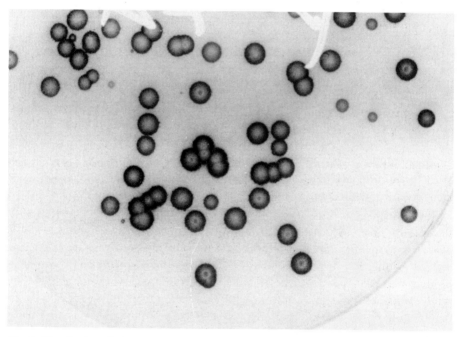

Fig. 4. Visualization of plaques without visible lawns by staining with ethidium bromide. Cells of the tryptophan auxotroph, *trp* A33, were grown overnight in maltose minimal medium supplemented with tryptophan (*trp*). The culture was washed twice with 10 mM $MgSO_4$ and starved for *trp* for 30 min. Approximately 10^8 cells were infected with λgt-λB (10^7) and $\lambda ptrp$ ABCD (500) and plated on glucose minimal plates at 37°C. After 2 days, plates were sprayed with 0.1 mg/ml ethidium bromide and visualized by fluorescence with ultraviolet light.

Lysogenic selection

Since λ is capable of integrating into the bacterial chromosome to form a stable lysogen, direct genetic selections can also be performed without killing the host. The establishment of lysogeny with the integration defective λgt-Sc phage is achieved by coinfection with an integration helper phage. We have isolated λgt-Sc phage which, when integrated into the chromosome of a auxotroph, allows this bacterium to grow in the absence of histidine [11]. The cell used is a non-reverting *his* B mutant of *E. coli* which lacks the enzyme imidazole glycerol phosphate (IGP) dehydratase. The *his* B complementation is dependent upon the presence of a λgt hybrid since all *his*+ colonies (200) contain a λ prophage, and removal of the prophage from the chromosome by curing results in the original *his* B− phenotype. The phage (λgt-Sc2601) which is responsible for the *his* B complementation has been cloned and contains a single inserted yeast *Eco*RI DNA fragment of 10.3 kb. This same *Eco*RI DNA fragment has been isolated from another independently generated λgt-Sc phage pool using an identical selection procedure. If λgt-Sc2601 is used in the selection with an integration helper, the frequency of obtaining *his*+ colonies is increased by 10^4 compared to using the hybrid pool with an integration helper. The inserted DNA fragment in λgt-Sc2601 is not a contaminant because labeled RNA made from it hybridizes to the same size *Eco*RI fragment of a different yeast DNA preparation. Therefore, we conclude that there is functional genetic expression of eukaryotic DNA in *E. coli* and that transcription is most probably initiated within the yeast DNA sequence since all known λ promotors are repressed or have been deleted.

Physical selection

Physical screening with complementary RNA

Clearly, not all genes can be obtained by genetic selection. For example, we are presently attempting to isolate the same yeast DNA fragment in λgt-Sc2601 from yeast mutants with no IGP dehydratase activity. A simple screening method has been developed in which ^{32}P RNA is hybridized to denatured DNA in plaques [12,13]. This involves forming plaques on bacterial lawns grown on nitrocellulose filters, and alkaline-denaturing the phage and the DNA to locally fix the single stranded DNA on the filter [13]. After washing the filter and drying, labeled RNA is added and hybridized. The filters are then washed. After RNase treatment, plaques containing DNA sequences complementary to the RNA are localized by autoradiography. An example is shown (fig. 5) using total and 5.8S yeast ribosomal RNA hybridized to randomly selected plaques from the hybrid

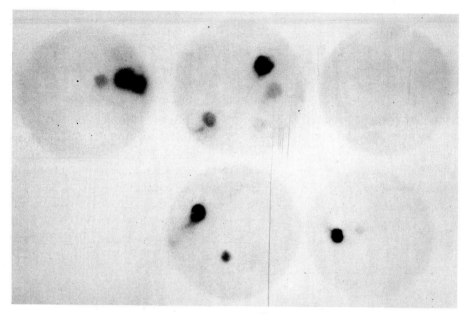

Fig. 5. Autoradiograph of in situ filter hybridization. ^{32}P-labeled RNA was extracted from yeast ribosomes and approximately 5×10^5 cpm applied to each nitrocellulose filter with denatured DNA from phage plaques fixed in situ. After hybridization and washing, the filters were placed against X-ray film to determine the location of phage which contained sequences homologous to rRNA (dark spots). The outline of the filters is visible due to a low level of non-specific binding of the RNA.

pool. Recently, we have found that tRNA can also be used for this screening procedure.

Physical selection with complementary RNA

Using the physical screening method, one can readily find a sequence in the hybrid pool at a frequency of about 10^{-3}, which is the approximate frequency of finding a single copy sequence in yeast. If a hybrid pool is constructed with DNA from higher organisms, the frequency of finding a single copy sequence drops to 10^{-6}. Because of this low frequency, a direct physical selection is desired. We are developing a selection procedure which is based on the observation [14] that RNA can be hybridized to double stranded DNA. A micrograph of such a molecule is shown in fig. 6 [15]. The DNA is from a λgt-Sc hybrid with an inserted yeast DNA segment complementary to yeast 26S rRNA [13]. This hybrid was selected by the physical screening procedure given in the previous section. The RNA displaces the identical DNA strand and forms a loop which has been designated an R-loop. It is formed because at high

Selection of eukaryotic genes cloned in E. coli

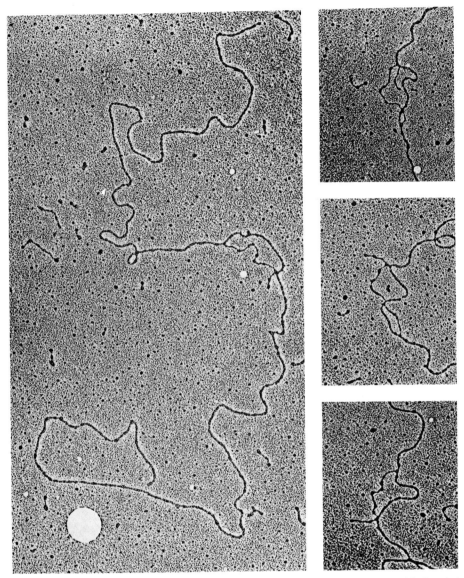

Fig. 6. The R-loops above were made by incubating 5 μg/ml λgt-Sc2056 DNA with 5 μg/ml total yeast rRNA in a solution containing 70% v/v formamide, 0.1 M PIPES, pH 7.8, and 0.01 M Na$_3$EDTA at 47°C for 20h. All of the 500 molecules examined by electron micrscopy contained an R-loop. Samples were mounted for electron microscopy by the basic protein film technique [2]. Grids were stained with uranyl acetate and shadowed with Pt/Pd.

formamide concentration, the RNA/DNA hybrid is more stable than the duplex DNA. The rate of R-loop formation is very dependent upon the incubation temperature (fig. 7), the rate being maximal at the denaturation temperature of the

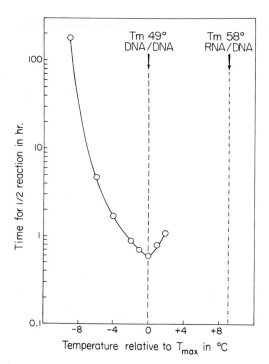

Fig. 7. Rate of R-loop formation as a function of temperature. R-loops were formed using the conditions given in fig. 6. The rate was determined by scoring the fraction of reacted DNA molecules in the electron microscope as a function of time. The melting temperatures (Tm) of the duplex DNA and of the duplex RNA/DNA hybrid were assessed by determining the temperature at which the duplex structures are converted to collapsed single strand structures in the electron microscope.

duplex DNA [15]. R-loops have considerable kinetic stability since the formamide can be removed by dialysis, and R-loop containing DNA can be cleaved with a restriction endonuclease without appreciable loss of R-loops (fig. 8). The ability to cleave R-loop containing DNA with restriction endonucleases should prove very useful in the detailed mapping of the location of the hybridized RNA.

R-loops may serve as the basis of a number of methods for physically selecting DNA sequences. One is based on the fact that the displaced DNA strand in the R-loop causes a DNA molecule containing an R-loop to be retained on a benzoylated DEAE cellulose (BD cellulose) column in the presence of high salt. This method has been used to enrich for the rDNA genes from *Dictyostelium discoideum* (fig. 9) and obtain λ clones containing these sequences [16]. Work is in progress to improve this technique so that it can be used to isolate single copy sequences from any organism.

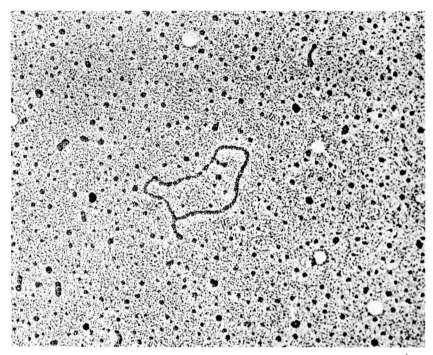

Fig. 8. DNA containing R-loops was dialyzed into 0.1 M NaCl, 0.05 M Tris, pH 7.5, 10^{-4} M EDTA, and subsequently cleaved with EcoRi endonuclease. Above is one example of such a molecule prepared from a DNA sample containing 100% R-loops. The digestion was done in the presence of 10 mM MgSO$_4$ at 37°C for 10 min. The sample was mounted for electron microscopy as described in fig. 6. The R-loop shown contains a fragment of rRNA.

References

[1] N.W. Murray and K. Murray, Manipulation of restriction targets in phage λ to form receptor chromosomes for DNA fragments. Nature 251, 476 (1974).

[2] M. Thomas, J.R. Cameron and R.W. Davis, Viable molecular hybrids of bacteriophage lambda and eukaryotic DNA. Proc. Nat. Acad. Sci. USA 71, 4579 (1974).

[3] J.R. Cameron, S.M. Panasenko, I.R. Lehman and R.W. Davis, In vitro construction of bacteriophage λ carrying segments of the Escherichia coli chromosome: selection of hybrids containing the gene for DNA ligase. Proc. Nat. Acad. Sci. USA 72, 3416 (1975).

[4] K. Murray and N.W. Murray, Phage lambda receptor chromosomes for DNA fragments made with restriction endonuclease III of Haemophilus influenzae and restriction endonuclease I of Escherichia coli. J. Mol. Biol. 98, 551 (1975).

[5] M. Mandel and A. Higa, Calcium-dependent bacteriophage DNA infection. J. Mol. Biol. 53, 159 (1970).

[6] C. Aaij and P. Borst, The gel electrophoresis of DNA. Biochim. Biophys. Acta 269, 192 (1972).

[7] P.A. Sharp, B. Sugden and J. Sambrook, Detection of two restriction endonuclease activities in Haemophilus parainfluenzae using analytical agarose-ethidium bromide electrophoresis. Biochemistry 12, 3055 (1973).

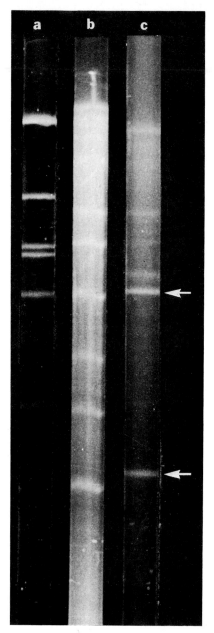

Fig. 9. Selection of rDNA cistrons from *Dictyostelium discoideum* by BD cellulose chromatograph of R-loops. R-loops were formed with 10 μg of *Eco*RI endonuclease cleaved *Dictyostelium discoideum* DNA using 5 μg 28S and 18S rRNA in 70% formamide, 0,5 M NaCl, 0,1 M PIPES, pH 7.8, 0.01 M Na$_3$EDTA and incubating at 52°C for 2 h. DNA molecules containing R-loops were separated from totally duplex DNA by selectively adsorbing the R-loop containing DNA to a BD cellulose [16] column in 1 M NaCl, 0.1 M Tris, pH 7.5, 0.01 M Na$_3$EDTA. The R-loop containing DNA was eluted by washing the column with 50% formamide, 1 M NaCl, 0.1 M Tris, pH 7.5, 0.01 M Na$_3$EDTA. The RNA was removed from the R-loop by RNase treatment. The resulting duplex DNA was electrophoresed on a 0.7% agarose gel as given in fig. 3. The arrows mark the bands that hybridize to 28S and 18S rRNA. a) *Eco*RI cleaved λ DNA. b) *Eco*RI cleaved *Dictyostelium discoideum* DNA. c) *Eco*RI cleaved *Dictyostelium discoideum* DNA with R-loops after adsorption and elution from BD cellulose.

[8] R.B. Helling, H.M. Goodman and H.W. Boyer, Analysis of endonuclease R. *Eco*RI fragments of DNA from lambdoid bacteriophages and other viruses by agarose gel electrophoresis. J. Virol. 14, 1235 (1974).

[9] M. Thomas and R.W. Davis, Studies on the cleavage of bacteriophage lambda DNA with *Eco*RI restriction endonuclease. J. Mol. Biol. 91, 315 (1974).

[10] J.R. Cameron and R.W. Davis, manuscript in preparation.

[11] K. Struhl, J.R. Cameron and R.W. Davis, Functional genetic expression of eukaryotic DNA in *E. coli*, Proc. Nat. Acad. Sci. USA 73, 1471 (1976).

[12] K.W. Jones and K. Murray, A procedure for detection of heterologous DNA sequences in lambdoid phage by in situ hybridization. J. Mol. Biol. 96, 455 (1975).

[13] R. Kramer, J.R. Cameron and R.W. Davis, Isolation of bacteriophage λ containing yeast ribosomal RNA genes: screening by in situ RNA hybridization to plaques. Cell 8, 227 (1976).

[14] R.L. White and D.S. Hogness, manuscript in preparation.

[15] M. Thomas R.L. White and R.W. Davis, Hybridization of RNA to double stranded DNA: formation of R-loops. Proc. Nat. Acad. Sci. USA. 73, 2294 (1976).

[16] M. Thomas, A. Jacobson and R.W. Davis, manuscript in preparation.

Chapter 3

DNA SEQUENCE ORGANIZATION AND EXPRESSION OF STRUCTURAL GENE SETS IN HIGHER ORGANISMS

R.C. ANGERER, W.R. CRAIN*, B.R. HOUGH-EVANS, G.A. GALAU, W.H. KLEIN, M.M. DAVIS, B.J. WOLD, M.J. SMITH**, R.J. BRITTEN*** and E.H. DAVIDSON

Division of Biology, California Institute of Technology, Pasadena, California 91125, U.S.A.

The two patterns of DNA sequence arrangement in eukaryotes

In the past several years the pattern of arrangement of single copy and repetitive DNA sequences has been determined for the genomes of a number of higher organisms. These studies show that a wide variety of phylogenetically diverse organisms have a strikingly similar pattern of DNA sequence organization. This pattern was first demonstrated for the DNA of *Xenopus laevis* [1,2] and we will refer to it as the '*Xenopus* pattern' of sequence interspersion. The basic feature of the *Xenopus* pattern is the interspersion of relatively short single copy and repetitive DNA sequences. In *Xenopus* the interspersed repeats average about 350 nucleotides in length. They terminate in single copy sequences which are about 700–900 nucleotides long. At least 70% of the single copy DNA in the *Xenopus* genome is organized in this manner. The remainder of the single copy DNA consists of longer stretches which are uninterrupted by repetitive sequences. The lengths of these stretches and the repeats which terminate them are at present unknown. About 75% of the repetitive DNA occurs in the short sequences. The remainder consists of repetitive DNA sequences at least 2000 nucleotides long. Quite a different pattern of sequence arrangement is present in the *Drosophila* genome. Manning et al. [3] concluded that the number average length of repetitive sequences in this genome is 5600 nucleotides. The data of Manning et al. indicate

---

* Present address: The Worcester Foundation for Experimental Biology, Shrewsbury, Massachusetts 01545, U.S.A.
** Present address: Department of Biological Sciences, Simon Fraser University, Burnaby, B.C., Canada V5A 1S6.
*** Also staff member. Carnegie Institution of Washington.

The Molecular Biology of the Mammalian Genetic Apparatus:
edited by P. Ts'o © 1977, Elsevier North-Holland Biomedical Press

that the single copy sequences average at least 13,000 and possibly as much as 30,000 nucleotides in length. We refer to this type of organization as the '*Drosophila* pattern' of sequence interspersion. Further experiments by Crain et al. [4] corroborated the pattern determined by Manning et al. and indicated that no more than 10% of the repetitive DNA occurs in short sequences.

TABLE 1

Higher organism DNA sequence organization.

Organism	Genome size	Single copy fraction of DNA	Repetitive sequence families (repetition frequency)	Fraction of repetitive DNA in short elements	Minimum fraction of single copy DNA interspersed with short repeats in fragments 2000–4000 nucleotides long	References
Dictyostelium discoideum slime mold	0.05 pg 4.5×10^7 NTP	~0.75	120–160	0.50	0.65	26
Aurelia aurita jelly fish	0.73 pg 6.7×10^8 NTP	0.70	180	0.60	0.80	12
Cerebratulus lacteus nemertean worm	1.4 pg 1.3×10^9 NTP	0.60	40 1200	0.55	0,70	12
Spisula solidissima surf clam	1.2 pg 1.1×10^9 NTP	0.75	30 3700	0.60	0.70	12
Crassostrea virginica oyster	0.69 pg 6.3×10^8 NTP	0.60	40	0.35	0.75	12
Aplysia californica sea hare	1.8 pg 1.7×10^9 NTP	~0.40	85	0.60	0.80	13
Loligo pealii squid	2.8 pg 2.6×10^9 NTP	0.75	100 4100	0.60	0.85	Galau, personal communication
Limulus polyphemus horseshoe crab	2.8 pg 2.6×10^9 NTP	0.70	50 2000	0.75	0.70	12
Antherea pernyi wild silk moth	1.0 pg 9.2×10^8 NTP	0.75	15 1600	0.75	0.65	27
Musca domestica house fly	0.89 pg 8.2×10^8 NTP	>0.50	200	0.70	0.70	28

DNA sequence in higher organisms

TABLE 1 (continued)

Higher organism DNA sequence organization.

Organism	Genome size	Single copy fraction of DNA	Repetitive sequence families (repetition frequency)	Fraction of repetitive DNA in short elements	Minimum fraction of single copy DNA interspersed with short repeats in fragments 2000–4000 nucleotides long	References
*Chironomus tentans midge	0.2 pg 1.8×10^8 NTP	0.90	90	<0.10	none	29
*Drosophila melanogaster fruit fly	0.12 pg 1.1×10^8 NTP	0.75	35	<0.10	none	3, 30
*Apis mellifera honey bee	0.35 pg 3.2×10^8 NTP	0.90	100	<0.10	none	28
Strongylocentrotus purpuratus sea urchin	0.89 pg 8.2×10^8 NTP	0.75	1500	0.75	0.70	7
Xenopus laevis clawed toad	2.7 pg 2.5×10^9 NTP	0.75	100 2100	0.75	0.70	1,2
Triturus carnifex salamander	∼24 pg 2.2×10^{10} NTP	∼0.65	10,000		∼0.70	35
Bos taurus cow	3.2 pg 2.9×10^9 NTP	0.65	60,000	0.55	0.65	31
Rattus norvegicus rat	3.2 pg 2.9×10^9 NTP	0.75	70 2000	∼0.55	∼0.65	32, 33
Homo sapiens human	3.5 pg 3.2×10^9 NTP	∼0.75	500	∼0.75	0.80	34
Triticum aestivum wheat	6 pg 5.5×10^9 NTP	0.25	4300	∼0.85	90	Flavell and Smith personal communication
Nicotiana tabacum tobacco	1.6 pg 1.5×10^9 NTP	∼0.43	250	>0.35	0.80	Zimmerman and Goldberg, personal communication
Gossypium herbaceum cotton	0.8 pg 7.4×10^8 NTP	0.65	300	0.80**	0.86	5

* Only long period interspersion is observed in these genomes.
** In cotton, the average length of the 'short' repetitive elements is 1250 NTP.

Our knowledge of the DNA sequence organization of a number of diverse higher organisms is summarized in table 1. The organisms surveyed are listed in the first column. A large majority of these organisms have the *Xenopus* pattern of genome organization. The phylogenetic diversity of this sample indicates that the *Xenopus* pattern is characteristic of most groups of higher organisms. The only exceptions so far reported are the three species of insect which are indicated in the table by asterisks. Haploid genome sizes are given in units of mass (picograms) and in nucleotide pairs in column 2 of the table. The range in genome size is from 0.05 pg for *Dictyostelium* (about 10 times the size of the *E. coli* genome) to about 24 pg for *Triturus*. With the exception of *Dictyostelium*, the organisms listed in the table with the smallest genomes are the insects, *Drosophila*, *Chironomus* and *Apis*. Only these three species exhibit the *Drosophila* pattern of interspersion. The fraction of the genome in single copy sequence is given in column 3. In most cases these values are the fraction of the reassociated DNA which is digested by single strand specific nuclease S1 at C_0ts where the repetitive DNA has reassociated. In a few cases the size of the single copy fraction has been determined from hydroxyapatite reassociation kinetics of short DNA fragments. This technique underestimates the fraction of single copy DNA in the genome due to inclusion of single copy sequences on repetitive duplexes. The values in column 3 have been corrected for this effect of interspersion using the fraction of repetitive DNA in short repeats, the length of the short repeats and the fragment length of the DNA used for the reassociation measurements.

Column 4 of table 1 lists the frequencies of repetition of the moderately repeated sequences in each genome. Minor components composed of sequences repeated more than 10^6 times, and satellite DNA sequences are omitted. The repetition frequencies are in most cases determined from the kinetics of reassociation of short DNA fragments by least squares analysis, and are likely to represent averages of what may be ranges of components. Most of the genomes examined contain a low frequency component repeated 15–200 times. In addition many of the genomes include sequences repeated a few thousand times. The only genome characterized by a major component of much higher repetition frequency (cow) has not been closely examined for the presence of lower frequency components. Sequences of low and moderate repetition frequency may thus be a general requirement of eukaryotic genomes. The fraction of the repetitive sequences present in the DNA in elements of 200–400 nucleotides long, interspersed with single copy sequences, is given in column 5 of table 1. Most of these estimates are based on chromatography on agarose A-50 of reassociated repetitive duplexes remaining after digestion of single stranded (single copy) DNA with nuclease S1. In several cases the lengths of repetitive sequences have also been determined by electron microscopy. In organisms exhibiting the short period interspersion pattern, usually 60–80% of the repetitive sequences are 200–400 nucleotides long. In the genomes of *Apis* and *Chironomus* as well as in *Drosophila*, less than 10%, if any, of the repetitive sequences are this short. An intermediate value for the length of repetitive sequences is observed for the cotton genome. The analysis of Walbot and Dure [5] showed that in this DNA the repetitive sequences average about 1250 nucleotides in length.

As mentioned earlier, in each genome of the *Xenopus* type a fraction of the repetitive sequences consists of duplex stretches 2000 nucleotides or longer. Eden et al. [6] showed that in sea urchin DNA both length classes of repetitive DNA, have the same distribution of repetition frequencies. In their experiments long and short repetitive sequence fractions were isolated from reassociated DNA by S1 treatment and agarose A-50 chromatography. Each fraction was then reassociated in the presence of a large excess of unlabeled sea urchin DNA. The kinetics of reassociation are similar for both the long and short repetitive sequence fractions. About 65% of each fraction consists of sequences repeated 1500 times and about 35% of sequences repeated 100 times per haploid genome.

The sixth column of table 1 lists the minimum fraction of the single copy DNA sequence which is interspersed with repetitive sequences at short fragment lengths. In some cases this minimum value can be determined from 'interspersion curves' which measure the fraction of the genome in molecules containing repetitive duplex as a function of DNA fragment length. (For a discussion of interspersion curves see Graham et al. [7].) In most cases the values listed here are based on the quantity of single copy DNA in fragments 2000–4000 nucleotides long which reassociate at a repetitive $C_0 t$. This probably represents the best approximation of the quantity of single copy DNA in the short interspersed pattern. In all the genomes which have the *Xenopus* pattern, at least 65% of the single copy sequence is interspersed with the short repeats.

Detailed investigation of sequence organization in the genomes of *Musca* and *Apis*

Here we review in more detail measurements of the sequence organization in the genomes of two insects. These insects, the honeybee (*Apis mellifera*, a Hymenopteran), and the housefly (*Musca domestica*, a Dipteran) were chosen to gain insight into the phylogenetic distribution of the two patterns. Honeybee DNA fragments 330 or 2200 nucleotides long were reassociated and the fraction of fragments containing duplex was assayed by hydroxyapatite chromatography. The results of this kinetic analysis are shown in fig. 1. The least squares solution for each set of data yields a single second order component which accounts for about 90% of the reassociation. The rate constants for the 330 and 2200 nucleotide fragments are 3.8×10^{-3} M^{-1} sec^{-1} and 1.3×10^{-2} M^{-1} sec^{-1}, respectively. These values are in the right proportion for fragments of these two lengths which consist entirely of single copy sequence. The rate of reassociation of such fragments is

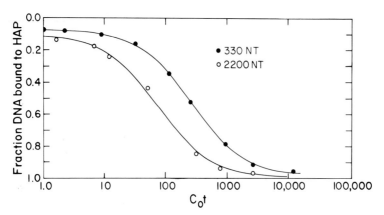

Fig. 1. Reassociation of honeybee DNA. Honeybee DNA fragments 330 nucleotides (●) and 2230 nucleotides (○) long were reassociated in 0.12 M phosphate buffer, pH 6.8 (PB) at 60°C. The fraction of fragments containing duplex regions was assayed by hydroxyapatite chromatography in 0.12 M PB at 60°C. The curves represent least squares solutions to the data for one second order component in each case. The fraction of fragments in the components and rate constants are: 330 nucleotide fragments 0.917 and 3.8×10^{-3} M^{-1} sec^{-1}; 2230 nucleotide fragments 0.897 and 1.3×10^{-2} M^{-1} sec^{-1}.

proportional to the square root of the fragment length [8]. This result shows that a large fraction of the single copy sequence in the honeybee genome occurs in long stretches rarely interrupted by repetitive sequences. Regions where the repetitive sequences occur contain little if any single copy sequence. We conclude that both the repetitive and nonrepetitive sequences are long compared to the 2200 nucleotide fragments.

Analysis of the reassociation kinetics of two different fragment lengths of housefly DNA gives a much different result. Fig. 2 shows the reassociation kinetics of fragments 250 and 2000 nucleotides long. The data for the 250 nucleotide fragments have been fit by least squares methods with three second order kinetic components. The slowest component accounts for 34% of the fragments and has a rate constant of 9.5×10^{-4} M^{-1} sec^{-1} which is very close to 9.3×10^{-4} M^{-1} sec^{-1}, the expected value for single copy fragments in a genome of 0.89 pg. In contrast, the reassociation of 2000 nucleotide fragments terminates at much lower values of $C_0 t$ (fig. 2b) than is expected for single copy sequence of this fragment length. This is a direct demonstration that a large fraction of the single copy sequence is interspersed with repeated sequences at some interval less than 2000 nucleotides.

In order to measure the size of the repetitive sequence elements of housefly DNA, fragments 2000 nucleotides long were reassociated to $C_0 t$ 9 and digested with S1 nuclease [10]. After digestion 34.4% of the DNA bound as duplex to hydroxyapatite. This fraction was eluted and chromatographed on Agarose A-50. As shown in fig. 3, the repetitive sequences display a bimodal length distribution.

DNA sequence in higher organisms

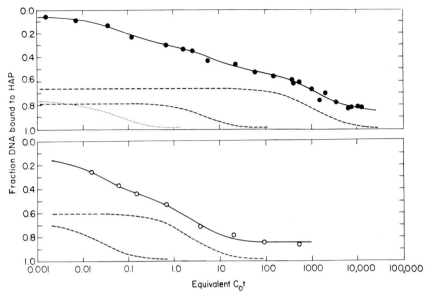

Fig. 2. Reassociation of housefly DNA. The solid circles represent the reassociation of tritium labeled housefly DNA fragments 250 nucleotides long. The solid line represents a least squares solution of the data with three second order components as indicated by the dashed lines. The fraction of fragments in each component and the rate constants are: fast, 0.23 and 17.2 M^{-1} sec^{-1}; slow, 0.22 and 0.192 M^{-1} sec^{-1}; and single copy, 0.34 and 9.5×10^{-4} M^{-1} sec^{-1}. The open circles represent the reassociation of unlabeled housefly DNA fragments 2000 nucleotides long. The solid line represents a least squares solution of the data with two second order components as indicated by the dashed lines. The fraction of fragments in each component and the rate constants are fast, 0.32 and 42.9 M^{-1} sec^{-1}; and slow 0.40 and 0.44 M^{-1} sec^{-1}.

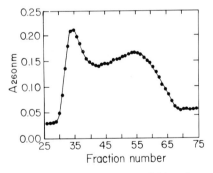

Fig. 3. Agarose A-50 profile of S1 resistant reassociated repetitive sequences of housefly DNA. Housefly DNA fragments 2000 nucleotides long were reassociated to C_0t 9 and the duplex regions excised with single strand specific S1 nuclease. Digestion was in 0.15 M NaCl, 0.005 M Pipes, 0.025 M Na acetate, pH 4.3, 0.1 mM $ZnSO_4$ at 37°C for 45 min. The S1 concentration used was one which has been shown not to digest moderately mismatched repetitive duplexes [10].

Those excluded from the column are about 2000 nucleotides or greater in length, while those which are included have a broad size distribution with a peak around 300 nucleotides. This distribution is qualitatively quite similar to that of other organisms which show a *Xenopus* type interspersion pattern [11,12,13]. As our knowledge grows we will likely learn of more subtle variations in the patterns of DNA sequence organization. However, at the moment the contrast between the *Xenopus* and *Drosophila* patterns stands out. It is not easy to visualize how an evolutionary transition of such magnitude could occur between two different patterns. Yet such an event has probably occurred at least twice since the *Drosophila* pattern is observed in some, but not all Diptera, and also in an Hymenopteran. The probable occurrence of evolutionary transitions between long and short period interspersion patterns suggests that they share some fundamental similarity which is obscured at the moment by our sensitivity to the differences. An example of such a similarity may be found in the internal sequence arrangement of long repetitive sequences. In the case of one such sequence in the *Drosophila* genome which has been examined as a recombinant DNA clone (Finnegan, Rubin and Hogness, personal communication), it appears that the long repetitive sequence is actually a tandem arrangement of shorter repetitive sequences which are members of different families. Perhaps most individual repetitive sequences are short, but in some organisms they are interspersed with single copy sequences while in others they are contiguous.

A possibly important point is that the three insect species with the long interspersion pattern all have small genomes. This fact is at least consistent with the concept of DNA loss in an evolutionary transition from the short to the long interspersion pattern. Other metazoan species have genome sizes in this range, but their DNA sequence organization has not yet been determined. All of the animal species whose DNA is known to have a short period pattern have genome sizes greater than 0.7 pg. All three insects without short period interspersion also have a small fraction of the genome as middle repetitive DNA, as compared to the 20–40% typically observed in metazoan genomes. Since the genome sizes of these three insects are also small, the absolute quantity of repetitive sequence is also very small, on the average less than 2×10^7 nucleotide pairs. These observations are consistent with the possibility that the evolution of genomes containing only long period interspersed sequences has involved DNA loss by the elimination from the genome of the short period interspersed sequences.

Functions of short interspersed repetitive sequences

The fact that the short pattern of interspersion is almost universally observed in the wide range of phyla studied implies that this aspect of genome structure is of some functional significance to a variety of organisms. We now know that two major classes of RNA are transcripts of DNA sequences which make up the short interspersion pattern. High molecular weight, rapidly turning over heterogeneous nuclear RNA has been shown to be transcribed from interspersed repetitive and non-repetitive sequences [14,15]. Most kinds of messenger RNA are transcripts of single copy DNA sequences [16,17,18,19]. Davidson et al. [20] demonstrated that at least 80% of the mRNA of sea urchin gastrula is transcribed from single copy DNA sequences which are adjacent to interspersed repetitive sequences. In addition, the duck hemoglobin genes have been reported to be adjacent to repetitive sequences [21]. These observations suggest that a part of the interspersion pattern represents a functional configuration of structural gene sequences with adjacent repetitive DNA.

Specificity of structural gene transcription

A key issue in considering the problem of gene regulation is the number of structural genes which are expressed in the various cell types of a complex animal. We have recently made direct measurements of the amount of sequence overlap in the mRNA populations present at various stages of sea urchin embryogenesis and in several adult tissues. Structural genes are here defined experimentally as single copy DNA sequences [16,22,23] which hybridize with mRNA released from purified polysomes. Previous work has provided considerable quantitative information about the structural genes of the sea urchin embryo. According to our earlier measurements [24], the complexity, or total length in nucleotide pairs of DNA sequences represented in the RNA, of the gastrula mRNA is about 17×10^6 nucleotide pairs. This means that about $1-1.5 \times 10^4$ diverse proteins are being translated in the polysomes. Some structural genes are far more extensively represented in sea urchin polysomal messenger RNA than are others. Galau et al. [24] calculated that about 90% of the polysomal mRNA in gastrula, the 'prevalent mRNA class' consists of species of relatively low total diversity, while the other 10% is made up of the $1-1.5 \times 10^4$ species referred to above. We term the latter the 'complex mRNA class'. Using a different technique, Nemer et al. [23] demonstrated that poly(A) mRNA in sea urchin embryos consists mainly of prevalent messages, and concluded that this class of mRNA includes about 1400 species.

The rationale for the current experiments [25] is as follows. When labeled single-copy DNA is hybridized to termination with mRNA from gastrulae at ratios such that mRNA is in excess, only about 1.35% of the DNA reacts. This is too small a fraction of the total single copy DNA to permit much resolution in comparing diverse mRNA populations. To increase the sensitivity of the measurements, we prepared a single copy DNA fraction which is highly enriched for sequences complementary to gastrula mRNA. Single copy DNA labeled in vitro was hybridized with excess gastrula polysomal mRNA to a RNA $C_0 t$ at which the reaction terminated. The hybrid fraction was collected on hydroxyapatite and again hybridized with excess mRNA. We refer to the duplex fraction isolated from this second hybridization as 'mDNA'. About 63% of the reactable DNA in this preparation hybridizes with gastrula mRNA as compared to the original 1.35%. This represents a purification by a factor of about 50, and provides more than an order of magnitude increase in sensitivity to differences in sequence content when hybridization reactions are carried out with other mRNAs. Reaction of a nongastrula mRNA preparation with mDNA from gastrula measures the fraction of the structural genes expressed in gastrula which is also shared in the test messenger RNA population.

A single copy DNA fraction depleted of the sequences represented in gastrula mRNA was also prepared, and is termed 'null mDNA'. This DNA was prepared from the single copy DNA which remained single stranded after the initial hybridization of single copy DNA with gastrula mRNA. The null mDNA preparation had no detectable reaction with gastrula mRNA. Reaction of a test RNA with null mDNA therefore measures the number of structural genes represented in the test mRNA which are different from those in gastrula mRNA.

The results of measurements of the reaction of mDNA and null mDNA with messenger RNA preparations from several embryonic stages and several adult tissues of the sea urchin are shown in fig. 4. Here the solid portion of each bar shows the complexity of those structural gene sets represented in the indicated mRNA population which are also represented in gastrula mRNA as assayed by hybridization with mDNA. Similarly, the open portion of each bar gives the complexity of the structural gene sets represented in each tissue which are excluded from the gastrula gene set as measured by saturation hybridization of the null mDNA. The sum of these two values is the total complexity of each mRNA population. As described in detail by Galau et al. [25] the total complexity values were verified in several cases by reaction of the RNAs with unfractionated single copy DNA. Complexity is expressed in three ways in fig. 4. The left-hand ordinate shows the total length of single copy nucleotide sequence, and the two right hand ordinates show the percentage of gastrula mRNA complexity and the percentage of total single copy sequences, respectively.

A number of interesting points emerge from these data. It is clear that each of the tissues investigated contains a different, although partially overlapping set of structural gene transcripts. Our observations on the three adult non-reproductive tissues (tubefoot, coelomocyte and intestine) suggest that these tissues share a common subset of mRNAs, expressed both in gastrula as well as in the three adult tissues. The number of such genes in the sea urchin genome appears to be no greater than 1000–1500.

The adult tissues we have studied consist of several different cell types, each present in significant concentration. In addition, there may be rare cell types

DNA sequence in higher organisms

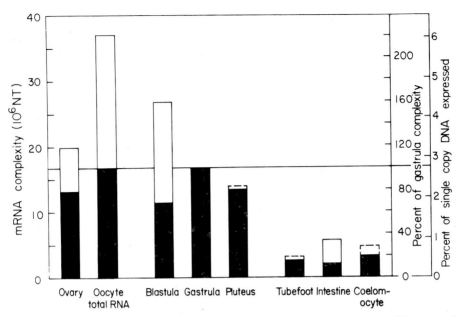

Fig. 4. Sets of structural genes active in sea urchin embryos and adult tissues. The solid portion of each bar indicates the amount of single-copy sequence shared between gastrula mRNA and the RNA preparations listed along the abcissa. These data are obtained from the mDNA reactions described in the text. The open portions show the amount of single copy complexity present in the various RNAs studied but absent from gastrula mRNA. These data derive from the null mDNA reactions described in the text. Dashed lines indicate the maximum amount of null mDNA reaction which could have been present and escaped detection, in terms of complexity, for cases where no apparent null mDNA reaction was observed. The total complexity of each RNA is indicated by the overall height of each bar. Complexity is calibrated in three ways along the three ordinates shown. From left to right these are: in nucleotides of single copy sequence; as percentage of gastrula mRNA complexity; as percentage of total single copy sequence.

whose specific mRNA populations do not contribute significantly to the total complexities measured. The overall complexity of the mRNA populations in these three adult nonreproductive tissues ranges from about 15 to 35% of that of gastrula mRNA. We cannot be certain that the measured complexities represent the total complexity of the mRNAs in the complete organ or tissue because of the unknown contribution of rare cell types. These measurements indicate that there are differentiated adult tissue cell types, if not complete organs, which require fewer than about 5000 structural genes to operate. This statement refers to the usual circumstance of tissue maintenance and to mRNAs present at a frequency within 1–2 orders of magnitude of the typical complex sequence class mRNAs.

The data for various embryonic stages show that, in comparison, the process of embryonic differentiation is relatively very expensive in terms of the number of structural genes required. To a first approximation, our data cover one phase of the sea urchin life cycle, that extending from oogenesis to feeding pluteus. The oocyte RNA may include most of the sequences in blastula mRNA which react with null mDNA, and we have shown that the oocyte RNA includes all of the gastrula mRNA sequences. Thus, as a crude maximum estimate, the complexity of oocyte RNA approximates the structural gene information needed to program and carry out development from oogenesis to the feeding pluteus stage. This amounts to about 6% of the total single copy sequence, some 37×10^6 nucleotides, or 20,000–30,000 diverse structural genes. Generally, the differences observed in mRNA sequence sets at the various developmental stages are quite large. For example, 56% of the mRNAs on the polysomes of blastula stage embryos are distinct from those of gastrulae, and 44% are homologous. These differences amount to many thousands of individual structural genes whose transcripts are present at detectable levels in one stage and absent in another.

Among the most surprising and significant findings in this work is that virtually all the polysomal gastrula mRNA sequences are found to be represented in mature oocyte RNA in per embryo concentrations higher than are on the polysomes at gastrula. Furthermore, a large fraction of the gastrula message set is present in pluteus and in blastula. Recent measurements of mRNA content, turnover and synthesis rates by Galau et al. (in preparation) demonstrate that most of the mRNAs of both the prevalent and the complex sequence classes turn over with a half life of approximately 5 h during gastrulation. The bulk of the messenger RNAs of all frequencies of representation in the polysomes are thus newly synthesized in the embryo. Since many of the same sequences are transcribed during oogenesis (fig. 4) these results imply that the embryo inherits a program for structural gene transcription similar to that used during oogenesis. The maternal messenger RNA molecules present in the egg are presumably used mainly to program protein synthesis early in development. The specific subsets of the gastrula mRNA set present in the pluteus and blastula thus appear to represent stage specific modulations in the embryonic transcriptional program, rather than in the sets of preserved maternal messages being translated.

Acknowledgments

This research was supported by NIH Grants HD-05753 and GM-20927 and by NSF Grant BMS 75-07359. RCA holds a postdoctoral fellowship from ACS, Calif. Div., Grant J-309. WRC was the recipient of an NIH postdoctoral fellowship (GM-55726). GAG was supported by NIH Training Grant GM-00086 and by the National Foundation March of Dimes Grant 1-404. WHK is supported by a postdoctoral fellowship from ACS, Calif. Div., Lievre Fellowship J-340. MMD and BJW are supported by NIH Training Grant GM-00086.

References

[1] E.H. Davidson, B.R. Hough, C.S. Amenson and R.J. Britten, General interspersion of repetitive with non-repetitive sequence elements in the DNA of *Xenopus*. J. Mol. Biol. 77, 1 (1973).
[2] M.E. Chamberlin, R.J. Britten and E.H. Davidson, Sequence organization in *Xenopus* DNA studied by the electron microscope. J. Mol. Biol. 96, 317 (1975).
[3] J.E. Manning, C.W. Schmid and N. Davidson, Interspersion of repetitive and nonrepetitive DNA sequences in the *Drosophila melanogaster* genome. Cell 4, 141 (1975).
[4] W.R. Crain, F.C. Eden, W.R. Pearson, E.H. Davidson and R.J. Britten, Absence of short period interspersion of repetitive and non-repetitive sequence in the DNA of *Drosophila melanogaster*. Chromosoma (Berl.) 56, 309 (1976).
[5] V. Walbot and L.S. Dure III, Developmental biochemistry of cotton seed embryogenesis and germination. VII. Characterization of the cotton genome. J. Mol. Biol. 101, 503 (1976).
[6] F.C. Eden, D.E. Graham, E.H. Davidson and R.J. Britten, Exploration of long and short repetitive sequence relationships in the sea urchin genome, Nucl. Acids Res., in press.
[7] D.E. Graham, B.R. Neufeld, E.H. Davidson and R.J. Britten, Interspercion of repetitive and nonrepetitive DNA sequences in the sea urchin genome, Cell 1, 127 (1974).
[8] J.G. Wetmur and N. Davidson, Kinetics of renaturation of DNA, J. Mol. Biol. 31, 349 (1968).
[9] K. Bier and W. Müller, DNS-Messungen bei insekten und eine hypothese über retardierte evolution und besonderen DNA-Reichtum in tierreich. Biol. Zbl. 88, 425 (1969).
[10] R.J. Britten, D.E. Graham, F.C. Eden, D.M. Painchaud and E.H. Davidson, Evolutionary divergence and length of repetitive sequences in sea urchin DNA. J. Mol. Evol. (in press).
[11] E.H. Davidson, D.E. Graham, B.R. Neufeld, M.E. Chamberlin, C.S. Amenson, B.R. Hough and R.J. Britten, Arrangement and characterization of repetitive sequence elements in animal DNAs. Cold Spring Harbor Symp. Quant. Biol. 38, 295 (1974).
[12] R.B. Goldberg, W.R. Crain, J.V. Ruderman, G.P. Moore, T.R. Barnett, R.C. Higgins, R.A. Gelfand, G.A. Galau, R.J. Britten and E.H. Davidson, DNA sequence organization in the genomes of five marine invertebrates. Chromosoma (Berl.) 51, 225 (1975).
[13] R.C. Angerer, E.H. Davidson and R.J. Britten, DNA sequence organization in the mollusc *Aplysia californica*. Cell 6, 29 (1975).
[14] D.S. Holmes and J. Bonner, Sequence composition of rat nuclear deoxyribonucleic acid and high molecular weight nuclear ribonucleic acid. Biochemistry 13, 841 (1974).
[15] M.J. Smith, B.R. Hough, M.E. Chamberlin and E.H. Davidson, Repetitive and non-repetitive sequence in sea urchin heterogenous nuclear RNA. J. Mol. Biol. 85, 103 (1974).
[16] R.B. Goldberg, G.A. Galau, R.J. Britten and E.H. Davidson, Non-repetitive DNA sequence representation in sea urchin embryo messenger RNA. Proc. Natl. Acad. Sci. U.S.A. 70, 3516 (1973).
[17] W.H. Klein, W. Murphy, G.A. Attardi, R.J. Britten and E.H. Davidson, Distribution of repetitive and non-repetitive sequence transcripts in HeLa cells. Proc. Natl. Acad. Sci. U.S.A. 71, 1785 (1974).
[18] A. Spradling, S. Penman, M.S. Campo and J.O. Bishop, Repetitious and unique sequences in the heterogeneous nuclear and cytoplasmic messenger RNA of mammalian and insect cells. Cell 3, 23 (1974).
[19] M.S. Campo and J.O. Bishop, Two classes of messenger RNA in cultured rat cells: repetitive sequence transcripts and unique sequence transcripts. J. Mol. Biol. 90, 649 (1974).
[20] E.H. Davidson, B.R. Hough, W.H. Klein and R.J. Britten, Structural genes adjacent to interspersed repetitive DNA sequences. Cell 4, 217 (1975).
[21] J.O. Bishop and K.B. Freeman DNA sequences neighboring the duck hemoglobin genes. Cold Spr. Harb. Symp. Quant. Biol. 38, 707 (1974).

[22] R.S. McColl and A.I. Aronson, Transcription from unique and redundant DNA sequences in sea urchin embryos. Biochem. Biophys. Res. Commun. 56, 47 (1974).
[23] M. Nemer, L.M. Dubroff and M. Graham, Properties of sea urchin embryo messenger RNA containing and lacking poly(A). Cell 6, 171 (1975).
[24] G.A. Galau, R.J. Britten and E.H. Davidson, A measurement of the sequence complexity of polysomal messenger RNA in sea urchin embryos. Cell 2, 9 (1974).
[25] G.A. Galau, W.H. Klein, M.M. Davis, B.J. Wold, R.J. Britten, and E.H. Davidson, Structural gene sets active in embryos and adult tissues of the sea urchin. Cell 7, 487 (1976).
[26] R.A. Firtel and K. Kindle, Structural organization of the genome of the cellular slime mold *Dictyostelium discoideum*; Interspersion of repetitive and single-copy DNA sequences. Cell 5, 401 (1975).
[27] A. Efstratiadis, W.R. Crain, R.J. Britten, E.H. Davidson and F. Kafatos, DNA sequence organization in the lepidopteran *Antheraea pernyi*. Proc. Natl. Acad. Sci. U.S.A. 73, 2289 (1976).
[28] W.R. Crain, E.H. Davidson and R.J. Britten, Contrasting patterns of DNA sequence arrangement in *Apis mellifera* (honeybee) and in *Musca domestica* (housefly). Chromosoma (Berl.) in press.
[29] R. Wells, H.D. Royer and C.P. Hollenberg, Non Xenopus-like DNA sequence organization in the *Chironomus tentans* genome. Mol. Gen. Genet., 147, 45 (1976).
[30] P.C. Wensink, D.J. Finnegan, J.E. Donelson and D.S. Hogness, A system for mapping DNA sequences in the chromosomes of *Drosophila melanogaster*. Cell 3, 315 (1974).
[31] R.J. Britten and J. Smith, A bovine genome. Carnegie Inst. Wash. Year Book 68, 378 (1970).
[32] J. Bonner, W.T. Garrard, D.S. Holmes, J.S. Sevall and M. Wilkes, Functional organization of the mammalian genome. Cold Spr. Harb. Symp. Quant. Biol. 38, 303 (1974).
[33] N. Rice, Some observations on interspersion of DNA sequences, Carnegie Inst. Wash. Year Book 69, 479 (1971).
[34] C.W. Schmid and P.L. Deininger, Sequence organization of the human genome. Cell 6, 345 (1975).
[35] J. Sommerville and P.B. Malcolm, Transcription of genetic information in amphibian oocytes. Chromosoma (Berl.) 55, 183 (1976).

Chapter 4

THE HIGHLY REPEATED DNA SEQUENCES OF *DROSOPHILA MELANOGASTER*

DOUGLAS L. BRUTLAG

Stanford University Medical Center, Stanford, California 94305, U.S.A.

The highly repeated DNA sequences of eukaryotes can usually be isolated as satellites in CsCl buoyant gradients. The tandem arrangement of short repeated sequences gives rise to the distinctive physical properties of satellites. These DNAs are usually restricted to centromeric heterochromatin [1,2] and, as might be expected, are neither transcribed nor translated. The function of centric heterochromatin is difficult to assess because, like satellite DNA itself, it is so variable between closely related species. Centric heterochromatin in *Drosophila* was thought to be genetically inert due to the lack of genes in this region and the viability of large heterochromatic deletions [3]. However, such deletions in the sex chromosomes do result in improper disjunction in the first meiotic division and cause a marked decrease in fertility [4]. This genetic evidence indicates that centric heterochromatin and perhaps satellite DNAs are involved in normal chromosome propagation in meiosis. We have, therefore, been studying the arrangement of satellite DNA in the heterochromatin in *Drosophila* in order to provide a molecular basis for such a proposed function.

We have found that the bulk of the heterochromatic DNA of *Drosophila* consists of four different satellite species [5]. These DNAs contain short nucleotide sequences tandemly repeated in arrays over 1,000,000 base pairs long [6]. Several of these satellites, although appearing homogeneous by many physical criteria, contain more than one distinct DNA with different repeating sequences [7]. The sequences in one satellite are so similar that the different DNAs cannot be separated by classical procedures. I will conclude this paper by summarizing the progress we have made in separating these components by cloning individual molecules in hybrid bacterial plasmids.

Drosophila is an ideal organism with which to work since most of the heterochromatic DNA can be isolated in the form of satellite DNA. Fig. 1 shows nuclear DNA centrifuged to equilibrium in a CsCl gradient containing the antibiotic actinomycin D. Three classes of highly repeated DNA do not bind

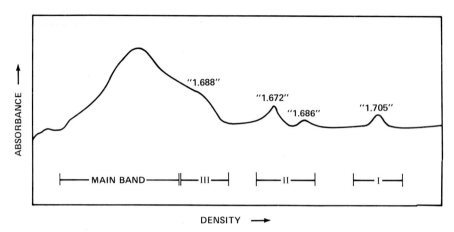

Fig. 1. Analytical actinomycin D-CsCl gradient of *Drosophila melanogaster* DNA. This DNA was isolated from diploid nuclei of developing embryos (9 h mean age).

actinomycin D, and their densities are not altered by the drug. The 1.705, 1.686 and 1.672 species are, therefore, well separated from the main band which is shifted from 1.701 g/cm^3 to 1.62 g/cm^3. A fourth satellite 1.688, which normally co-bands with the 1.686 g/cm^3 species does bind actinomycin D, allowing the resolution of these two satellites. The 1.688 g/cm^3 material can be purified to near homogeneity by repeated centrifugation in CsCl gradients.

Repurification of each species results in DNA which appears homogeneous in both neutral and alkaline CsCl gradients. In alkaline CsCl each satellite separates into its complementary strands (fig. 2). Note the extreme separation of heavy and light strands of the 1.672 and 1.686 species which has facilitated sequence studies.

These four classes of DNA constitute 18% of the *Drosophila* genome and are located exclusively in centromeric heterochromatin as determined by in situ hybridization [5]. Three of these satellites have been isolated in pure form as molecules 225,000 base pairs in length [6]. From the yield of this isolation, and assuming that satellite regions have a uniform size, we can calculate the minimum length of such a region to be 1,000,000 base pairs. Since the heterochromatin of the X chromosome contains perhaps 8,000,000 base pairs, a block of 1,000,000 satellite base pairs could constitute a major cytological subdivision.

We have carried out partial nucleotide sequence analysis of these species to determine the basic repeated segment (fig. 3). The lightest satellite contains two different repeating sequences [7]. Evidence for this is three-fold. First, pancreatic RNase digests of RNA complementary to the T rich strand give AAU and AU as major products. Partial ribonuclease digests show the major repeating units are these five and seven nucleotide sequences. Longer oligonucleotides with two and three of these five nucleotide repeats in a row have been found. Similarly,

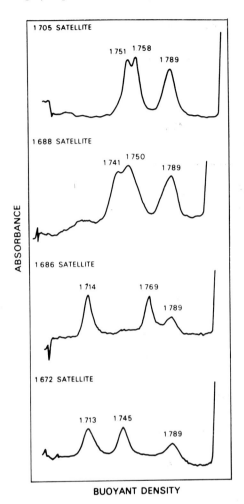

Fig. 2. Alkaline CsCl gradients of purified satellite DNA fractions. *M. luteus* DNA was added as a density marker in each gradient ($\rho = 1.789$ g/cm^3):

oligonucleotides containing two and three repeats of the seven nucleotide sequence have been found.

A second line of evidence for two distinct DNAs comes from thermal denaturation studies on the separated complementary strands. Since each sequence is almost an alternating AT, we predicted that both the heavy and light strands might undergo base pairing with itself in the form of a hairpin. One sequence could pair with itself but with one TT mismatch every five nucleotides. Likewise, the other sequence would pair with itself with a TT mismatch every seven nucleotides. Electron microscopy of the separated strands showed that they were double helical and thermal denaturation of each showed a biphasic melt. These transitions occured 15° and 23° below the T_m for a perfectly paired

SEQUENCES OF HIGHLY REPEATED DNA

SATELLITE	SEQUENCE	COMPOSITION
1.672	A A T A T T T A T A	60%
	A A T A T A T T T A T A T A	40%
1.686	A A T A A C A T A G T T A T T G T A T C	80%
	$\begin{pmatrix} A_5 & T_3 & C_1 & G_1 \\ T_5 & A_3 & G_1 & C_1 \end{pmatrix}$	10%
	$\begin{pmatrix} A_7 & T_1 & C_1 & G_1 \\ T_7 & A_1 & G_1 & C_1 \end{pmatrix}$	10%
1.688	Hae III SITE $\begin{pmatrix} G & G & C & C \\ C & C & G & G \end{pmatrix}$ DISTRIBUTED REGULARLY, OFTEN 365 NUCLEOTIDES APART	
1.705	A A G A G T T C T C	40%
	A A G A A G A G A G T T C T T C T C T C	~25%

Fig. 3. Sequences detected in satellite DNA [5,7,9,10].

d(AT), consistent with two DNA species, one with a mismatch every five base pairs and one with a mismatch every seven.

A third experiment indicates that these sequences are on different molecules. If the A rich strand of this satellite is used as a template for DNA polymerase, pyrimidine tract analysis of the product showed both sequences were copied faithfully and confirmed the RNA sequencing results. If, however, dGTP and dCTP were left out of the DNA polymerase reaction, the amount of product was reduced 50% and pyrimidine tract analysis showed only the five nucleotide repeat was synthesized. We conclude that molecules containing the five nucleotide repeat are highly uniform, but the seven nucleotide repeat may have a rare G or C residue distributed throughout. Base composition indicates G and C are present at the 0.5% level.

Sequence analysis of the 1.686 satellite shows three repeating sequences. When RNA complementary to the heavy strand of this satellite is cleaved with T_1 ribonuclease, three oligonucleotides ten long are produced. The major component has the sequence 5′ AAUAACAUAG 3′ determined by sequential digestion with spleen phosphodiesterase. The two minor sequences have differing base compositions. We do not know if these different sequences are tandemly arranged or if they are present as distinct DNA species.

The 1.688 DNA is the most difficult to purify and sequence analysis has only begun. Manteuil et al. [8] have shown that a particular restriction site (the HaeIII sequence 5′GGCC 3′) is distributed in a regular fashion in this satellite. Work in our laboratory confirms this and preliminary sequence analysis indicates the basic repeating sequence of this DNA is much more complex than the others.

The sequence of the 1.705 satellite has been reported in three labs. The sequence AAGAG has been reported by Sederoff et al. [9] and by Endow et al. [10]. These workers used only ATP and GTP to transcribe complementary RNA. We have used all four triphosphates and have detected a second repeating sequence in the RNA produced. These sequences were analyzed by partial T_1 ribonuclease followed by sequential digestion of the partial products.

It is clear that separation of such closely related DNAs would be difficult, if not impossible, by classical physical techniques. Two repeating sequences such as those present in 1.705 g/cm^3 species have identical base compositions, nearest neighbors and nucleotide strand bias. Moreover, for the study of the arrangement of these sequences in the centric heterochromatin, it is essential to have all sequences separated and in pure form. We have, therefore, made synthetic hybrid bacteria plasmids [11], each containing a single molecule of satellite DNA. These hybrids will allow us to separate each buoyant class into its various molecular components in order to simplify sequencing and to provide hybridization probes to locate each sequence in situ.

Insertion of satellite DNA into bacterial plasmids is not straightforward. Since none of the Drosophila satellites contain sites for R1 endonuclease, one must synthesize complementary terminal sequences on the plasmid and the satellite. This is normally accomplished by shearing the DNA and then treating it with a 5′ exonuclease to expose single-stranded 3′ termini to act as primers for terminal transferase [12]. Treating satellite DNA with a 5′ exonuclease exposes complementary 3′ sequences, and the molecules immediately cyclize and the ends become unavailable for terminal addition. We have, therefore, found conditions for terminal addition to double-stranded DNA. Basically, by lowering the ionic strength of the terminal transferase reaction, we have favored fraying at the end of a helix which permits the enzyme to add dA tails to double-stranded satellite DNA. Similar conditions allow addition of dT tails directly to R1 cleaved plasmid DNA without 5' exonuclease treatment (fig. 4).

We have then annealed the satellite and plasmid DNA in the normal way and transfected bacteria deficient for recombination, restriction and modification (fig. 4). Colonies resistant to tetracycline, a gene carried on the plasmid DNA, were selected. Among these we have further selected colonies containing satellite DNA sequences by hybridization of RNA complementary to the satellite (1.705 in this case) to the plasmid DNA of the bacterial cells [13].

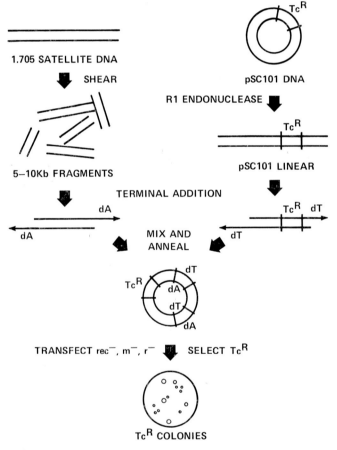

Fig. 4. Scheme for synthesis of hybrid bacterial plasmids containing satellite DNA. Basically this procedure is as described in [11] with the exception of the 5' exonuclease step, which was omitted as described in the text.

Only 30% of all the colonies tested contained 1.705 DNA sequences by this hybridization criteria. This was unexpected since the presence of satellite DNA with A tails was absolutely required to obtain Tc^R colonies. The pSC DNA by itself was not infectious. The second unexpected result was the length of the inserted sequence. DNA of 5–10 kb average length was used and was present in annealed molecules, but the average insert in the plasmids was usually less than 1 kb. Occasionally, individual clones contained plasmid DNA heterogeneous in size. Upon subcloning, stable and homogeneous plasmids are obtained but with small 1 Kb satellite inserts. Based on the instability of large satellite inserts and the stability of shorter satellite regions, we have concluded that intramolecular recombination is occurring within tandemly repeated regions. We have used both recA$^-$ and recB$^-$C$^-$ double mutants. This instability may be due to small amounts of residual recombination in these strains, combined with the traditional instability of tandemly repeated sequences in *E. coli*, or it may be due to a strong selection against the presence of such unusual sequences.

With plasmids which have probably undergone deletion and rearrangement, it is clear that one cannot study long range sequence regularities such as the organization of restriction sites in satellite DNA. However, since each plasmid has derived from a single molecule, we are now in a position to isolate various molecular components of each buoyant class by hybridization. We will also be able to determine the basic repeating units contained within each molecule and to detect other closely related sequences in the satellites which may have escaped our sequence analysis. By using complementary RNA prepared from these plasmids as probes for in situ hybridization, we will test the hypothesis of a chromosome specific location of each of the different repeated sequences [7].

Acknowledgements

This work was supported in part by the National Institutes of Health (GM 21498-01) and the National Foundation March of Dimes.

The sequence work reported in this paper was performed in collaboration with Dr. W.J. Peacock in his laboratory in Canberra, Australia. The work on the plasmids was in association with Dr. Kirk Fry, Timothy Nelson, Marian Carlson and Peggy Hung.

References

[1] M.L. Pardue and J.G. Gall, Chromosomal localization of mouse satellite DNA. Science 168, 1356 (1970).
[2] K.W. Jones, Chromosomal and nuclear location of mouse satellite DNA in individual cells. Nature 225, 912 (1970).

[3] S. Gershenson, Studies on the genetically inert region of the X-chromosome of *Drosophila*. J. Genetics 28, 297 (1933).
[4] W.J. Peacock and G.L.G. Miklos, Meiotic drive in *Drosophila*: New interpretations of segregation distorter and sex chromosome systems. Adv. Genetics 17, 361 (1973).
[5] W.J. Peacock, D. Brutlag, E. Goldring, R. Appels, C.W. Hinton and D.L. Lindsley, The organization of highly repeated DNA sequences in *Drosophila melanogaster* chromosomes. Cold Spring Harbor Symp. Quant. Biol. 38, 405 (1973).
[6] E.S. Goldring, D.L. Brutlag and W.J. Peacock, Arrangement of highly repeated DNA of *Drosophila melanogaster*, in: The Eukaryote Chromosome (W.J. Peacock and R.D. Brock, eds.), Australian National University Press, Canberra, Australia, 1975, p. 47.
[7] D.L. Brutlag and W.J. Peacock, Sequences of highly repeated DNA in *Drosophila melanogaster*, in: The Eukaryote Chromosome (W.J. Peacock and R.D. Brock, eds.), Australian National University Press, Canberra, Australia, 1975, p. 35.
[8] S. Manteuil, D.H. Hamer and C.A. Thomas, Jr., Regular arrangement of restriction sites in *Drosophila* DNA. Cell 5, 413 (1975).
[9] R. Sederoff, L. Lowenstein and H.C. Birnboim, Polypyrimidine segments in *Drosophila melanogaster* DNA: II Chromosome location and nucleotide sequence. Cell 5, 183 (1975).
[10] S.A. Endow, M.L. Polan and J.G. Gall, Satellite DNA sequences of *Drosophila melanogaster*. J. Mol. Biol. 96, 665 (1975).
[11] P.C. Wensink, D.J. Finnegan, J.E. Donelson and D.S. Hogness, A system for mapping DNA sequences in the chromosome of *Drosophila melanogaster*. Cell 3, 315 (1974).
[12] P.E. Lobban and A.D. Kaiser, Enzymatic end to end joining of DNA molecules. J. Mol. Biol. 78, 453 (1973).
[13] M. Grunstein and D.S. Hogness, Colony hybridization: A method for the isolation of cloned DNAs that contain a specific gene. Proc. Nat. Acad. Sci. U.S.A. 72, 3961 (1975).

Chapter 5

ANALYSIS OF SEQUENCE STRUCTURE OF THE RAT GENOME

JUNG-RUNG WU, WILLIAM R. PEARSON, MAHLON WILKES and JAMES BONNER

Division of Biology, California Institute of Technology, Pasadena, California 91125, U.S.A.

One structural approach to the problem of gene regulation in higher organisms is to characterize the DNA sequences which may serve as control elements in gene expression. Studies on the length and organization of repetitive DNA sequences have shown highly ordered patterns of sequence organization which are characterized by a repetitive DNA sequence length of 200 to 400 nucleotides separated by single copy sequences from 500 to 3000 nucleotides in length for more than 65% of the genome [1,2]. This organization was one of the predictions of a model for sequence specific gene expression presented by Britten and Davidson [3–5], which suggested that repetitive DNA sequences coordinately control the expression of adjacent single copy genes.

In this paper, we will summarize the results of some experiments to determine the structural parameters of sequence organization in a mammal, the rat. Two classes of sequences have been examined: the foldback or most rapidly reannealing sequences, and the sequences repeated from 100–100,000 times in the genome. Three types of measurements have been made to characterize these two classes: 1) the length distribution for the class; 2) the separation or interspersion period for the sequence; and 3) the fraction of the genome associated with the class.

These measurements have been made using a variety of experimental techniques. Length measurements have been made using digestion with the single strand nuclease S-1 and visualization in the electron microscope. Interspersion distances have been determined by binding to hydroxyapatite and inferred from electron microscopic data. All measurements give values for the fraction of the genome displaying the particular arrangement. Visualization of structures in the electron microscope has provided additional insight into the real distribution of sequences giving rise to the measurements made by more classical techniques (hydroxyapatite binding and nuclease digestion [6,7]) which offer an important control for the observations obtained from microscopy.

The Molecular Biology of the Mammalian Genetic Apparatus:
edited by P. Ts'o 1977, Elsevier North-Holland Biomedical Press

The renaturation kinetic classes of rat DNA

To provide a basis for this discussion of the rat genome, we include fig. 1, a renaturation curve of rat DNA fragments 350 nucleotides in length, compiled over a number of years by members of the laboratory. The solid lines show the components and sum for a fit of three second order reactions. The actual numbers for the fraction, rate and repetition frequency for each component derived from that fit are shown in table 1. About 60% of the DNA in this curve is non-repeated single copy DNA, while about 10% is repeated about 10 to 100-fold and 20% repeated about 3000 times. Another 10% has renatured by the earliest time on this curve, C_0t 0.01.

The numbers in table 1 reflect the fraction of repetitive and foldback DNA contained on fragments averaging 350 nucleotides in length. The actual numbers for foldback and repetitive sequences must be determined by digestion with

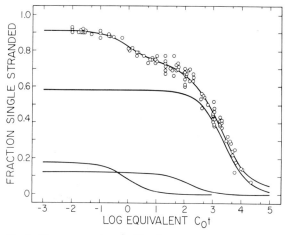

Fig. 1. The renaturation profile of rat DNA fragments 350 nucleotides long assayed with hydroxy-apatite. The upper curve represents the best root mean square (RMS = 2.95%) fit for three combined second order components. The lower curves represent the three separate components of that fit.

TABLE 1

Renaturation kinetic components of rat DNA.

Kinetic component	Fraction of DNA in component	C_0t 1/2 observed	Repetition frequency
1	0.176	0.93	2660
2	0.122	145	17
3	0.581	2460	1.0

TABLE 2

Duplex content of renatured 800 nucleotide fragments.

C_0t	0.05	0.5	5.0	50
Fraction bound to HAP	0.22	0.32	0.30	0.44
		0.28	0.25	0.36
			0.30	0.41
Fraction S-1 nuclease resistant	0.065	0.087	0.217	0.253
	0.060	0.125	0.157	0.203
			0.164	0.204
Fraction of HAP duplexes single-stranded	0.70	0.66	0.45	0.50

nuclease so that single strand contributions are not included. When this measurement is made (table 2), 6% of the DNA is in duplex due to foldbacks at C_0t 0.05, and 20–25% due to foldbacks and repetitive sequences at C_0t 50. Therefore, about 70–75% of rat DNA is unique sequence, 15–20% is repeated 100 to 100,000 times and 6% contains foldback or more highly repetitive sequences.

We will discuss the length and interspersion of the foldback and moderately repetitive DNA (3000-fold on the average) in this paper; no experiments have been done on the slightly (10–100-fold) repetitive DNA sequences and they will be considered as part of the single copy population for the rest of the discussion.

Foldback DNA

The binding of DNA to hydroxyapatite (HAP) after short incubation times is due to foldback and highly repetitive sequences. To study the structure of the foldback sequences, 20 to 30 kb DNA (1 kb = 1000 nucleotides) was alkaline denatured, allowed to renature to C_0t 0.005 and spread from 50% formamide for electron microscopy. Double-stranded foldbacks can be visualized as simple rods, or rods terminated by loops when the complementary sequences are not contiguous. Some examples of these structures are shown in fig. 2. The duplex lengths of the foldback rods and spacing between foldbacks of the same strand were measured. Fig. 3a shows the distribution of foldback lengths; the number average (mean) length of all foldback duplexes is about 700 nucleotides. If the distribution is divided at the length of 1000 nucleotides for the classification of the long and short foldbacks, the shorter class contains 1/3 of the DNA but 90% of the fragments with a number average length of 300 nucleotides.

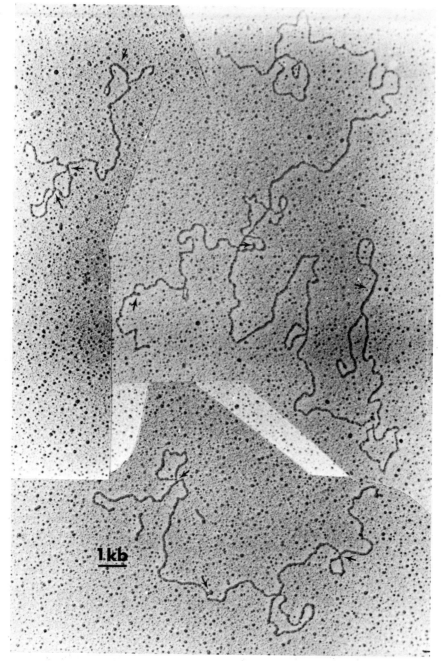

Fig. 2. Multiple foldback duplexes on long single strands. Arrows on the micrograph point to the foldback duplexes. A bar indicates the equivalent length of 1000 nucleotides (1 kb).

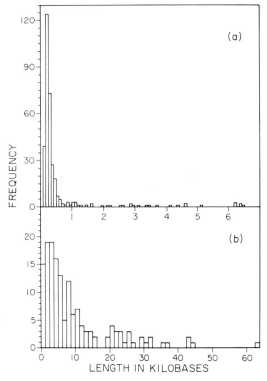

Fig. 3. (a) The distribution of foldback duplex lengths. The measurements are of the duplex length of the foldbacks, not of the combined length of the two complementary sequences. (b) The distribution of spacings between foldback duplexes. Spacing measurements were between the proximal ends of two adjacent foldback duplexes.

The distribution of spacings between foldback pairs on the same molecule is shown in fig. 3b. The average spacing is 9700 nucleotides of single strand DNA with a modal value between 2000 and 3000 nucleotides. Five percent of the rat DNA was found to be in duplex form at C_0t 0.005 by electron microscopy, in good agreement with the figure of 6% determined by digestion with the single- strand specific endonuclease S-1.

Size distribution of repetitive sequences

Nuclease digestion and electron microscopy have also been used to determine the length of the middle repetitive DNA sequences. Two types of measurement have been made using the electron microscope. In one set of experiments, DNA from 1000 to 2500 nucleotides in length was stripped of highly repetitive as well as unique sequences and then allowed to renature with itself. The length

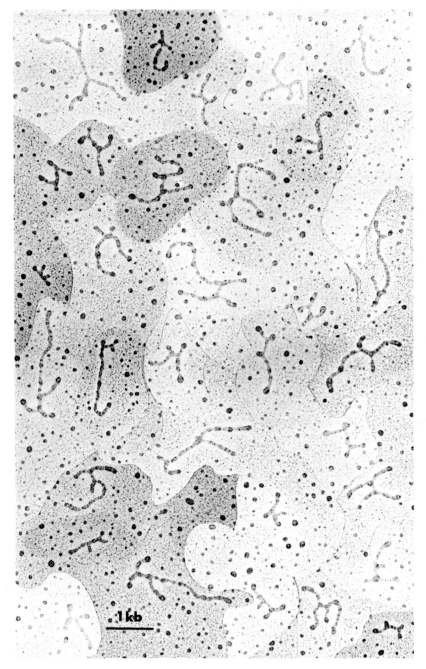

Fig. 4. 'H' structures formed among the repetitive DNA strands 1000 nucleotides long. A marker in the micrograph indicates the equivalent length of 1000 nucleotides (1 kb).

of the repetitive sequences, as shown in fig. 4, in simple structures consisting of a duplex with four single strand tails can be measured and the average is about 250 to 400 nucleotides (fig. 5).

A second type of experiment allows visualization of both the repetitive duplex lengths and the single strand lengths separating the duplexes. In these experiments, DNA fragments 1000 to 2500 nucleotides long containing middle repetitive sequences were used to drive trace amounts of very long (7.5–30 kb) DNA. In these experiments, identification of duplexes is more ambiguous but more information about the spacing between duplex regions can be obtained. The average duplex length determined from this data was again 250 to 400 nucleotides, while the distance between two duplex regions averaged 1400 to 2400 nucleotides.

The length of repetitive sequences has also been estimated by digestion with endonuclease S-1. DNA fragments 3000 to 4000 nucleotides long were renatured to different repetitive $C_o t$s, digested with the nuclease and chromatographed on hydroxyapatite to separate duplexes from nucleotides and single strands, and then separated by size on Biogel A-50. Fig. 6 shows the results of such fractionation. About 40% of the DNA is excluded, indicating a length greater than 2000 nucleotides. The remainder is included with a length of about 300 nucleotides. Apparently there are two size classes of rat repetitive sequences: one short, around 300 nucleotides, and the other much longer, at least 2000 to 3000 nucleotides. This distribution of repetitive sequence lengths has also been reported for a number of other organisms [7,8]. The size fractionation has been carried out at a number of different $C_o t$s with the results shown in table 3. At $C_o t$ 5 and $C_o t$ 50 the repetitive DNA distributions are about the same; 60% of the DNA is included in the column with a length of about 300 nucleotides. At $C_o t$ 0.05 the distribution is different, with less than 40% of the DNA included.

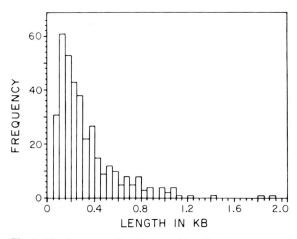

Fig. 5. The frequency distribution of lengths of renatured double-stranded duplexes of repetitive DNA.

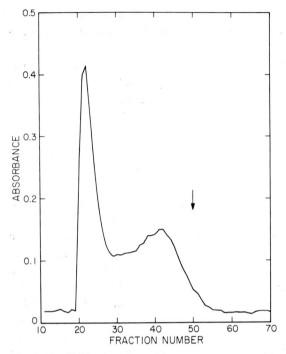

Fig. 6. Size distribution of repetitive sequences separated by Biogel A-50 agarose chromatography. ^{32}P-labelled phosphate is used as an inclusion marker and is indicated with an arrow.

TABLE 3

Size distribution of S-1 nuclease resisted duplex fragments after biogel A-50 chromatography.

$C_o t$	0.05	5.0	50
Excluded fraction	0.683	0.385	0.452
		0.306	0.408
Included fraction	0.317	0.614	0.548
		0.694	0.592

This difference may be due to very rapidly reannealing sequences and is reflected in the longer average length of foldback sequences measured by microscopy (see fig. 3a).

Interspersion of repetitive with single copy sequences

The interspersion of repetitive with single copy sequences in the rat indicated by the microscopy data in the previous section has also been measured by

Analysis of sequence structure of the rat genome

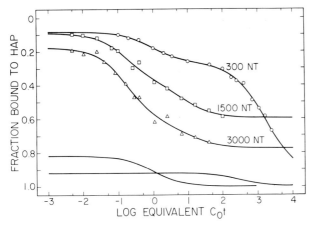

Fig. 7. The self-renaturation profiles of fragments 300, 1500 and 3000 nucleotides long assayed with hydroxyapatite.

chemical techniques. The physical contiguity of repetitive and single copy sequences can be demonstrated by the self-reassociation of successively longer DNA fragments [6]. Fig. 7 shows the self-reannealing of DNA fragments 300, 1500 and 3000 nucleotides in length. While DNA 300 nucleotides in length shows the repetitive and single copy fractions of DNA seen in fig. 1, as the fragment length is increased, the DNA appears to reanneal more rapidly and much of the slowly reannealing fraction is lost. These experiments demonstrate the interspersion of single copy DNA with repetitive DNA in the rat. At C_ot 50, 30% of the fragments 300 nucleotides in length are bound to HAP but 55% of the fragments 1500 nucleotides in length and 70% of the fragments 3000 nucleotides in length are bound. Thus, while only 30% of DNA fragments 300 nucleotides long contain a repetitive sequence, 55% and 70% of strands with lengths of 1500 and 3000 nucleotides contain a repetitive sequence. The increase in the fraction of the DNA bound to HAP is due to single copy tails linked to the bound repetitive sequence.

In order to measure the length of the single copy sequence separating the interspersed repetitive sequences, labeled DNA fragments of various lengths were used as tracers in renaturations with large amounts of sheared whole rat DNA. The excess DNA was renatured with the tracer to C_ot 5 and C_ot 50 and the fraction of the tracer bound to HAP determined. As the length of the DNA tracer is increased, the fraction of the tracer bound to HAP increases due to single strand non-repetitive tails. No more increase in bound DNA is found when the average DNA strand contains two repetitive sequences. At that point, the distance between two repetitive sequences has been found [7].

Fig. 8 shows the results of this experiment. The amount of labeled DNA bound to HAP corrected for binding in the absence of driver (zero-time binding) is

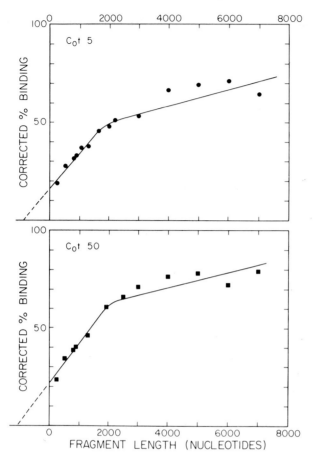

Fig. 8. The fraction of tracer DNA containing repetitive sequence segments bound to HAP at both C_0t 5 and 50 as a function of fragment length.

plotted on the Y-axis, while the length of the labeled tracer is shown on the X-axis. The curves have two important features. First, there is a transition around 2000 nucleotides when the slope of the plotted data changes. This fragment length is the interspersion distance or distance between two repetitive sequences for sequences renaturing at C_0t 5 and C_0t 50. At C_0t 5, 50% of the fragments contain two repeat sequences within 2000 nucleotides, while by C_0t 50, when more of the repetitive sequences have renatured, 65% of fragments 2000 nucleotides long contain two repeats. The continuing but decreased slope past the 2000 fragments length is due to repetitive sequences separated by a longer period.

Similar short period interspersion has been measured in *Xenopus* [6], sea urchin [7] and human DNA [9]. The 2000 nucleotide, short period interspersion distance is also similar to the average length of messenger RNA found in the rat [5].

A second feature of these curves is the fraction of fragments of zero length which would be bound to HAP. This number, the Y-intercept, is the fraction of the genome containing repetitive sequence. At C_0t 5, 16% of the genome contains repetitive duplex, while at C_0t 50, 21% is renatured. These numbers agree with the nuclease digestion studies reported in table 2.

Discussion

The data presented above indicate that at least 65% of rat DNA is organized in a way similar to the classic *'Xenopus'* pattern of repeats of 400 nucleotides in length separated by single copy sequences 750 to 2500 nucleotides in length. This pattern has also been measured in human and sea urchin DNA and can be inferred from qualitative data presented for a survey of organisms [8]. Our electron microscopic measurements will allow us to search for higher order sequence organization which may shed more light on the evolutionary origin and functional significance of this widespread pattern of organization.

The repetitive DNA population displays two distinct size classes with 60% of the repetitive DNA mass 200 to 500 nucleotides in length and the remainder longer than 2000 nucleotides. While it is clear that the short sequences are responsible for the interspersion patterns seen in both the DNA and the nuclear RNA transcribed from it [10–12], the distribution and function of the longer repetitive sequences is unclear. Some of this DNA must contain the known repetitive genes such as ribosomal and histone genes [13,14] and thus must be distinct from the short sequences. The long repetitive DNA may contain other sequences which are shared with short repetitive DNA. Some repetitive sequences may sometimes be bounded by single copy sequences and at other times be bounded by other repetitive sequences as part of a long repetitive sequence. Measurement of the sequence overlap between the long and short repetitive DNA is made more difficult by the ease with which short repeat sequences can be generated by fragmentation of long sequences. If sequence overlap can be unambiguously demonstrated, this may provide evidence for the 'integrator' gene sequences postulated by Britten and Davidson [3].

We have also studied the foldback sequences in the rat genome. Six percent of the DNA is organized in foldback sequences, the majority of which are 300 nucleotides in length. These sequences seem to be uniformly distributed throughout the genome with an average spacing of 10,000 nucleotides, similar to the length of many nuclear RNA transcripts. A large number of foldback sequences can be found spaced 2000–3000 nucleotides apart, however, in common with the spacing of repetitive sequences and the length of messenger RNA in the rat.

These measurements, along with further analysis of sequence organization by electron microscopy and measurements of sequence overlap between long and short repetitive populations, should provide more insight into the structure of highly ordered repetitive DNA and its functional significance.

Acknowledgements

We wish to thank Dr. Francine C. Eden for generously supplying us with endonuclease S-1. This work was supported in part by the United States Public Health Service Grants GM 13762 and GM 20927.

References

[1] J. Bonner, W.T. Garrard, J. Gottesfeld, D.S. Holmes, J.S. Sevall and M. Wilkes, Functional organization of the mammalian genome. Cold Spring Harbor Symp. Quant. Biol. 38, 303 (1974).
[2] E.H. Davidson, G.A. Galau, R.C. Angerer and R.J. Britten, Comparative aspects of DNA sequence organization in metazoa. Chromosoma (Berl.) 51, 253 (1975).
[3] R.J. Britten and E.H. Davidson, Gene regulation for higher cells: a theory. Science 165, 349 (1969).
[4] R.J. Britten and E.H. Davidson, Repetitive and nonrepetitive DNA sequences and a speculation on the origins of evolutionary novelty. Quart. Rev. Biol. 46, 111 (1971).
[5] E.H. Davidson and R.J. Britten, Organization, transcription, and regulation in the animal genome. Quart. Rev. Biol. 48, 565 (1973).
[6] E.H. Davidson, B.R. Hough, C.S. Amenson and R.J. Britten, General interspersion of repetitive with nonrepetitive sequence elements in the DNA of *Xenopus*. J. Mol. Biol. 77, 1 (1973).
[7] D.E. Graham, B.R. Neufeld, E.H. Davidson and R.J. Britten, Interspersion of repetitive and nonrepetitive DNA sequences in the sea urchin genome. Cell 1, 127 (1974).
[8] R.B. Goldberg, W.R. Crain, J.V. Ruderman, G.P. Moore, T.R. Barnett, R.C. Higgins, R.A. Gelfand, G.A. Galau, R.J. Britten and E.H. Davidson, Sequence organization in the genomes of five marine invertebrates. Chromosoma (Berl.) 51, 225 (1975).
[9] C.W. Schmid and P.L. Deininger, Sequence organization of the human genome, Cell 6, 345 (1975).
[10] D.S. Holmes and J. Bonner, Sequence composition of rat nuclear DNA and high molecular weight nuclear RNA. Biochemistry 13, 841 (1974).
[11] M.J. Smith, B.R. Hough, M.E. Chamberlin and E.H. Davidson, Repetitive and non-repetitive sequences in sea urchin heterogeneous nuclear RNA. J. Mol. Biol. 85, 103 (1974).
[12] A. Spradling, S. Penman, M.S. Campo and J.O. Bishop, Repetitious and unique sequences in the heterogeneous nuclear and cytoplasmic messenger RNA of mammalian and insect cells. Cell 3, 23 (1974).
[13] D.D. Brown, P.C. Wensink and E. Jordan, A comparison of the ribosomal DNA's of *Xenopus laevis* and *Xenopus mulleri*: the evolution of tandem genes. J. Mol. Biol. 63, 57 (1972).
[14] M. Birnstiel, J. Telford, E. Weinberg and D. Stafford, Isolation and properties of the genes coding for histone proteins. Proc. Nat. Acad. Sci. 71, 2900 (1974).
[15] M.S. Campo and J.O. Bishop, Two classes of messenger RNA in cultured rat cells: Repetitive sequence transcripts and unique sequence transcripts. J. Mol. Biol. 90, 649 (1974).

Chapter 6

DISPERSAL OF SATELLITE DNA SEQUENCES THROUGHOUT THE MUSKMELON GENOME AND THE NATURE OF FAMILIES OF REPEATED DNA SEQUENCES IN PLANTS

ARNOLD J. BENDICH

Departments of Botany and Genetics, University of Washington, Seattle, Washington 98195, U.S.A.

It is likely that all eukaryotes contain in their nuclei DNA base sequences which are represented many times; this feature of the genome undoubtedly has profound biological significance. Two classes of reiterated sequences which are often considered separately are satellite, or simple-sequence DNA and 'intermediately repetitive' DNA. Though the function of neither is as yet understood, satellite sequences are thought to be tandemly arranged, to be concentrated in centromeric and heterochromatic regions of chromosomes [1,2] and possibly to serve a noncoding function in chromosome folding, pairing and movement [3]. Intermediately repetitive sequences are less highly reiterated, and are in part clustered and in part dispersively scattered throughout the genome [4,5]. Those repeated sequences which are interspersed with non-repetitious DNA have been considered as likely to serve in a gene regulatory capacity [5,6]. The terms 'repeated', 'reiterated', and the like are somewhat misleading since only rarely do the sequences so designated approach complete base sequence repetition. This has not been sufficiently appreciated, and I will return to this important issue later.

From theoretical considerations, Britten and Kohne [7] proposed that families of repeated DNA sequences were created in rare 'saltatory events' (bursts of DNA synthesis) leading to clustered repeats often detectable as satellites in CsCl gradients. With time, the repeats accumulate base changes and are translocated throughout the genome until we observe them today as dispersed and partially mismatched intermediately repetitive sequences. The proposal is an attractive one, but the only observation which can be taken as support for the idea derives from a low level of homology of mouse satellite DNA with mouse main band DNA [8]. This result was not interpreted as support for the above translocation proposal and later efforts failed to confirm the observation [9]. There is apparently no other work which deals with the possible satellite origin for dispersed repeats.

The Molecular Biology of the Mammalian Genetic Apparatus:
edited by P. Ts'o © 1977, Elsevier North-Holland Biomedical Press

One reason for this lack may be that most, if not all, of the characterized animal satellite DNA's have simple sequences with a basic repeat of less than 20 nucleotide pairs (NTP) (discussed in 10), whereas the repeat length of intermediately repeated sequences in many animals was thought to be at least 300 NTP [4]. However, recent work from Hogness' group with cloned repetitive sequences (described elsewhere in this volume) indicates that long repeat lengths in the *Drosophila* genome are composed of short, tandem arrays of members of different families of intermediately repeated sequences. The actual length of the reiterated unit will become important when considering the biological role of functionally repeated sequences.

The work described below concerns both satellite and dispersed repeated DNA sequences in plants. I have studied the major frequency component of DNA from four vascular plants in some detail in order to describe sequence repetition more quantitatively. In addition, sequences complementary to two satellite DNAs from the muskmelon have been detected in its main band DNA. These observations will be used to support the idea that repeats dispersed throughout the genome originated as satellites. In the concluding section of this paper, I shall offer an appraisal of the current direction of repeated sequence research and an assessment of whether this direction may lead us to an understanding of functionally repetitive DNA.

How do we measure DNA base sequence repetition?

Surprisingly little is known about the relationship among the nominally 'repeated' sequences, despite a decade of research since the discovery of the phenomenon by Britten and his coworkers [11]. Let us examine the commonly used operational definition of repeated sequences and the shortcomings of this definition.

Repeated sequences are recognized by the fact that their rate of reassociation is greater than would be anticipated if each sequence were present once per haploid genome. The procedure used to assess base sequence similarity in almost all previous work was to allow single DNA strands to form a duplex structure (reassociate) at 25° below the T_m of native DNA. According to Ullman and McCarthy [12], this permits sequences to reassociate if their average complementarity is at least 85%. They estimate that 0.6% mispaired bases lowers the T_m of reassociated DNA by about 1°. An estimate of 1% mispairing lowering the T_m by 1° has also been made [13]. Consequently, those sequences less than 85% (or 75%) related are classified as 'non-repeated'. Furthermore, most of the previous work employed hydroxyapatite (HA) chromatography to separate single from double stranded structures. Difficulties of interpretation which have been associated with HA-based measurements include the following. 1) Short repeated sequence elements may go undetected. In the usual assay conditions (0.12 M sodium phosphate buffer at 60°) the minimal stable length of duplex can be calculated to be about 20 NTP, but a run of at least 40 NTP is required for quantitative retention by HA [14]. In bacteria and viruses, operator sites and sequences in promoters responsible for the formation of a tight binary complex with RNA polymerase are about 20 NTP or less [15–17]. 2) Unreassociated sequences contiguous to reassociated ones will be retained by HA, altering the shape of the kinetic curve and increasing the apparent extent of reassociation. The effect is striking and is a function of strand length [18–20]. Since we may expect the interspersion pattern of repeated and non-repeated sequences to vary with the organism in question, HA-based species comparisons of percent repetitious DNA are difficult to interpret. 3) At any convenient strand length, unpaired loops and ends of strands may be present in duplex structures which are retained by HA. Thus, we do not know – for most of the organisms reported – what fraction of a duplex structure participates in base-pairing. 4) Low-melting regions in a duplex-containing structure would not elute from HA until the most stable region melts. 5) Molecules eluted from HA at high temperature under certain conditions may not be melted [21] – a possibility not discussed in most reports which do include melting profiles.

Thus, for the eukaryotes investigated to date, we know that there exist families of sequences related by at least 85 (or 75%) complementarity. We know also that in some organisms a larger fraction of the DNA meets this criterion than in others. But we know almost nothing of less closely related families of sequences. Perhaps most important of all, we do not know for any eukaryote what degree of complementarity constitutes a functional repeat (the 'biological criterion' for repeated sequences). Consequently, we do not yet know what fraction of any eukaryotic DNA is functionally repeated.

Homogeneous families of sequences

Two important parameters for assessing the reassociation of DNA strands are the fraction of bases involved in the duplex and the degree of mispairing in the duplex. The first may be determined by measuring the hypochromic effect as strands come together or the hyperchromic effect as they dissociate. The second is determined by thermal stability measurement. In order to avoid some of the difficulties mentioned above, reassociation kinetics were measured at several temperatures in the spectrophotometer.

In our approach to the characterization of repeated sequences, a survey was made of 20 vascular plants. All the plants exhibited DNA which reassociated with kinetics characteristic of intermediately repetitive sequences when assayed at $T_m-25°$ (25° below the T_m of native DNA). Four species were then chosen for further analysis because each lacked (contained less than a few percent) extremely fast sequences but did contain a DNA component whose reassociation followed a single ideal second order curve. I will designate such a component as 'early'. These early components comprised a fraction of total DNA which was sufficiently large to permit the analysis on unfractionated DNA. The presence of a single early component is usually interpreted to indicate a single family of related sequences, or possibly several different families with the same number of members. A family is defined as a group of sequences of sufficient similarity that they may reassociate with one another at a given criterion of stringency set by temperature and salt. The many family members would be of varying similarity since the size of the family decreases with increasing stringency used for reassociation. Such a family will be termed a *heterogenous family*. However, a single early component may also indicate the presence of several or many families of a different type. These families would differ from one another in degree of similarity within their membership, but each family would contain members of the same similarity. In one family the members are all related by, for example, 90% sequence homology. Such a family will be termed a *homogeneous family*. Thus, for homogeneous families, as the stringency is raised (by increasing the reassociation temperature, for example), entire families are operationally moved to the non-repeated sequence class, since their members are no longer similar enough to reassociate at the higher temperature. The size of each family would have to be the same in order to show a single early component. Were the size to differ among the several families, no single early component would be observable. The two alternatives, heterogeneous versus homogeneous families, may be distinguished experimentally as shown below.

For heterogeneous families, the observed rate of reassociation should decrease as the family size decreases. Thus, if the $C_0t_{1/2}$ [7] were, say, equal to 1 for a 60% component at $T_m-25°$, it would increase to 3 at that higher temperature at which the component size becomes 20% of the total DNA. This follows since there are now only 1/3 the number of sequences (family members) which can interact. For homogeneous families, the observed rate of reassociation should remain constant as the families comprised of less well-matched members are eliminated. The $C_0t_{1/2}$ of the 60% component would remain equal to 1 at the higher temperature at which the component size becomes 20% of total DNA. This follows since the size, and therefore the reassociation rate, of each family is independent of the presence of other families with dissimilar sequences. In order to determine whether families are homogeneous or heterogeneous, reassociation kinetics were measured as a function of increasingly stringent conditions from 35° to about 5° below the T_m (fig. 1). Data for the two monocotyledons and two ferns are summarized in fig. 2. For all 4 species the size of the repeated component (to the nearest 5% in all cases) is unchanged between $T_m-35°$ and $T_m-25°$ and then decreases with further increasing stringency. The shape of these curves indicates a

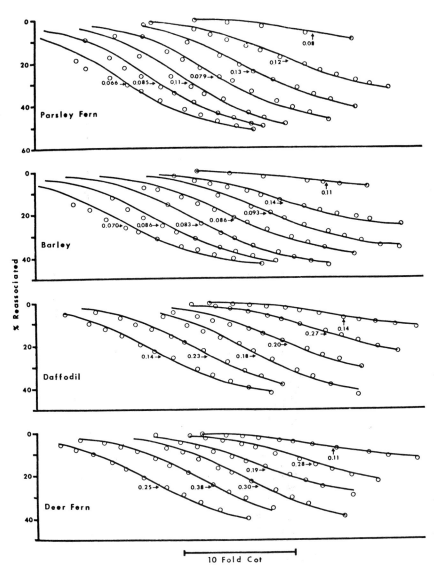

Fig. 1. Reassociation kinetics of plant DNAs. Hypochromicity was measured at from 35.1 to 5.1° (Parsley fern and Barley) or 34.6 to 7.6° (Daffodil and Deer fern) below the T_m in 1 M $NaClO_4$, 0.03 M Tris buffer pH 8.0 with DNA at 77–100 μg/ml using fragments of about 1100 nucleotide pairs. Data have been standardized against *B. subtilis* DNA analyzed simultaneously in each case to account for temperature effects on rate and for thermal damage, and have been corrected for the effect of base mispairing (ΔT_m) on rate [22]. Lines are ideal second order curves for components (to the nearest 5%) best fitting the data points which are arbitrarily displaced laterally. Arrows indicate the $C_0 t_{1/2}$ for each curve. The $C_0 t_{1/2}$ for *B. subtilis* DNA was about 0.5 at $T_m-25°$. With DNA from *Bacillus megaterium*, *B subtilis* and bacteriophages T4 and PS8, the decrease in absorbance at the completion of reassociation approaches only 90–91% of the hyperchromic effect of melted DNA. Therefore, 90% hypochromicity was taken as the value for complete reassociation of DNA. All data for each plant were obtained by serially melting and reassociating a single sample in a Gilford 2400 recording spectrophotometer equipped with the thermal programmer and reference compensator accessories.

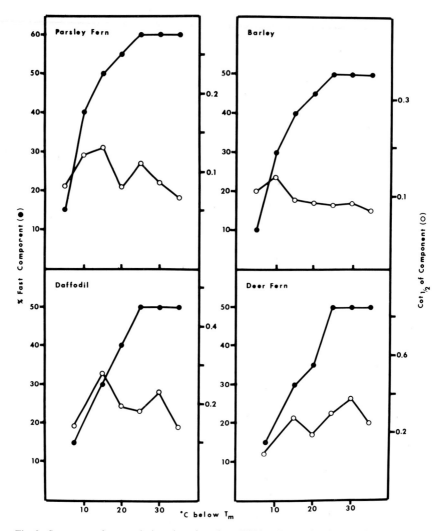

Fig. 2. Summary of reassociation data for plant DNAs. From the data in fig. 1, the size of the second order component (to the nearest 5%) and the $C_0t_{1/2}$ of each component are plotted against degrees below the T_m for reassociation.

distribution of relatedness classes among the repeats which is continuous after the plateau. The $C_0t_{1/2}$ does not increase with decreasing repeated component size; rates have been corrected for the effect of base mispairing [22]. With parsley fern, for example, the heterogeneous family alternative predicts an increase in $C_0t_{1/2}$ of 60/15 = 4-fold as the component size is reduced from 60% to 15% by increased stringency. The fold increases predicted for barley, daffodil and deer fern would be 5, 3.3 and 3.3, respectively. The homogeneous families alternative predicts a constant $C_0t_{1/2}$ regardless of component size. Since for none of the

4 plants tested does the $C_0t_{1/2}$ increase appreciably, the data favor this alternative.

Data consistent with the homogeneous family alternative may be found in the classic paper of Britten and Kohne [7]. Reassociated repetitious salmon DNA was fractionated with HA on the basis of its thermal stability. Both low and high melting duplexes reformed those duplexes which were predominantly low and high melting, respectively, upon a second cycle of reassociation. These and other experiments with calf DNA indicated little sequence homology between the well-matched and less well-matched sets of repetitive DNA. However, such measurements could not themselves lead to the homogeneous family alternative because a single frequency class of sequences was not used in that work.

The data in figs. 1 and 2 do not rule out the presence of a minor fraction of DNA in heterogeneous families which could cause a small increase in $C_0t_{1/2}$ for the lower temperature range. However, the data do indicate that most if not all the sequences which reassociated at high temperature did not interact with more distantly related sequences at low temperature. Most or all of the families containing poorly matched sequences cease to exist at high temperature; their DNA is now recognized as 'non-repeated'.

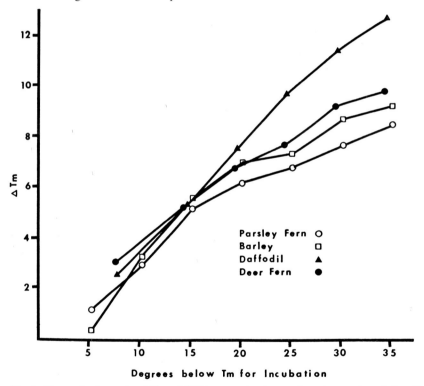

Fig. 3. Change in thermal stability of DNA reassociated at various temperatures below the T_m of native DNA. The difference between the T_m of native and reassociated DNA is plotted against the difference between the native T_m and the incubation temperature for reassociation.

When DNA devoid of partially related base sequences (such as viral or bacterial DNA) is melted, reassociated and remelted, the difference (ΔT_m) between the first and second T_m is small. This, of course, is evidence for the absence of higher organism-type 'repeated' sequences. Fig. 3 shows the ΔT_m as a function of increasing stringency of the conditions used in reassociation; this represents a relatedness profile for the various homogeneous families. The shape of the four curves is similar. This indicates a similar, continuous gradation in the degree of relatedness among the various families of repeated sequences in the four plants.

It should appear obvious that a more complete understanding of the relationships among related sequences may be achieved when measurements are made at several temperatures rather than at a single temperature.

Muskmelon satellite DNA sequences interspersed with main band DNA

Muskmelon DNA contains two satellites which have been characterized [10,23]. Satellites I and II represent about 10 and 20% of total DNA and have kinetic complexities of about 560 NTP and 1.7×10^6 NTP, respectively. Although the two satellites band at the same density in neutral CsCl, I melts about 8° higher than II and is not linked to satellite II. Satellite I occurs in blocks of at least 67 tandemly repeating units of the 560 NTP sequence. The 1.8×10^5 and 72 copies per haploid genome of satellites I and II, respectively, are essentially identical replicas of each other, for the ΔT_m of each is near 0°.

Muskmelon DNA with main band density contains short sequences that are homologous to satellite DNA [23]. This conclusion was reached from experiments in which denatured satellite DNA strands were incorporated into hyperpolymeric structures called networks [11,24] or aggregates [25,26]. Britten first described the aggregation phenomenon with mouse DNA and suggested that in order to form such networks, there probably is a requirement for an average of more than two regions per strand that are capable of pairing with complementary regions on other strands [11]. When aggregates are formed with eukaryotic DNA incubated to low C_0t, it seems likely that the interacting strands contain an average of two or more repeated sequence elements. Fig. 4 shows an electron micrograph of a pea DNA aggregate. Both single and double stranded regions on interlocking strands can be seen. The entire aggregate contains more than 1000 strands.

Tritium-labeled satellites I and II were separated in an Ag^+-Cs_2SO_4 gradient [23] and then sheared to about 1100 NTP. A trace of each satellite was incubated with excess long main band DNA and aggregation assays were performed (table 1). We have shown that the amount of satellite-density material in the main band preparation used here was < 1%, a level of contamination which had no significant effect on the amount of ^3H in the aggregate pellet [23]. After incubation to a

Dispersal of satellite DNA sequences 71

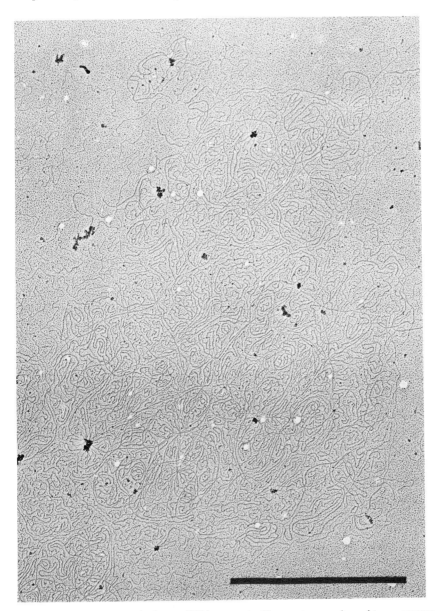

Fig. 4. Electron micrograph of a pea DNA aggregate. Fragments were sheared to an average length of 4900 nucleotide pairs, reassociated at 25° below the T_m in 1 M NaClO$_4$, 0.03 M Tris buffer pH 8.0, to a C_0t of 1.8, and spread in 60% formamide by a modified Kleinschmidt method [35]. The bar equals 4900 nucleotide pairs (1 micron).

C_0t of 32, 36% of Satellite I, and only 9% of Satellite II, is found in the main band aggregate. Upon more extensive incubation, the amount of Satellite II in the aggregate approaches that for I. The more rapid aggregation for I is consistent

TABLE 1

Aggregation of sheared ^{3}H-DNA with excess long DNA.

	Long DNA	^{3}H-DNA	%A_{260} in Agg	%^{3}H in Agg
Low C_0t	Main band	I	41, 40	37, 35
	Calf	I	55, 55	3, 3
	Main band	II	40, 40	9, 8
	Calf	II	50, 53	2, 2
	Main band	PS8	46, 50	1, 1
High C_0t	Main band	I	48, 56	55, 55
	Calf	I	70, 69	6, 5
	Main band	II	53, 54	30, 32
	Calf	II	70, 71	5, 5

Main band DNA was incubated to C_0t of 32 or equivalent C_0t of 310. Calf DNA was incubated to C_0t of 46 or equivalent C_0t of 440.

Tritium-labeled DNA (0.7–0.8 ng, 300–380 cpm for Satellites I and II; 4.6 ng, 640 cpm for PS8) mixed with excess unlabeled DNA (1.5 µg main band or 2.1 µg calf) was denatured at 103° for 3 min. and incubated in 0.2 M NaCl, 5 mM Tris pH 8 at 56° (low C_0t) or in 1 M NaClO$_4$, 0.03 M Tris pH 8 at 61° (high, equivalent C_0t). After incubation, the 10 µl samples were quickly cooled to 0°, diluted to 0.2 ml with incubation buffer, reheated for 5 min. at incubation temperature, again quickly cooled at 0° and centrifuged at 4° for 30 min. at 27,000 g. After removing the supernatant, the aggregate pellet was resuspended in 0.195 ml incubation buffer and both fractions were again heated to 100° and quickly cooled to 0° before absorbance and radioactivity were determined. The C_0t and equivalent C_0t values for ^{3}H-labeled Satellites I and II were 0.017, 0.018 and 0.14, 0.17, respectively; that for ^{3}H-PS8 was 0.10. Sheared and long DNA were about 1100 and 4900 nucleotide pairs, respectively. Satellite and main band preparations were obtained by rebanding the dense and light portions from the satellite and main band peaks in CsCl, respectively. Satellite I and II were then resolved in a Ag$^+$-Cs$_2$SO$_4$ gradient [10].

with its lower kinetic complexity. Neither Satellite I nor Satellite II join the calf DNA aggregate; DNA of bacteriophage PS8 is not incorporated into the muskmelon DNA aggregate. Thus, short sequences complementary to both satellites are interspersed in the main band of muskmelon DNA. These sequences must be distributed such that they comprise a minor fraction of any given 4900-nucleotide main band fragment. Any main band fragment containing more than a minor fraction of satellite sequences would be found in the dense side of the main band peak which was discarded. The most likely situation is that a short stretch of Satellite I sequence is present on some 4900-nucleotide main band fragments, which contain at least two repeated sequence elements unrelated to Satellite I; this situation also applies to Satellite II. These conclusions were previously reached for Satellite I using a mixture of I and II (not resolved in Ag$^+$-Cs$_2$SO$_4$) with main band DNA incubated to C_0t 58 where most of the ^{3}H in the aggregate would be from Satellite I [23]. We estimated that up to 1% of aggregated main band DNA (half the DNA in that case) was homologous to Satellite I. This represented the equivalent of some 5000 units of the 560 NTP sequence, or 3% of Satellite I. Roughly the same degree of dispersal for Satellite II can be assumed since the amount of II in the aggregate approaches that of I (table 1). Since the specific radioactivity of satellite DNA in the present study is much greater than in the former [23], the fraction of main band DNA homologous to satellite DNA is even further below the saturation level than in the former study, and thus we use the indirect estimate for Satellite II.

The evolutionary history of homogeneous families

For four plants surveyed here (fig. 1) there is probably more DNA which reassociates at $T_m-25°$ with ideal second order kinetics in the single early component than exists in the entire mammalian genome (unpublished results). The single early component implies that one-half or more of the genome is composed of sequences which are each represented the same number of times. If some sequences were represented more frequently than others, no single component would be observed. How did this come about? Let us consider an event in antiqiuty in which all DNA in the component arose from one sequence in a single saltatory replication or in saltations occurring over an evolutionarily short period of time. The length of the sequence taken for replication would be its kinetic complexity. The number of families just after saltation would be one. However, the number of families could be considered as many, if the sequence were subdivided many times, either conceptually or physically, as by translocation. In a recent example [5], the 'fast' component of DNA from the mollusc *Aplysia californica* analyzed at $T_m-25°$ has a kinetic complexity of 5×10^4 NTP, exists as dispersed sequences 300 NTP in length, and could be considered to contain $(5 \times 10^4) \div 300 = 170$ families. Dispersal per se does not affect intrinsic complexity. It is likely, as Britten and Kohne have reasoned [7], that the nucleotide sequences of the members of the families are not conserved by severe selection. The members of all families may then change slowly and independently of each other, leading, after a long period of time, to families with divergent members such as are observed. This process should result in heterogeneous families comprising the single early component. But the data indicate homogeneous families. To be consistent with the data and retain the idea of a single saltation, we must invoke an unusual selection. The thousands of members of each family must be restricted to a specific rate of change and members of different families must change at different and widely varying rates in order to generate homogeneous families. Furthermore, the changes must not alter the constant number of members in each family. Clearly, this is an unlikely possibility and we should abandon the concept that a single event created the homogeneous families we observe.

Let us now assume a constant rate of accumulation of base changes in members of all families. The families composed of closely related members would then be identified as recently created families, while those with distantly related members would be older. Thus, the genesis of a set of homogeneous families is most likely an extended process involving many minor saltatory events over a long period of time. The length of sequence taken for replication could be variable and would be much shorter than the kinetic complexity measured for the single early component. This measurement represents the sum of the lengths of all sequences which have been copied the same number of times during the series of saltatory events. Therefore, we can only place an approximate upper limit on the total length of sequence taken for replication in the most recent saltations by using data in fig. 2 at the most stringent criteria. The $C_0t_{1/2}$ for Parsley Fern is about 0.1 and the size of the component is 15% at $T_m-5°$. If not diluted in total DNA, its $C_0t_{1/2}$ would be $0.1 \times 0.15 = 0.015$. Within a factor of 2, the corrected $C_0t_{1/2}$ values for the other species in fig. 2 are the same. The DNA of *Bacillus subtilis* has a $C_0t_{1/2}$ of about 0.5 under the same conditions and has a kinetic complexity of 2×10^9 daltons or 3×10^6 NTP [27]. A $C_0t_{1/2}$ of 0.015 would then represent a kinetic complexity of about 10^5 NTP. If the basic length of a family member in these plants is about 300 NTP, as has been estimated for the repeat length interspersed with single-copy DNA in many animals [4], the plant DNA components observed at high stringency would each contain about 300 families. However, the 300 NTP estimate for animals was made from data obtained at only one condition of stringency: about $T_m-25°$. It is possible that a shorter repeat length may be observed at higher stringency. The fact that the length of the parental sequence of a family is not known precludes an accurate estimate of the number of families or saltations in the plants.

I suggest that the homogeneous families are the result of saltatory replications occurring at many different times with a constant rate of sequence change in all families. In order to account for the single early component we observe, one of two requirements must be met. Either the parental sequence was copied the same number of times in each saltation, or family members were discarded subsequent to their synthesis until a constant family size remained. It is not obvious why such a sequence amplification to a constant level should exist. Nor is it easy to formulate a mechanism by which the constancy is maintained. Nevertheless, a phenomenon termed 'gene compensation' has been described which appears relevant to the question of constancy of sequence amplification. In *Drosophila* the number of genes for ribosomal RNA [28] and for 5S RNA [29] is sometimes returned to a wild-type level after individual flies with various deletions for some of these genes are crossed. Perhaps it is not unreasonable to accept a controlled amplification of sequences whose function we do not understand.

In my view the creation of each of the two muskmelon satellites represented the initial step in a process which leads to a family whose sequences are scattered throughout the genome. That the satellites were created recently can be inferred from the fact that the ΔT_m for each is near $0°$. The short sequences of main band density which are homologous to the satellites represent the first translocation events which will ultimately convert each satellite into a single family of dispersed repetitive sequences. As the families grow older, their member sequences will gradually diverge from one another. It is the families composed of dispersed and divergent member sequences in animals that have been extensively studied as possible regulatory elements for the control of gene expression [4,6]. I shall consider the regulatory potential of such sequences in the concluding section.

On the value of sequence repetition for the investigator and for the cell

In this section I will consider the recent direction of inquiry concerning sequence repetition, and whether it may lead us to an understanding of the biological role of repetitive DNA.

Let us assume for the moment that there are DNA sequences – other than those which specify the stable RNAs – that can function in concert because of their similarity and that such sequences are needed for orderly functioning of the organism. What degree of sequence homology would the cell require for such functional equivalence? Would the cell consider two stretches of DNA to be unrelated if they differ in sequence by 10%? By 1%? I have seen no discussion of this issue in the literature which deals with sequence repetition in eukaryotes. Investigators in this area usually set an operational definition for repetition by choosing $T_m-25°$ as the single criterion for their reassociation experiments, although there have been some notable exceptions [30–34]. But we do not know what level of discrimination is used by any eukaryote to distinguish repeat from non-repeat. It would therefore be fortuitous if the properties of sequences classified as repetitious at $T_m-25°$ were to be representative of functionally repetitious DNA.

The data indicate that several to many homogeneous families comprise the single rapidly reassociating components in the plants studied here. Which of the families might be utilized for function? Since there is such a large proportion and amount of DNA in these families, it seems unlikely that all this DNA would function in a sequence-dependent manner. The families composed of more divergent members might be useful, but those with closely similar members should afford the advantage of more precise control over a postulated network of functions. I will assume for this argument that to be considered as reiterated by a cell, sequences must be nearly perfect replicas. The short, dispersed sequences homologous to the muskmelon satellites might be examples of such sequences. Just how similar the sequences need be is conjectural, but there are examples in prokaryotic systems in which a single base change can alter the recognition of a regulatory protein for its binding sequence [15,16]. In accordance with my assumption, nearly all the families would be unsuitable for purposes which require functional repetition. Perhaps only some of the sequences which reassociated at $T_m-5°$ $(0-1° \Delta T_m)$ would be sufficiently similar. Data for many organisms have been gathered from reassociation experiments conducted at $T_m-25°$, the criterion of stringency adopted as standard by most investigators. These data have led to estimates of the length, spacing, number and complexity of DNA sequences which appear to be repetitious at the standard criterion, as well as to assignments of the fraction of total DNA in such sequences. However, these properties and the generalizations derived from them may not be characteristic of functionally repetitious sequences. If functional repeats exist as a small number of nearly perfect replicas, and if their properties are different from those for the much larger amount of DNA which reassociates rapidly at the standard criterion, then the properties of functional repeats have been *obscured* by those of sequences regarded as reiterated by the investigator but not by the cell. The current of research may, in fact, be carrying us away from our goal of understanding the biological utility of repeated DNA sequences.

At present the manner in which families of repeated sequences are important to the phenotype remains elusive. A solution to this central problem will require a more quantitative description of sequence repetition than the one usually provided. For example, properties such as length, spacing and distribution should be derived from measurements employing several criteria, rather than a single criterion. We should determine whether a single kinetic component is comprised of heterogeneous or homogeneous families. After we obtain such information for a variety of organisms, we will be better able to formulate generalizations concerning sequence organization. Without this basic analysis it seems unlikely that we will understand the putative role of repeated sequences in gene regulation, chromosome structure, evolution or any other process.

Summary

Reassociation kinetics were measured at several temperatures for DNA from barley, daffodil, deer fern and parsley fern. The data indicate that several to many families of related DNA base sequences comprise half or more of the genome in these plants. The various families are not related to one another. They are *homogeneous* because each family contains member sequences related by the same degree of similarity. In one family the members would all be related by, for example, 95% sequence homology. Sequences homologous to each of two satellite DNAs from muskmelon have been detected in its main band DNA. The observations provide the first experimental evidence to support the idea that families of repeated sequences are created as clustered repeats which are subsequently dispersed throughout the genome. Thus, sequences recognized as 'intermediately repetitive' may exist in homogeneous families which originated as satellites or as cryptic satellites.

The question is considered: are sequences designated by investigators as 'repeated', in fact, accepted by the cell as repetitious in some functional sense? I conclude that before we can understand the functional and phenotypic significance of repeated sequences, we may need to ascertain the physical parameters of sequence organization by methods different from those in common use.

Acknowledgements

I thank M.-D. Chilton and W.C. Taylor for their helpful consultation, W.C. Taylor for the electron microscopy and Aaron Bendich for fatherly advice in the preparation of the manuscript. This work was supported by a grant from the National Science Foundation.

References

[1] M.L. Pardue and J.G. Gall, Chromosomal localization of mouse satellite DNA. Science 168, 1356 (1970).
[2] K.W. Jones, Chromosomal and nuclear location of mouse satellite DNA in individual cells. Nature, 225, 912 (1970).
[3] P.M.B. Walker, W.F. Flamm and A. McLaren, Highly repetitive DNA in rodents, in: Handbook of Molecular Cytology (A. Lima-de-Faria, ed.) N. Holland, Amsterdam, 1969, p. 53.
[4] E.H. Davidson, G.G. Galau, R.C. Angerer and R.J. Britten, Comparative aspects of DNA organization in metazoa. Chromosoma 51, 253 (1975).
[5] R.C. Angerer, E.H. Davidson and R.J. Britten, DNA sequence organization in the mollusc *Aplysia californica*. Cell 6, 29 (1975).
[6] R.J. Britten and E.H. Davidson, Gene regulation for higher cells: a theory. Science 165, 349 (1969).
[7] R.J. Britten and D.E. Kohne, Repeated sequences in DNA. Science 161, 529 (1968).
[8] W.G. Flamm, P.M.B. Walker and M. McCallum, Some properties of the single strands isolated from the DNA of the nuclear satellite of the mouse (*Mus musculus*). J. Mol. Biol. 40, 423 (1969).
[9] T.R. Cech, A. Rosenfeld and J.E. Hearst, Characterization of the most rapidly renaturing sequences in mouse main-band DNA. J. Mol. Biol. 81, 299 (1973).
[10] A.J. Bendich and R.S. Anderson, Novel properties of satellite DNA from muskmelon. Proc. Nat. Acad. Sci. U.S. 71, 1511 (1974).
[11] E.T. Bolton, R.J. Britten, D.B. Cowie, R.B. Roberts, P. Szafranski and M.J. Waring, 'Renaturation' of the DNA of higher organisms, in: Carnegie Inst. Wash. Yrbk. 64, 1965, p. 316.
[12] J.S. Ullman and B.J. McCarthy, The relationship between mismatched base pairs and the thermal stability of DNA duplexes II: effects of deamination of cytosine. Biochim. Biophys. Acta 294, 416 (1973).
[13] T.I. Bonner, D.J. Brenner, B.R. Neufeld and R.J. Britten, Reduction in the rate of DNA reassociation by sequence divergence. J. Mol. Biol. 81, 123 (1973).
[14] D.A. Wilson and C.A. Thomas, Jr., Hydroxyapatite chromatography of short double-helical DNA. Biochim. Biophys. Acta 331, 333 (1973).

[15] W. Gilbert, N. Maizels, and A. Maxam, Sequences of controlling regions of the lactose operon. Cold Spring Harbor Symp. Quant. Biol. 38, 845 (1974).
[16] T. Maniatis, M. Ptashne, K. Backman, D. Kleid, S. Flashman, A. Jeffrey, and R. Maurer, Recognition sequences of repressor and polymerase in the operators of bacteriophage lambda. Cell 5, 109 (1975).
[17] D. Prinbow, Nucleotide sequence of an RNA polymerase binding site at an early T7 promotor. Proc. Nat. Acid. Sci. U.S. 72, 784 (1975).
[18] L.M. Grouse, M. Chilton and B.J. McCarthy, Hybridization of ribonucleic acid with unique sequences of mouse deoxyribonucleic acid. Biochemistry 11, 798 (1972).
[19] E. Pays and A. Ronsse, Interspersion of repetitive sequences in rat liver DNA. Biochem. Biophys. Res. Commun. 62, 862 (1975).
[20] R.B. Goldberg, W.R. Crain, J.V. Ruderman, G.P. Moore, T.R. Barnett, R.C. Higgins, R.A. Gelfand, G.A. Galau, R.J. Britten and E.H. Davidson, DNA sequence organization in the genomes of five marine invertebrates. Chromosoma 51, 225 (1975).
[21] H.G. Martinson, E.B. Wagenaar, Thermal elution chromatography and the resolution of nucleic acids on hydroxyapatite. Anal. Biochem. 61, 144 (1974).
[22] J.L. Marsh and B.J. McCarthy, Effect of reaction conditions on the reassociation of divergent deoxyribonucleic acid sequences. Biochemistry 13, 3382 (1974).
[23] A.J. Bendich and W.C. Taylor, Sequence arrangement in satellite DNA from the muskmelon. Plant Physiol., in press.
[24] M.J. Waring and R.J. Britten, Nucleotide sequence repetition: a rapidly reassociating fraction of mouse DNA. Science 154, 791 (1966).
[25] A.J. Bendich and E.T. Bolton, Relatedness among plants as measured by the DNA-agar technique. Plant Physiol. 42, 959 (1967).
[26] W.F. Thompson, Aggregate formation from short fragments of plant DNA. Plant Physiol, 57, 617 (1976).
[27] M. Gilles and J. DeLey, Determination of the molecular complexity of double-stranded phage genome DNA from initial renaturation rates. The effect of DNA base composition. J. Mol. Biol. 98, 447 (1975).
[28] K.D. Tartof, Increasing the multiplicity of ribosomal RNA genes in *Drosophila melanogaster*. Science 171, 294 (1971).
[29] J.D. Procunier and K.D. Tartof, Genetic analysis of the 5S RNA genes in *Drosophila melanogaster*. Genetics 81, 515 (1975).
[30] P.M.B. Walker and A. McLaren, Specific duplex formation in vitro of mammalian DNA. J. Mol. Biol. 12, 394 (1965).
[31] M.A. Martin and B.H. Hoyer, Thermal stabilities and species specificities of reannealed animal deoxyribonucleic acids. Biochemistry 5, 2706 (1966).
[32] A.J. Bendich and B.J. McCarthy, DNA comparisons among barley, oats, rye, and wheat. Genetics 65, 545 (1970).
[33] B.J. McCarthy and M.N. Farquhar, The rate of change of DNA in evolution, in: Evolution of Genetic Systems (H.H. Smith, ed.), Gordon and Breach, New York, 1972, p. 1.
[34] J.L. Marsh, Characterization of repeated DNA sequences in eukaryotic genomes, Ph. D. thesis, Univ. of Washington, Seattle, 1974.
[35] R.W. Davis, M. Simon and N. Davidson, Electron microscope heteroduplex methods for mapping regions of base sequence homology in nucleic acids, in: Methods in Enzymology, Vol. 21D (L. Grossman and K. Moldave, eds.), Academic Press, New York, 1971, p. 413.

Chapter 7

THE RIBOSOMAL DNA OF *TETRAHYMENA*

JOSEPH G. GALL, KATHLEEN KARRER and MENG-CHAO YAO

Department of Biology, Yale University, New Haven, Conn., U.S.A.

The genes coding for ribosomal RNA have been studied extensively in several organisms. Typically, several hundred to thousands of rRNA genes are present in the haploid genome. These copies are serially repeated at one or a few nucleolar organizer loci. Exceptions to the chromosomal location of rDNA are known (reviewed in 1), the best studied case being that of the oocyte nucleus of Amphibia. In the toad, *Xenopus*, up to 2 million extrachromosomal copies of the rRNA genes are located in approximately 1000 free nucleoli just inside the nuclear envelope. Recently the rDNA in the macronucleus of the protozoan, *Tetrahymena pyriformis* has been shown to be extrachromosomal and the same is true for two other protists, the slime mold, *Physarum* [2,3] and the alga, *Acetabularia* [4]. We shall review briefly some pertinent characteristics of the rDNA in *Tetrahymena*.

Extrachromosomal location

Among the first observations to suggest independent behavior of the rDNA in *Tetrahymena* were the autoradiographic studies of Charret [5]. She showed that [^{3}H]thymidine can be incorporated into the multiple nucleoli of the macronucleus at a time when no other nuclear DNA synthesis is under way. That Charret's 'nucleolar DNA' was indeed rDNA was shown by Engberg and co-workers [6,7,8]. They found that starved organisms, upon refeeding, preferentially synthesized rDNA and mitochondrial DNA for up to 90 min. before generalized DNA synthesis resumed. They identified the rDNA by its buoyant density in CsCl and by its specific hybridization to rRNA. They also centrifuged newly synthesized rDNA from starved and refed organisms and showed that it sedimented more slowly on a sucrose gradient than the bulk macronuclear DNA. Gall [9] showed by electron microscopy that the low molecular weight rDNA

The Molecular Biology of the Mammalian Genetic Apparatus:
edited by P. Ts'o © 1977, Elsevier North-Holland Biomedical Press

consisted of a population of identical or nearly identical molecules. In the amicronucleate strain GL about 90% of the macronuclear rDNA consisted of linear molecules 12.6×10^6 daltons in mass. Most of the remaining 10% consisted of circles of the same size along with a small fraction of dimers and more complex molecules (e.g. lariats). Partial denaturation showed a common pattern among the linear molecules, implying that they all might have the same sequence. Karrer and Gall [10,11] found that the rDNA from another strain (syngen 1) was similar, although the molecular weight was slightly higher (13.5×10^6 daltons) and the denaturation pattern was different.

The uniform and small size of the isolated rDNA molecules, their cytological localization in multiple nucleoli beneath the nuclear envelope, and their independent replication, during both the normal vegetative cycle and after starvation, imply that the macronuclear rDNA of *Tetrahymena* is extrachromosomal.

The bulk of the macronuclear DNA exists as larger molecules [7,8,12], but very little else is known about its physical organization. It does not consist of 'gene-sized' pieces as in the hypotrich ciliate *Stylonychia* [13], but whether there are chromosome-sized pieces of DNA in the *Tetrahymena* macronucleus remains to be determined.

Detailed structure of the rDNA

Detailed physical maps of the rDNA are being developed using information from electron microscopy, partial denaturation, enzymatic cleavage with restriction endonucleases, and RNA-DNA hybridization. The most remarkable feature so far revealed is the palindromic nature of the rDNA molecule; that is, the center of the molecule is a point of rotational symmetry: the right and left halves are identical. Thus, the two complementary polynucleotide strands of the molecule are identical, and each is self-complementary about its center (fig. 1). These relationships were first suggested by the symmetrical partial denaturation map and were conclusively demonstrated by the behavior of the rDNA upon melting [10,11]. When rDNA was heated to 100°C under low salt conditions, cooled rapidly, and then examined by electron microscopy, double stranded molecules were found having half the length of the native rDNA. These half-length molecules were the products of 'snapback' renaturation. This was shown

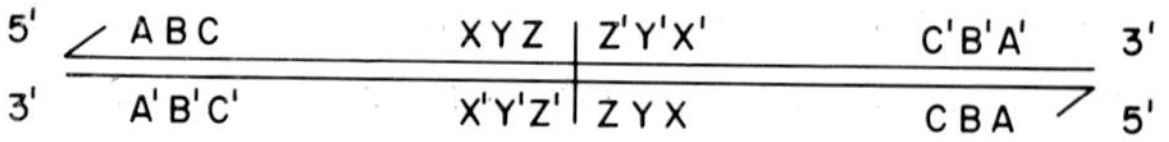

```
5'   A B C          X Y Z | Z'Y'X'        C'B'A'    3'
3'     A'B'C'      X'Y'Z' | Z Y X          C B A    5'
```

Fig. 1. Diagram of the structure of a linear rDNA molecule from *Tetrahymena*. The molecule has an axis of rotational symmetry: the two polynucleotide strands are identical and each is self-complementary about its center.

by spreading the molecules for electron microscopy under partially denaturing conditions, whereupon one 'end' of the molecule opened into a loop. The loop was an easily denaturable, high A + T region originally at the center of the native molecule (fig. 2). Various other structures were generated from the palindromic molecule, including spectacular cruciforms (figs. 2 and 3).

Additional evidence for symmetry came from analysis of the fragments produced by restriction enzyme digestion. If the rDNA is a perfect palindrome, any cuts by specific endonucleases should occur in pairs symmetrically disposed about the center of the molecule. These relationships have been demonstrated for the enzymes *Eco* RI and *Bam* on rDNA of syngen 1 and for *Eco* RI on strain GL [11,14,15].

The coding regions of the rDNA have been studied by RNA-DNA hybridization. In an earlier report [9] it was concluded that each rDNA molecule hybridized with only one 17S and one 25S rRNA molecule. We now believe that estimate to be in error, because the rDNA attached to the filters in the hybridization assay was partially renatured by snapback. The most recent hybridization data, obtained by immobilizing rRNA and measuring the amount of rDNA with which it can be hybridized, indicate two coding regions per molecule for each rRNA species. Furthermore, by analyzing the hybridization capacity of the *Eco* RI fragments of rDNA from strain GL, Engberg et al. [15] have established the order of the genes: the two 17S sequences are near the center of the molecule and the two 25S regions are more distal. The existence of spacers can be inferred from the relative sizes of the rDNA and rRNA molecules, but their exact distribution is still under study. For instance, the two 17S coding regions in the center of the molecule could be immediately adjacent, or they could be separated by a short spacer [11,15]. The hybridization data indicate that the distal segments are non-coding, and preliminary sequencing information

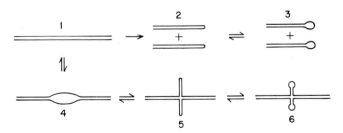

Fig. 2. Various configurations of the rDNA of *Tetrahymena*. The native molecule (1) upon denaturation yields two single strands, which immediately 'snap-back' upon themselves because of self-complementarity (2). Partial denaturation of the native molecule reveals a high A + T region in the middle (4). If this region is allowed to renature, the molecule may return to its original state or it may yield a cruciform (5). Partial denaturation of the 'snap back' molecules and the cruciforms yields molecules (3 and 6) with loops corresponding to the high A + T region. A molecule in configuration 6 is shown in fig. 3.

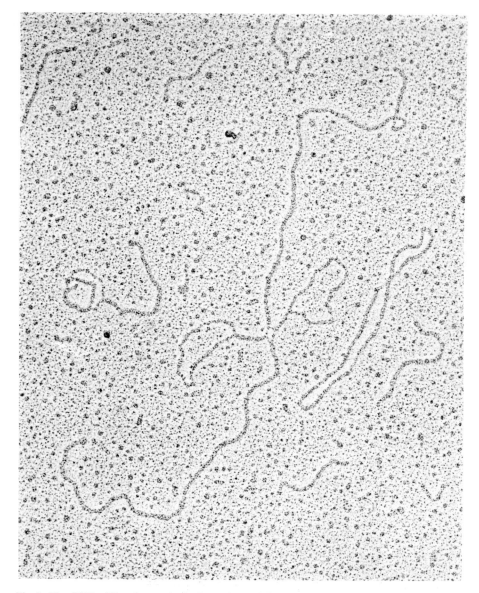

Fig. 3. The rDNA of *Tetrahymena* in the form of a partially denatured cruciform molecule (cf. fig. 2). The adjacent circular molecule is ϕ X 174 with a mass of 3.4×10^6 daltons.

(E. Blackburn) suggests that these regions contain short repeated sequences in both syngen 1 and strain GL.

The structure of the ends of the molecule should be of importance in understanding the origin and replication of the rDNA as well as the relationship between the linear and circular molecules. Because the frequency of circular molecules is lower in samples of rDNA spread in 50–70% formamide than in samples spread under aqueous conditions [11], some of the circles may arise by pairing of complementary single-stranded ends ('sticky ends').

Amplification

The macronuclear rDNA of *Tetrahymena* shares certain properties with the amplified rDNA of *Xenopus* oocytes. Both are extrachromosomal and both are found in multiple nucleoli closely associated with the nuclear envelope. They differ in that the rDNA of *Xenopus* consists of circles containing one to many coding units [16], whereas that of *Tetrahymena* is largely in the form of the palindromic dimers just described.

The oocyte nucleus and the protozoan macronucleus are similar in another important manner – in both cases rDNA is amplified relative to its amount in the haploid genome. This was shown by Tao et al. [17] for *Tetrahymena* by comparing the hybridization values of micronuclei and macronuclei separately. In *Tetrahymena*, as in other ciliated Protozoa, the micronucleus is the germinal or genetic nucleus, responsible for gene transfer at the time of conjugation. The macronucleus, by contrast, is destroyed during conjugation. Nevertheless, the macronucleus controls essentially all gene expression and RNA synthesis during normal vegetative growth [18]. The macronucleus of *Tetrahymena* is polyploid, and its DNA sequence complexity is similar to that of the micronucleus [19].

Yao et al. [17] found that total macronuclear DNA of syngen 1 hybridized with 0.3% of its mass of rRNA, similar to the value obtained previously on strain GL by Engberg and Pearlman [20]. Yet micronuclear DNA from the same organisms hybridized with only 0.014–0.030% of its mass of rRNA. These values correspond to about 200 rDNA coding units per genome for macronuclei but less than 20 per genome for micronuclei. In this sense, then, it is appropriate to say that the macronuclear rDNA is not only extrachromosomal, but is amplified relative to the rDNA of the basic haploid genome.

A plausible interpretation of these data, although one not yet tested, is that the haploid nucleus contains one or a few rDNA coding regions integrated into one or more chromosomes. During conjugation, when the macronucleus breaks down and is replaced by one of the division products of the diploid zygote nucleus [21], one or more extrachromosomal copies of rDNA are produced by some as yet unknown mechanism. By independent replication the number of rDNA copies is then amplified to the level found in the vegetative macronucleus.

An equally plausible scheme would be a wholly extrachromosomal transmission of the rDNA (episome model). If alternative forms of rDNA were

Fig. 4. A possible configuration of chromosomally integrated rDNA. The rDNA is represented by a straight line, the adjacent chromosome segments by wavy lines. The two symmetrically placed cuts in the rDNA produced by the restriction endonuclease *Eco* RI are indicated. *Eco* RI cuts made outside the rDNA define connecting segments (brackets) that could be identified by their size and hybridization properties (see text).

available, so that appropriate genetic crosses could be made, the chromosome *vs* episome models could be distinguished. So far our search for alleles distinguishable by restriction enzyme cutting or other features has been unsuccessful. We are encouraged in our search by the obvious differences between the rDNAs of syngen 1 and GL; unfortunately these two strains cannot be crossed since GL is amicronucleate.

In principle one should be able to use restriction enzymes to distinguish free rDNA molecules from rDNA sequences integrated into a chromosome. For instance, the enzyme *Eco* RI cuts the extrachromosomal rDNA of syngen 1 into three sections, two end pieces of 1.6×10^6 daltons and a middle piece of 9.8×10^6 daltons. If the rDNA of the micronucleus is integrated into a chromosome, normal-sized end pieces should not be produced by *Eco* RI digestion. Instead, the end sequences should be part of longer fragments whose length depends upon the position of the first *Eco* RI site outside the rDNA (fig. 4). It should be possible to find such connecting segments, if they exist, by hybridization experiments. Total micronuclear DNA would be digested with *Eco* RI enzyme and the resulting fragments separated on an agarose gel. Using radioactive RNA complementary to the end sequences, one could search for such connecting segments. If the segments were identified and subsequently cloned in an appropriate plasmid, a detailed sequence analysis of the DNA adjacent to the integrated genes would then be possible. Such an analysis would hopefully give new insight into the mechanism of gene amplification.

We thank Ms. C. Barney and Ms. M. Truett for expert assistance. Parts of this study were supported by grant GM 12427 from the National Institute of General Medical Sciences and grant VC 85 from the American Cancer Society.

References

[1] H. Tobler, Occurrence and developmental significance of gene amplification. Biochemistry of Animal Development, Vol. 3 Academic Press, N.Y. 1975, p. 91.
[2] H.-J. Bohnert, B. Schiller, R. Böhme and H.W. Sauer, Circular DNA and rolling circles in nucleolar rDNA from mitotic nuclei of *Physarum polycephalum*. Eur. J. Biochem. 57, 361 (1975).

[3] V. Vogt and R. Braun, The structure of ribosomal DNA in *Physarum polycephalum*. J. Mol. Biol. 106, 567 (1976).

[4] M. Trendelenburg, H. Spring, U. Scheer and W.F. Franke, Morphology of nucleolar cistrons in a plant cell, *Acetabularia mediterranea*. Proc. Nat. Acad. Sci. 71, 3626 (1974).

[5] R. Charret, L'ADN nucleolaire chez *Tetrahymena pyriformis*: chronologie de sa replication. Exp. Cell Res. 54, 353 (1969).

[6] J. Engberg, D. Mowat and R. Pearlman, Preferential replication of the ribosomal RNA genes during a nutritional shift-up in *Tetrahymena pyriformis*. Biochim. Biophys. Acta 272, 312 (1972).

[7] J. Engberg, G. Christiansen and V. Leick, Autonomous rDNA molecules containing single copies of the ribosomal RNA genes in the macronucleus of *Tetrahymena pyriformis*. Biochem. Biophys. Res. Comm. 59, 1356 (1974).

[8] J. Engberg, J. Nilsson, R. Pearlman and V. Leick, Induction of nucleolar and mitochondrial DNA replication in *Tetrahymena pyriformis*. Proc. Nat. Acad. Sci. 71, 894 (1974).

[9] J. Gall, Free ribosomal RNA genes in the macronucleus of *Tetrahymena*. Proc. Nat. Acad. Sci. 71, 3078 (1974).

[10] K. Karrer and J. Gall, The macronuclear ribosomal DNA of *Tetrahymena* is a palindrome. J. Cell Biol. 67, 202a (1975).

[11] K. Karrer and J. Gall, The macronuclear ribosomal DNA of *Tetrahymena pyriformis* is a palindrome. J. Mol. Biol. 104, 421 (1976).

[12] A. Miyagishi and T. Andoh, The DNA of *Tetrahymena pyriformis* GL strain. A mild method for preparation and its characterization. Biochim. Biophys. Acta 299, 507 (1973).

[13] D. Prescott, K. Murti and C. Bostock, The genetic apparatus of *Stylonychia sp*. Nature 242, 576 (1973).

[14] M.-C. Yao and M. Gorovsky, Length homogeneity of rDNA of the macronuclei of *Tetrahymena pyriformis*. J. Cell Biol. 67, 467a (1975).

[15] J. Engberg, P. Andersson, V. Leick and J. Collins. The free rDNA molecules from *Tetrahymena pyriformis* GL are giant palindromes. J. Mol. Biol. 104, 455 (1976).

[16] D. Hourcade, D. Dressler and J. Wolfson, The amplification of ribosomal RNA genes involves a rolling circle intermediate. Proc. Nat. Acad. Sci. 70, 2926 (1973).

[17] M.-C. Yao, A. Kimmel and M. Gorovsky, A small number of cistrons for ribosomal RNA in the germinal nucleus of a eukaryote, *Tetrahymena pyriformis*. Proc. Nat. Acad. Sci. 71, 3082 (1974).

[18] M. Gorovsky, Macro- and micronuclei of *Tetrahymena pyriformis*: a model system for studying the structure and function of eukaryotic nuclei. J. Protozool. 20, 19 (1973).

[19] M.-C. Yao and M. Gorovsky, Comparison of the sequences of macro- and micronuclear DNA of *Tetrahymena pyriformis*. Chromosoma 48, 1, (1974).

[20] J. Engberg and R. Pearlman, The amount of ribosomal RNA genes in *Tetrahymena pyriformis* in different physiological states. Eur. J. Biochem. 26, 393 (1972).

[21] A.M. Elliott, Life cycle and distribution of *Tetrahymena*, Biology of *Tetrahymena* (Ed. A.M. Elliott), Dowden, Hutchinson and Ross, Stroudsburg, Pa., 1973, p. 259.

Chapter 8

ANATOMY OF THE GENE CLUSTER CODING FOR THE FIVE HISTONE PROTEINS OF THE SEA URCHIN, A PROGRESS REPORT

M.L. BIRNSTIEL, K. GROSS, W. SCHAFFNER, R. PORTMANN and E. PROBST

Institut für Molekularbiologie II der Universität Zürich, Winterthurerstr. 266A, 8057 Zürich, Switzerland

Introduction

Three interesting aspects of gene regulation involving the histone coding sequences have been discovered, which require explanations at the molecular level.

Regulation of histone mRNA production

The synthesis of histone mRNA is clearly regulated in the cell cycle of dividing cells. In HeLa tissue culture cells, histone messenger appears in the cytoplasm at the onset of DNA synthesis [1]. Prior to this, in G^1, no histone mRNA can be detected [2]. In mouse L-cells histone mRNA, once synthesized, persists for about 11 h. (a time span also taken in by the S-phase), directs the synthesis of histones in step with DNA synthesis and is then destroyed rapidly [3]. Regulation of the histone transcription is thought to be under control of non-histone chromosomal proteins [4].

The abundance of histone mRNA can also be seen to be regulated in development. The sea urchin egg already possesses a pool of maternal histone mRNA which is increased about five fold during the early cleavage stages [5,6] through rapid transcription of the many hundred fold repeated histone genes [7,8]. At gastrula, when cell replication has slowed down, a low level of histone mRNA is found [5,6]. Although cleavage proceeds rapidly in the fertilized egg of *Xenopus laevis* and there is a superficial resemblance to the early embryological events, histone mRNA synthesis becomes detectable in *Xenopus* only at gastrulation

(Malacinski and Jacob, unpublished observation). The rapid synthesis of histone proteins in this species is supported by large maternal pools of histone mRNA ([9]; Destrée, unpublished). During early development of the *Xenopus* embryo there is clearly no coordination between histone protein and DNA production [9].

Coordination of histone protein production

Histones H1, H2A, H2B, H3 and H4 are present in chromatin in a molar ratio of $1/2:1:1:1:1$ [10]. There are two main mechanisms possible by which the synthesis of these molar ratios of histones might be regulated, a) at a transcriptional level through control of the amounts of histone mRNA; this would be most easily achieved by tandem transcription of closely linked genes, as in the case of ribosomal DNA; b) at a post-transcriptional level, by regulation of the rates of nuclear processing, of translation in the cytoplasm, or of mRNA turn-over. In the latter three cases the propinquity of the histone genes would serve some functions other than that of gene regulation.

Production of histone variants

H1 [11], as well as other histone proteins [12], exhibit variation in primary structure within one and the same organism. The synthesis of histone variants appears to be specific to developmental stages and tissues [13]. This sequence divergence is consistent with the general tenet that repeated genes are also likely to be polymorphic. The finding that histone variants are sometimes tissue specific [12] is interesting also from the standpoint of gene regulation because this may mean that rather subtle mechanisms may exist within the cell which allow selective transcription of the genes coding for the histone variants from the repeated gene cluster.

Information pertinent to our understanding of the molecular mechanism of histone gene regulation is: the arrangement of the histone genes relative to one another on the chromosome, their relationship to spacer DNA and to sequences coding for the histone variants, the location of the initiation sites for transcription and the characterization of the initial transcription product. We have made a start in the analysis of histone gene regulation by studying the molecular structure of the genes on the chromosome. Here we review the evidence that the five histone coding sequences in the sea urchin *Psammechinus miliaris* are clustered together in a basic repeat of 5.6 kilo bases. We shall give the exact order of the five histone coding sequences as well as the polarity of the gene arrangement. This work involves the use of restriction enzymes and the use of 3′ and 5′ exonucleases as well as hybridization of the five highly purified histone mRNAs to restricted and resected DNA fragments. Finally, partial denaturation mapping of cloned histone DNA has been used to reveal the distribution of spacers and genes.

Identification of the repeat unit of histone DNA

That the histone DNA consists of DNA repeat units of (nominally) 4×10^6 daltons was demonstrated by cleaving isolated histone DNA [14] with either *Eco*RI or *Hin*dIII which yield in either case a prominent 5.6 kilo base DNA fragment [15,16]. This DNA was shown to hybridize to all five histone mRNAs [15,17] and hence clearly harbor the coding sequences for the H4, H3, H2A, H2B and H1 histone proteins.

When *Psammechinus* DNA was only partially digested with *Eco*RI, the fragments separated on agarose gels by electrophoresis and the histone DNA complements detected by hybridization to histone mRNA [18], higher order fragments consisting of multiples of the 4×10^6 dalton (~ 5.6 kilo bases) unit were detected at 8 and 12×10^6 daltons (fig. 1). This demonstrates that histone DNA consists of tandem repeats of 4×10^6 daltons (determined from its electrophoretic mobility) or 5.6 kilo bases (determined in the electron microscope by comparison with T7 molecules [25]), which are joined together on the chromosomal DNA without any detectable DNA intervening between them [15].

Isolation of the five messengers coding for the five kinds of histone

To determine the exact order of the histone coding sequences it was first necessary to isolate and identify the radioactively labelled mRNAs corresponding to the five histone proteins, since these RNAs constituted the probes for analysis of the gene clusters.

During the early cleavage stage of sea urchin development there is extensive synthesis of histone mRNAs and they comprise the major species of mRNA appearing in the cytoplasm at this time. The histone mRNAs can be labelled with great ease by incubating developing sea urchin embryos with [^{3}H]uridine. Since five histone proteins are produced, one might reasonably expect that five corresponding mRNAs could be isolated by gel electrophoresis. There are, however, two major difficulties to be overcome. Firstly, while the mRNAs of the histone H1 and H4 are clearly distinct in MW, corresponding to the smallest and largest histone proteins respectively, the histone proteins H2A, H2B and H3 are closely similar in MW and correspondingly have mRNAs of closely similar electrophoretic mobilities.

Secondly, it seems that histone mRNAs may be folded in many different ways, so that if they are fractionated under relaxed conditions of temperature and ionic strength, the pattern obtained is very complex and diffuse. This is best illustrated by the difficulties encountered by a group working on *Strongylocentrotus*, where 3 major sets of mRNAs separated into 11 bands upon gel electrophoresis [19], an unsatisfactory situation for the mapping of five coding sequences.

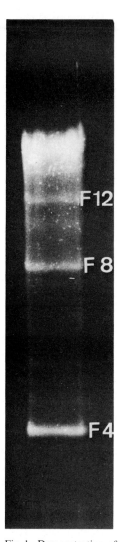

Fig. 1. Demonstration of multimers of the basic repeat unit of hDNA. 2 µg Psammechinus DNA of high molecular weight (about $5 \times 10^7 \Delta$) was enriched for histone sequences about $7 \times$ by density gradient centrifugation and restricted with sufficiently diluted enzyme such that a large portion of the DNA remained uncut. The sample was electrophoresed through a 0.8% agarose slab gel. The DNA was denatured inside the gel and was transferred onto a Millipore membrane filter [18]. ^{3}H RNA from *Psammechinus* embryos representing mainly histone coding sequences was hybridized in $2 \times$ SSC, 0.2% SDS at 65° for 4 h to the filter-bound DNA, followed by RNase treatment. The membrane filter was dried, soaked with 7% PPO in ether [26] and exposed to an 'activated' x-ray film [27]. The developed film 'fluorogram' demonstrates the presence of a dimer (molecular weight $8 \times 10^6 \Delta$) and a trimer ($12 \times 10^6 \Delta$) of the basic repeat unit. Molecular weight assignments were made by comparison with *Eco*RI digests of adenovirus DNA and phage ϕ 29 DNA run in parallel slots. The standards yielded fragments of 13.7×10^6, 6.1×10^6 and $3.5 \times 10^6 \Delta$ as well as smaller fragments.

Fractionation of these complex histone mRNA mixtures in *Psammechinus* was made possible by separating the mRNAs by gel electrophoresis under stringent conditions, in which most, but not all, of the secondary structure of the mRNAs was abolished [20].

In this way mRNAs of unimodel distribution were isolated by preparative gel electrophoresis, including those of H2A, H2B and H3 where the slight degree of secondary structure remaining played an important part in the separation of these mRNAs.

Once purified, the mRNAs were translated in a wheat germ system. Each fraction promoted the synthesis of just one major histone protein, and on this basis each mRNA was assigned to one of the five kinds of histone protein [20].

Dissection of the 5.6 kilo base unit and ordering of the coding sequences

Since it has already been established in our laboratory that the 5.6 kilo base unit of *Psammechinus* histone DNA [15] contains all five histone coding sequences, the histone DNA was dissected with restriction enzymes in such a way that each DNA segment, delimited by restriction sites, contained just one coding sequence [17]. Identification of the coding sequences was accomplished by hybridization of the individual fragments to each of the five purified mRNAs.

For these experiments histone DNA was enriched by actinomycin-CsCl gradient centrifugation [14] and restricted with either *Eco*RI and *Hin*dII, or *Hin*dIII and *Hin*dII. Each of these two treatments yielded four restriction fragments which could then be ordered from the overlap existing between them [17]. Each set of four fragments was hybridized to each of the five individual histone mRNAs. The restriction map shown in fig. 2 was constructed from these results and the five coding sequences were assigned to the DNA segments [17]. Accordingly, the order of the sequences starting at the *Hin*dIII cleavage site is H4, H2B, H3, H2A, H1 (fig. 2).

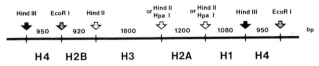

Fig. 2. Restriction map of *Psammechinus miliaris* DNA and assignments of coding sequences. The upper diagram depicts the five DNA segments delimited by the cleavage sites of several restriction enzymes. The numbers denote the lengths in kilo bases totalling a repeat length of (nominally) 6 kilo bases. In the lower diagram the five coding sequences for the five histone proteins have been assigned to the DNA segments. (See text for experimental design.)

Polarity and asymmetry of the coding sequences in *Psammechinus miliaris* [21]

It may be easily demonstrated that in histone DNA all five coding sequences reside on the same DNA strand and therefore show the same polarity. This is possible because the recombinant phage λ $b538$ red B113 imm^{434} (Pm histone)–22 [22] containing the inserted 5.6 kb unit can be conveniently strand separated by electrophoresis of the denatured DNA. When both slow and fast-moving single strands are challenged individually with each of the five mRNAs, all five messengers anneal to the slow-moving DNA strand of the recombinant phage [21].

The polarity of the gene cluster was determined by λ exonuclease digestion of the 5′ ends of the *Hin*dIII or *Eco*RI generated 5.6 kb histone DNA units. The rationale underlying the experiments may be explained with the aid of fig. 3, where the two possible orientations of the coding sequence are given. If the coding sequence for H4 mRNA lies near the 5′ end of the 5.6 kilo base fragment, then the removal of 1000 bases from the 5′ end by means of 5′ exonuclease digestion will destroy this sequence. Limited 5′ exonuclease digestion of the *Hin*dIII fragment did not remove this coding sequence since, after denaturation, the resected DNA still hybridized to mRNA H4. In this fragment, therefore, the H4 coding sequence must reside near the 3′ end of the molecule. If true, then the exonuclease treatment must have removed the sequences paired with the H4 coding sequence. That this had indeed happened was demonstrated by the successful annealing of H4 mRNA to resected but *not* denatured 5.6 kilo base DNA fragments.

The *Eco*RI restriction fragments did not hybridize with H4 mRNA after 5′ exonuclease treatment, and from this it may be concluded that the H4 coding

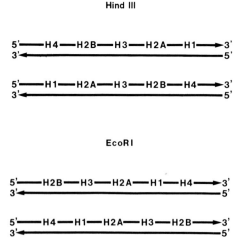

Fig. 3. Alternative polarities of the histone gene arrangement within the *Hin*dIII and *Eco*RI restriction fragments of histone DNA.

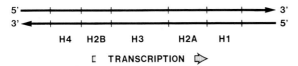

Fig. 4. Polarity and asymmetry of the histone gene cluster.

sequence lies near the 5′ end. This is consistent with the results from the *Hin*dIII fragment and shows that the polarity of the coding strand runs 5′ H1-H2A-H3-H2B-H4 3′ [21]. Thus, the pattern we obtain is that of a gene cluster containing five coding sequences for the five histone proteins (fig. 4). These all lie on the same DNA strand and are transcribed in the direction H4 → H1 although the site (or sites) of initiation of transcription are not as yet known.

The organization of the histone genes of the sea urchin *S. purpuratus* has been investigated by the Stanford group [23]. RNA-DNA hybridization between chimeric plasmid DNA and purified individual histone mRNA was used to assign coding sequences to eiter the 2.2 or the 4.8 kilo base region of the histone DNA repeat unit. The 2.2 kilo bases segment was found to hybridize with the mRNAs B1, B2 and B3, while the 4.8 kilo base unit contained the sequences complementary to mRNA A, B4 and C. Hence a total of six mRNAs were mapped, for a total length of 7 kilo bases. According to the mRNA assignments made by these workers (table 2 ref. 23), the order in that species would be (H2A; H2B) – (H4; H3; H1). This conflicts with our data in that H2A and H2B are clearly not adjacent to one another in our map.

So far, we have shown that all five histone genes are linked on the same DNA strand and hence possess the same polarity. Thus, they are arranged in a manner which would allow co-ordinate tandem transcription and regulation of all five genes; for example, the production of the rRNA 28S, 18S and 5.8S fractions are co-ordinated at the transcriptional level. Such co-ordination by transcription would provide the cell with a simple mechanism to ensure production of equimolar quantities of the different histone mRNAs, which in turn could produce the equimolar amounts of histone proteins found in the cell nucleus, at least in the case of histones H4, H3, H2A and H2B. There is at the moment no conclusive evidence for the tandem transcription of the histone genes in sea urchins, but in HeLa cells a large size ($\sim$50S) nuclear precursor of a cytoplasmic monocistronic histone mRNA has been found [24].

Distribution of spacer and genes [25]

Taken all together the five histone proteins comprise some 710 amino acids, so that the corresponding coding sequences must consist of 2.1 kilo bases of DNA. However, the mRNA of H4 protein is known to contain an additional

100 bases as untranscribed, leading and trailing sequences. Assuming that similar sequences are found in the other histone mRNAs, this means that the mRNA coding sequences extend to about 2.7 kilo bases.

From the restriction map it can be seen that the coding sequences occur rather regularly along the whole length of the repeat unit of 5.6 kilo bases. This therefore suggests a rather regular intermixing of coding sequences and spacer sequences of about equal lengths.

That this is indeed the case can be demonstrated by partial denaturation mapping in the electron microscope [25]. This mapping is facilitated by the fortunate circumstance that the spacer and coding sequences of *Psammechinus* histone DNA have distinctive and widely disparate base compositions.

In all sea urchins investigated so far, the histone mRNAs have been shown to possess a G-C content of 50–56%. By contrast, buoyant density measurements on isolated histone DNA indicate that its overall G-C content is 44%. Clearly the spacer component of the histone DNA must be AT-rich. This prediction is clearly borne out out by the thermal denaturation profile of the 5.6 kilo base unit reclaimed from the recombinant λ phage. One half of the DNA melts at high temperature and from its T_m has a G-C content of about 52% G-C, i.e., that predicted for the mRNA coding sequences; the other half melts in two steps with components of 34% and 40% G-C content, respectively. Furthermore, it is clear from the melting curve that these two components are separated by a plateau which extends over 5°C or so [25].

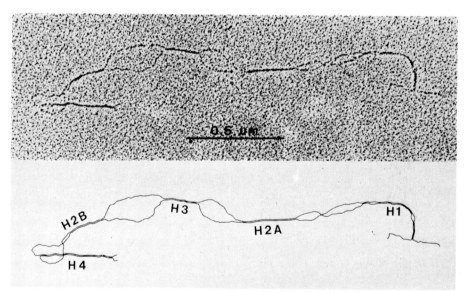

Fig. 5. Electronmicrograph of partially denatured 5.6 kilo base histone DNA repeat unit. The 5.6 kilo bases histone DNA reclaimed from recombinant λ phage was melted and examined in the electronmicroscope to determine the distribution of spacer DNA [25].

This is therefore an ideal situation for determining the topologies of low and high-melting DNA by melting the DNA up to this temperature plateau and inspection of the partially denatured DNA in the electron microscope. The denaturation patterns obtained are shown in fig. 5.

A typical molecule starts with a small fork, followed by a series of native high G-C regions and AT-rich blisters. At the other end there is a large open fork so that the molecule starts and ends with AT-rich spacer DNA (fig. 5). Because of this striking small fork and long fork arrangement the molecule is clearly asymmetrical, and all molecules used for statistical evaluation were oriented in this way. The double-stranded and single-stranded regions of 118 molecules were measured and a histogram of their distribution was constructed with a resolution of 28 bp (fig. 6). In this diagram the block areas denote the positions and lengths of the blisters. As can be seen, the black (single-stranded) and white (double-stranded) areas are about equal in size, so that the desired 50% denaturation was in fact attained.

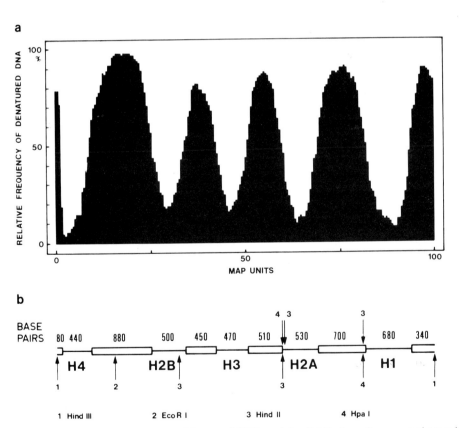

Fig. 6. Histogram of the frequency of denatured DNA, and the distribution of spacers and genes in the 5.6 kilo base histone DNA repeat unit [25]. The diagram is drawn at a resolution of 28 bp.

TABLE 1

Number of base pairs in histone and histone mRNA coding sequences. Values for H3, H2A and H1 are those for mammals.

Histones	Number of amino acids in protein	Amino acid coding sequence; bp	mRNA coding sequence; bp	GC-rich DNA segments; bp
H4	102	306	400	440
H2B	125	375	n.d.	500
H3	135	405	n.d.	470
H2A	128	384	n.d.	530
H1	heterogeneous, approx. 220	660*	n.d.	680
All five	710	2.13 kilo bases		2.62 kilo bases

*There is some circumstantial evidence [28] that the H1 of the sea urchin may be smaller than that of mammals.

The number of nucleotides in the native and denatured DNA stretches can then be estimated from a line dissecting the histogram at half-height, since this represents their distribution in the 'average' molecule. The relevant numbers are given in fig. 6 and table 1.

In order to fit the biochemical data to the blister map it is necessary to know which is 'left' and which is 'right' in the map, i.e., which end of the molecule contains the H4 and which the H1 sequence. This can be determined unequivocally by partial denaturation and blister mapping of the histone DNA within the recombinant λ phage into which the 5.6 kilo base unit was inserted in a unique and known orientation [25]. From this work it is clear [25] that the H4 gene is near the small fork and the H1 gene near the large fork. This leads to the assignment of the coding sequences to the native DNA regions shown in fig. 5.

The very close correlation between the biochemical and electron microscopic data can be clearly seen from a study of the quantitative aspects of the blister map. Thus, the DNA assigned to the smallest histone constitutes the smallest unmelted region; with 440 bp it includes an excess of some 140 bp over the amino-acid coding sequence (table 1). The DNA assigned to the largest of the histones is also the largest unmelted region of the DNA. The middle-sized histones are allotted to middle-sized DNA sequences which are larger by some 80–150 bp than the actual amino-acid coding sequences.

Conclusion

We have now established that in *Psammechinus miliaris* the histone DNA consists of a 5.6 kilo base unit many times repeated. Within the repeat unit the five histone genes are arranged in the direction of transcription H4 →H2B → H3 →H2A →H1. They lie on the same DNA strand, and hence possess the same polarity and are separated from one another by spacer DNA.

The arrangement of the histone genes is compatible with the view that their regulation occurs at the transcriptional level. Thus, the clustering of multiple repeat units would facilitate the switching on of a large battery of histone genes, as required for the large-scale synthesis of histone mRNA observed in early sea urchin development. Similarly, tandem transcription of all five coding sequences present per unit, possibly into a single polycistronic primary transcript, could ensure the simultaneous production of equimolar quantities of the five histone mRNAs. However, before a definite conclusion can be reached, it is first necessary to learn more about the sites of initiation of transcription and the nature of the primary transcription product.

Acknowledgements

This research was supported in part by a grant from the Kanton of Zürich and by the Schweiz. Nationalfonds, No. 3.8630.72 SR. Dr. K. Gross is a fellow of the Helen Hay Whitney Foundation. We thank Dr. S. Clarkson and Prof. H.O. Smith for the gift of a battery of restriction enzymes.

References

[1] M. Breindl and D. Gallwitz, Identification of histone mRNA from HeLa cells. Eur. J. Biochem. 32, 381 (1973).
[2] T.W. Borun, F. Gabrielli, K. Ajiro, A. Zweidler and C. Baglioni, Further evidence of transcriptional and translational control of histone mRNA during the HeLa S-3 cell cycle. Cell 4, 59 (1975).
[3] R.P. Perry and D.E. Kelly, mRNA Turnover in Mouse L-Cells. J. Mol. Biol. 79, 681 (1973).
[4] G. Stein, W. Park, C. Thrall, R. Mans and J. Stein, Regulation of cell cycle stage-specific transcription of histone genes from chromatin by non-histone chromosomal proteins. Nature 257, 764 (1975).
[5] M.N. Farquhar and B.J. McCarthy, Histone mRNA in eggs and embryos of *Strongylocentrotus purpuratus*. Biochem. Biophys. Res. Comm. 53, 515 (1973).
[6] P.R. Gross, K.W. Gross, A.I. Skoultchi and J.V. Ruderman, Maternal mRNA and protein synthesis in the embryo, *6th* Symposium: Protein synthesis in reproductive tissue, Karolinska Symposia on Research Methods in Reproductive Endocrinology (1973).

[7] L.H. Kedes and M.L. Birnstiel, Reiteration and clustering of DNA sequences complementary to histone mRNA. Nature New Biol. 230, 165 (1971).

[8] E.S. Weinberg, M.L. Birnstiel, I.F. Purdom and R. Williamson, Genes coding for polysomal 9S RNA of sea urchins: conservation and divergence. Nature 240, 225 (1972).

[9] E.D. Adamson and H.R. Woodland, Histone synthesis in early amplification development: histone and DNA synthesis are not coordinated. J. Mol. Biol. 88, 263 (1974).

[10] R.D. Kornberg, Chromatin structure: A repeating unit of histones and DNA. Science 184, 868 (1974).

[11] J.M. Kinkade and R.D. Cole, A structural comparison of different lysine-rich histones of calf thymus. J. Biol. Chem. 241, 5798 (1966).
M. Bustin and R.D. Cole, A study of the multiplicity of lysine-rich histones. J. Biol. Chem. 244, 5286 (1969).

[12] A. Zweidler and L. Cohen, Histone variants and new histone species isolated from mammalian tissues. J. Cell Biol. 59, 378a (1973).

[13] L. Cohen, K. Newrock and A. Zweidler, Stage-specific switches in histone synthesis during embryogenesis of the sea urchin. Science, 190, 994 (1975).

[14] M.L. Birnstiel, J.L. Telford, E.S. Weinberg and D. Stafford, Isolation and some properties of the genes coding for histone proteins. Proc. Nat. Acad. Sci. US 71, 2900 (1974).

[15] M.L. Birnstiel, K. Gross, W. Schaffner and J. Telford, Biochemical dissection of the histone gene cluster of sea urchin, in: FEBS Proceedings 10th Paris Meeting 38, 3, (1975).

[16] M.L. Birnstiel, E. Southern, J. Telford, M. Wilson, W. Schaffner and K. Gross, Molecular structure of the histone polygenes. Experientia 31, 735 (1975).

[17] W. Schaffner, K. Gross, J. Telford and M. Birnstiel, Molecular analysis of the histone gene cluster of *Psammechinus miliaris*: II. The arrangement of the five histone coding and spacer sequences. Cell 8, 471 (1976).

[18] E.M. Southern, Detection of specific sequences among DNA fragments separated by gel electrophoresis. J. Mol. Biol. 98, 503 (1975).

[19] M. Grunstein, S. Levi, P. Schedl and L. Kedes, Messenger RNAs for individual histone proteins: fingerprint analysis and in vitro translation. Cold Spring Harb. Symp. Q. Biol. 18, 717 (1973).

[20] K. Gross, E. Probst, W. Schaffner and M. Birnstiel, Molecular analysis of the histone gene cluster of *Psammechinus miliaris*: I. Fractionation and identification of five individual histone mRNAs. Cell 8, 455 (1976).

[21] K. Gross, W. Schaffner, J. Telford and M. Birnstiel, Molecular analysis of the histone gene cluster of *Psammechinus miliaris*: III. Polarity and asymmetry of the histone coding sequences. Cell 8, 479 (1976).

[22] S.G. Clarkson, H.O. Smith, W. Schaffner, K.W. Gross and M.L. Birnstiel, Integration of eukaryotic genes for 5S RNA and histone proteins into a phage λ receptor. Nucl. Acids Res. 3, 2617 (1976).

[23] L.H. Kedes, R.H. Cohen, J.C. Lowry, A.C.Y. Chang and S.N. Cohen, The organization of sea urchin histone genes. Cell 6, 359 (1975).

[24] M.L. Melli, H. Wyssling and G. Spinelli, in preparation.

[25] R. Portmann, W. Schaffner and M. Birnstiel, Partial denaturation mapping of cloned histone DNA from the sea urchin *Psammechinus miliaris*. Nature 264, 31 (1976).

[26] K. Randerath, An evaluation of film detection methods for weak β-emitters, particularly tritium. Anal. Biochem. 34, 188 (1970).

[27] R.A. Laskey and A.D. Mills, Quantitative film detection of ^{3}H and ^{14}C in polyacrylamide gels by fluorography. Eur. J. Biochem. 56, 335 (1975).

[28] L.H. Cohen and V.B. Gotchel, Histones of polytene and nonpolytene nuclei of *Drosophila melanogaster*. J. Biol. Chem. 246, 1841 (1971).

Chapter 9

SOME ASPECTS OF THE INTERACTION BETWEEN THE MOUSE (LA9) NICKING-CLOSING ENZYME AND CLOSED CIRCULAR DNA

J. VINOGRAD*

I am going to talk about a protein found in the nucleus of eukaryotic cells that has a particular enzymatic activity. This activity was discovered by James Wang [1] four or five years ago in extracts of *E. coli*, and has the property of nicking a phosphodiester bond in duplex DNA, allowing rotation to occur and then closing the bond. Only two assays for this enzyme have been suggested. The first is a fairly easy one using closed circular DNA and its properties which I will explain below. The other is to use oxygen 18 in the hopes of detecting an exchange reaction. As far as I know this has not been tried.

Our interest in this enzyme comes from the assay as well as from its potential function in the replication of DNA where there is an unwinding problem. The DNA must be unwound at the replicating fork, generating positive supercoils ahead of the replicating fork which have to be relieved. Either a nicking-closing enzyme or a combination of an endonuclease and a ligase are required. I should also say that after Wang's work on the *E. coli* system, where everything was lovely except for the fact that the enzyme works only on negative supercoils and not on positive supercoils (leaving its function a worrisome matter), Champoux and Dulbecco succeeded in demonstrating a similar activity in extracts of mouse cell nuclei [2]. The mouse activity works on both positive and negative supercoils. Other people have worked in this field, and similar enzymes have been partially purified from other eukaryotic cell nuclei by Keller and Wendel at Cold Spring Harbor [3], by Pulleyblank and Morgan at Edmonton [4], and by Baase and Wang at Berkeley [5]. I understand that Champoux is also in the process of purifying this enzyme.

---

* Based on a talk delivered by Jerome Vinograd at the Symposium and completed after his death by David Pulleyblank, Mavis Shure and David Tang.

The Molecular Biology of the Mammalian Genetic Apparatus:
edited by P. Ts'o © 1977, Elsevier North-Holland Biomedical Press

The first figure illustrates the closed circle problem. This is an old illustration many of you have already seen. If two single stranded circles are interwound once, twice or ten times, a structure resembling DNA emerges. This diagram illustrates a very fundamental statement which is that the number of times one strand is interwound with the other is invariant once both strands are closed. The only way to alter the number of times the two circles are interwound is to break one of them, rotate one with respect to the other, and then reseal the break. If the pair of circles are made to lie in a plane, one can count the number of times one strand winds around the other to get the topological winding number, which is called

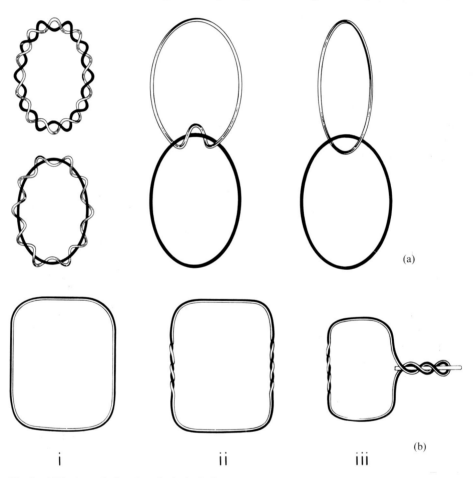

Fig. 1. a) The interwinding (topological winding number, α) of two strands is invariant once they are closed. Shown here are locked circles interwound once, twice, and ten times. b) Introduction of superhelical turns into a closed duplex of two noninterwound strands. In all three circles $0 = \alpha = \tau + \beta$. (i) $0 = 0 + 0$, two noninterwound strands; (ii) $0 = 0 + 0$, three right-handed duplex turns cancel three left-handed duplex turns; (iii) $0 = -3 + 3$, three right-handed duplex turns and three right-handed interwound superhelical turns.

α. If an attempt is made to wind the molecule up by changing the number of duplex turns (β), then one has to compensate by putting in an equal number of superhelical turns (τ) of the opposite sense so that the sum of τ and β is always equal to α.

The enzyme that I am going to talk about has the property of introducing a single stranded break or nick into a closed circular duplex DNA, allowing rotation to occur, and then sealing the nick to form a molecule with a new α. These steps are illustrated in fig. 2, which is taken from a paper which has appeared in the last issue of the *Proceedings* [6] and is accompanied by a paper by Depew and Wang [7] on another aspect of this problem. A closed circular DNA can be treated with a nicking-closing enzyme and a hypothetical enzyme–DNA complex is obtained in which free rotation about a swivel relieves the supercoiling. The DNA is then covalently closed to give a set of molecules having various degrees of supercoiling, differing by integral values such as $+1$, -1, and so on. Under the conditions of the experiment, the lowest energy state is not necessarily the state in which the cleaved phosphodiester bond was originally in register. To align the two ends in the hypothetical nicked intermediate, the DNA will have to be twisted through a small angle that we have called ε; at this point the bond may be resealed. There is another way to proceed with this experiment, which has been done by James Wang and ourselves. This is to take the same closed circular DNA, treat it with pancreatic DNase to introduce a nick and then close it with ligase. As we will show, this generates a set of molecules of similar variety. I should mention that the work I am going to talk about was performed by two

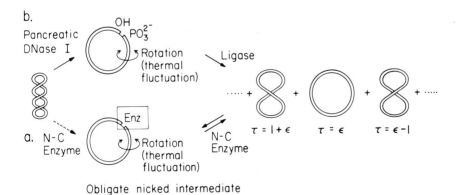

Fig. 2. The formation of a Boltzmann distribution of topological isomers of closed circular DNA. a) The action of N-C enzyme on a closed circular DNA substrate. The diagram is not intended to specify the mechanism for N-C action. The obligate elementary steps include nicking, releasing preexisting supercoils, random rotation at the swivel, and closure. The quantity ε is the difference between the duplex winding number (β) of a nicked DNA and the duplex winding number of a similar hypothetical molecule with $\tau = 0$. The fractional turn ε is not illustrated in this fig. b) The closure of nicked circular DNA by polynucleotide ligase.

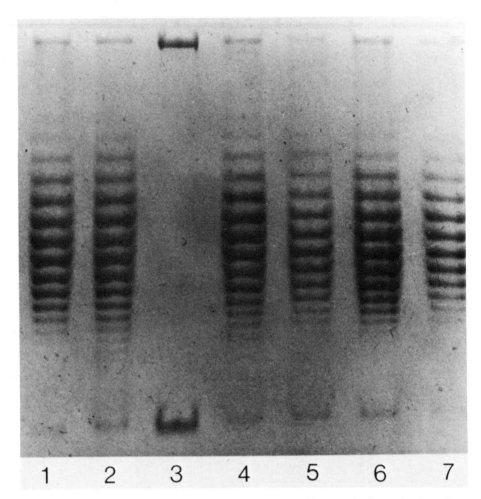

Fig. 3. The limit product of the N-C enzyme is indistinguishable from the ligase-closed product. 1) PM2 II + ligase; 2) PM2 I + N-C enzyme; 3) I, II; 4) mixture of products in (1) and (2); 5) PM2 I, II + N-C enzyme + ligase; 6) PM2 II + N-C enzyme + ligase; 7) ligase closed PM2 + N-C enzyme. Reactions were carried out in 0.2 M NaCl, 3 mM $MgCl_2$, 5 mM $(NH_4)_2SO_4$, 33 μM NAD, 0,2 mM EDTA, 0.1 mM spermidine, 20 μg/ml of bovine serum albumin, 30 mM Tris–HCl at pH 7.8 for 8 h at 37°C; 5 μg of DNA, 140 units of N-C enzyme, 2×10^{-3} units of ligase were used in the reactions. Nicked and linear DNA were removed by EtdBr-CsCl buoyant centrifugation. PM2 I (10 ng) was added to each channel.

postdoctoral collaborators, David Pulleyblank and Peter Vosberg and two graduate students, Mavis Shure and David Tang, in my laboratory.

Fig. 3 shows a gel of a type suggested by the work of Keller in which one particular DNA, PM2 DNA, separates into a series of bands depending on the number of supercoils, and each of these bands differs from the next one in having one more or one less supercoil. Nicked PM2 DNA appears at the top of the figure and native supercoiled PM2 DNA is at the bottom. By adding magnesium to the running buffer and lowering the temperature, the electrophoresis conditions have been adjusted so that the whole set of species generated either by the action of the nicking-closing enzyme on the closed circular DNA, or by the action of ligase on the nicked circular counterpart, are resolved as negatively supercoiled species. This figure contains quite a bit of information. The first piece of information is that the limit product of the nicking-closing enzyme is a set of bands or species. We can count thirteen to fourteen bands in this distribution of PM2 DNA. When the nicking-closing reaction is conducted at 37°C in the presence of 0.2 M salt, 3 mM magnesium ion and 33 μM NAD (Ch. 2), we get exactly the same set of bands as those generated by closing nicked PM2 DNA with ligase under the identical conditions (Ch. 1). We also get the same set of bands when nicked DNA is closed in the presence of both enzymes (Ch. 6) or

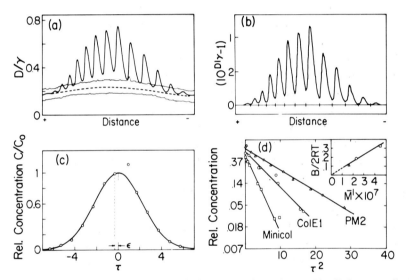

Fig. 4. Processing of data for the relative masses in each set. The Boltzmann distribution. a) Absorbance trace of a photograph of an ethidium stained gel of relaxed PM2 DNA with background traces from both sides of the sample channel. Sample background (---) was calculated by averaging the background traces. The ordinate is plotted in units of D/γ. b) Plot of fluorescence intensity against distance, generated from the optical density trace shown in 4a. The curve was calculated by the equation $J = 10^{D/\gamma} - 1$, where D is the optical density above background. The peaks were integrated between the indicated limits, to determine the relative masses of the species present. c) The concentration of species present in solution are normalized by C_0, the concentration of a theoretical species lacking supercoils. The curve through the points is the calculated Gaussian curve with the best least squares parameters, 0.105 and 0.33 for $B/2RT$ and ϵ, respectively. The point lying above the curve at $\tau = 0.67$ was not used in the least squares fit because of contamination of this species by linear PM2 DNA. d) The natural logarithm of the relative masses of species in the limit products of three closed DNAs treated with N-C enzyme, plotted against the square of the superhelical winding number. The intercepts of the traces have been displaced for clarity of presentation.

when native closed circular DNA is relaxed in the presence of both enzymes (Ch. 7). Hence, the distribution of species is independent of the enzymology involved, but reflects something about the rotation of DNA about the swivel.

Tracings of this gel were quantitated (fig. 4) after appropriate photography with calibration wedges, and with appropriate attention to baseline variation along the channels, and to the logarithmic nature of the photographic response. Fig. 4b shows a profile of fluorescence intensity calculated from fig. 4a, which is a tracing of the optical density on the film. Fluorescence intensity is proportional to the amount of DNA. From the areas under this particular profile (fig. 4b) one can find out how much DNA there is in each of the bands.

If one plots the relative amount of DNA in any particular band as a function of the degree of supercoiling of the band, and the most intense band is taken to have ε superhelical turns, then one can see that a Gaussian distribution in τ is obtained (fig. 4c). Notice that there is one maximal band, but the system is not symmetrical about the band. The center of symmetry lies a short distance away, separated from the maximal band by the ε value previously mentioned. In fig. 4d we have changed the plot to a semilogarithmic form, with the logarithm of the intensity of a species on the ordinate and the square of the number of superhelical turns on the abscissa. Distributions have been evaluated for three different N-C enzyme relaxed DNAs: PM2, Col E1 and Minicol. In all cases we obtain straight lines, the slopes of which can be related to a Boltzmann distribution of the states of the molecules in these equilibrium systems. The important point is that the molecule can rotate around the swivel. As it rotates it meets a resistance and the energy required to overcome that resistance is proportional to τ, and to the reciprocal of the molecular weight of the DNA (fig. 4d, upper right panel).

Fig. 5 shows the results of an experiment performed by Mavis Shure in which the following question was asked: Is supercoiling necessary for the action of the nicking-closing enzyme? The question was answered by preparing a limit product set of species by the action of an excess of N-C enzyme on Minicol DNA (fig. 5, Ch. 2, 5, 8). The product was then resolved into its constituent species by electrophoresis on a preparative gel and the individual species isolated by extracting them from the sliced gel. Three such species were electrophoresed again and are present in channels 3, 6 and 9 of fig. 5. Each of these three species, having approximately $+1$, 0 and -1 superhelical turns under the reaction conditions, was retreated with the enzyme to generate new sets of species entirely comparable to the original distribution (fig. 5, Ch. 4, 7, 10). The sets of bands migrating behind the minicol sets are Col E1 DNA added as a control to ensure that the enzyme was active. The important point here is that Mavis has demonstrated that supercoiling is not necessary for the action of the nicking-closing enzyme and that the distributions are of thermal origin. The species shown in fig. 5, channel 6 has zero supercoiling and it regenerated the whole set, showing that the enzymatic activity did not depend on the superhelix free energy. Apparently, in solution

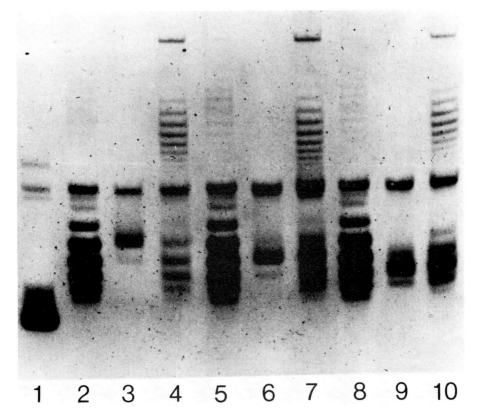

Fig. 5. Demonstration that non-supercoiled DNA is a substrate for the action of N-C enzyme and that N-C enzyme reacts with a homogeneously supercoiled species of closed circular DNA ($\tau \approx 0$) to form a thermal distribution of species. (1) Minicol I and II; (2, 5, and 8) limit products of the initial reaction of Minicol I with N-C enzyme; (3, 6, and 9) purified species with $\tau \approx +1, 0, -1$, respectively; (4, 7, and 10) limit products regenerated after reaction of purified species with N-C enzyme. Bands migrating slower than Minicol II in channels 4, 7, and 10 are sets of relaxed ColE1 DNA. Native ColE1 DNA was added to the reaction mixtures to monitor the activity of the enzyme.

DNA is subject to local torsional fluctuations which can be released by the enzyme. These fluctuations must certainly be a property of DNA in general. If a closed circle with zero supercoils is a substrate for the enzyme, we reason that a nicked circle would also be a substrate; and if a nicked circle is a substrate, we reason that a linear piece of DNA would also be a substrate. However, in neither of the latter two cases do we have an assay for the activity.

A final point in connection with these experiments is shown in fig. 6. This shows the effect of temperature on the limit product of the enzyme. The reactions were performed between 13° and 42°, and a progression on the same gel of the positions of the centers of the sets of bands can be seen. From this progression

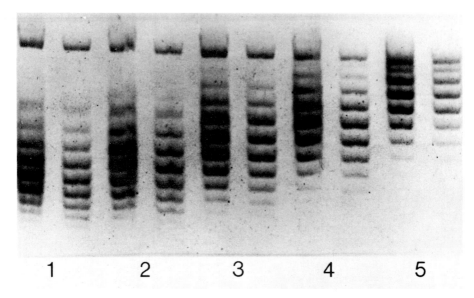

Fig. 6. The effect of the incubation temperature on the position of limit products in a slab-gel electrophoresis experiment. The paired samples contained 100 and 200 ng of DNA. The incubation temperatures were 41.5°, 37.2°, 29.6°, 22.5°, and 13.4° in (1) to (5).

the temperature dependence of the rotation angle between the base pairs of duplex DNA can be calculated. Some years ago James Wang measured this using ethidium bromide titrations with the assumption that the angle of unwinding of ethidium was 12° [8]. In more recent work he presented strong evidence that the 12° angle was wrong and that 26° was better [9]. By comparing the apparent temperature dependence of the duplex rotation angle obtained in his previous work with the present value measured in the experiment shown in fig. 6, we obtain a similar new value for the unwinding angle due to ethidium. It is for this reason, among others [10,11], that we are now entirely in agreement with Wang that the unwinding angle due to ethidium intercalation was mis-estimated several years ago, and that in general supercoiled molecules are supercoiled to about twice the extent that we had thought before.

In the presentation of the action of an enzyme, I would like to describe its purification. We published this a year ago [12] and have since modified the procedure, and I would now like to give you the method of purification of the enzyme, because we would hesitate in doing work like this unless we thought we had a pure enzyme [13]. The procedure for purifying the enzyme was worked out basically by Peter Vosberg from the following point of view. He had learned, as a result of a lot of hard work, that there were many things that one can do and consequently lose the enzyme activity. Dialysis was something to be avoided, phosphocellulose was something to be avoided, and a good many resins normally

TABLE 1

Purification of nicking-closing enzyme*.

Fraction	Vol. (ml)	Total protein (mg)	Total activity (units)	Specific activity (units/mg)	Purification
0 Sonicated nuclei	100	370	15×10^6	4×10^4	$\times 1$
I Supernatant of sonicated nuclei	98	277	14.7×10^6	5.3×10^4	$\times 1.3$
II PEG	94	183	12.5×10^6	6.9×10^4	$\times 1.7$
III Hydroxyapatite I	98.5	5.14	7.2×10^6	1.4×10^6	$\times 40$
IV DEAE cellulose	990	2.52	4.8×10^6	1.9×10^6	$\times 47.5$
V Hydroxyapatite II	31.8	0.80	1.85×10^6	2.31×10^6	$\times 58$

* Prepared from 4×10^9 cells.

used to purify enzymes were to be avoided. So he finally settled on a fairly rapid, very reproducible and very elegant procedure.

The activities at the various stages of purification are shown in table 1. From 4×10^9 cells we obtain 1.5×10^7 units of activity, which means that in each cell the amount of enzyme present is enough to carry out 10^{10} nicking-closing events. This number is estimated conservatively; the actual number could be as high as 10^{11} such events.

He purified nuclei, sonicated them in 1 M salt, centrifuged and took the supernatant. The DNA was largely removed by treatment with polyethyleneglycol in step 2 with only a very small loss of activity. The material was then loaded onto hydroxyapatite and washed with 1 M KCl and the activity sticks tightly. The column is then developed with a gradient from 0.2 to 1 M phosphate. The enzyme is eluted in 0.55 M phosphate. This material is then diluted and layered onto a DEAE cellulose column, where much of the protein is adsorbed and the enzyme passes through. The enzyme is finally concentrated on hydroxyapatite giving a recovery of about 15% of the original activity. The overall purification was not high. The results of this procedure are shown in the gel in fig. 7. Channel 2 is the starting material and Channel 1 is the final product with an estimated molecular weight of 37,000. Channels 3–5 contain the molecular weight markers.

After completing the work that I have just described, we embarked on a study of the protein chemistry of the purified enzyme. It has become clear that the purification procedure just described provides in a reproducible way a protein which is most likely histone H1, or a subfraction thereof. The protein which Vosberg was isolating in milligram quantities was shown to coelectrophorese with H1 samples from mouse cells, calf thymus and rat liver in SDS gels, in acid-urea gels and in gels containing a pH 9 running buffer. The last experiment showed that the protein was highly basic. Amino acid analyses of the sample were also consistent with the predominant polypeptide being histone H1. Arguing

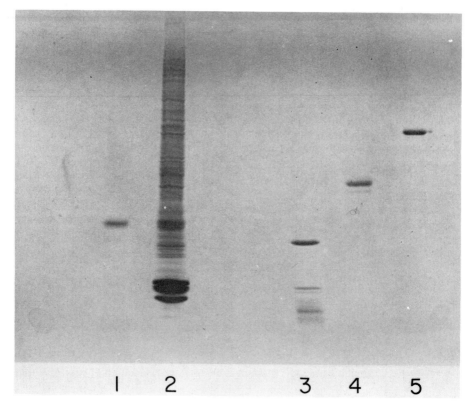

Fig. 7. SDS polyacrylamide gel electrophoresis was performed as Laemmli [14]. (1) fraction V, the final product (2) fraction I, the starting material (3) chymotrypsinogen (4) ovalbumin (5) albumin.

against the interpretation that the enzyme is H1 are the following experimental results: The activity is inhibited by N-ethylmaleimide, a sulfhydryl reagent. H1 does not contain cysteine. In addition, a modification of the extraction procedure resulted in a partial purification of the enzyme in high yield. The sample prepared by this procedure contained substantially no H1.

Cole, our first speaker, pointed out earlier in this conference that H1 is capable of reacting with other proteins as well as with DNA. The problem here may very well be a case of such an interaction. The foregoing ideas are obviously tentative in nature and we recognize that further work will be required to answer the question posed: Is the presence of H1 in our purified preparations trivial or significant?

In conclusion, we have shown that the limit product of the action of mouse cell N-C enzyme on closed circular DNA is an equilibrium distribution of isomers that differ by single superhelical turns. A comparison of these products with those produced by ligase closure of the nicked circular DNA shows that, providing the two reactions are performed under the same conditions, the products generated are indistinguishable. A further experiment demonstrated the regeneration of the equilibrium distribution from a single species. This experiment confirms the thermal origins of these distributions and shows that the N-C enzyme does not require supercoiling of its substrate. In subsequent work we have shown that the enzyme used in these experiments contained a large amount of histone H1. Regardless of whether H1 has any functional relationship to the enzyme, its presence did not effect the results of the above experiments, as evidenced by the indistinguishability of the limit products generated by ligase and by N-C enzyme.

References

[1] Wang, J.C., Interaction between DNA and an *Escherichia coli* Protein ω. J. Mol. Biol. 55, 523–533 (1971).
[2] Champoux, J.J. and Dulbecco, R. An activity from mammalian cells that untwists superhelical DNA – A possible swivel for DNA replication. Proc. Nat. Acad. Sci. 69, 143–146 (1972).
[3] Keller, W. and Wendel I. Stepwise relaxation of supercoiled SV40 DNA. Cold Spring Harbor Symp. Quant. Biol. 39, 199–208 (1974).
[4] Pulleyblank, D.E. and Morgan, A.R. Partial purification of 'ω' protein from calf thymus. Biochemistry 14, 5205–5209 (1975).
[5] Baase, W.A. and Wang, J.C. An ω protein from *Drosophila melanogaster*. Biochemistry 13, 4299–4303 (1974).
[6] Pulleyblank, D.E., Shure, M., Tang, D., Vinograd, J. and Vosberg, H.-P. Action of nicking-closing enzyme on supercoiled and nonsupercoiled closed circular DNA: formation of a Boltzmann distribution of topological isomers. Proc. Nat. Acad. Sci. 72, 4280–4284 (1975).
[7] Depew, R.E. and Wang, J.C. Conformational fluctuations of DNA helix. Proc. Nat. Acad. Sci. 72, 4275–4279 (1975).
[8] Wang, J.C. Variation of the average rotation angle of the DNA helix and the superhelical turns of covalently closed cyclic λ DNA. J. Mol. Biol. 43, 25–39 (1969).
[9] Wang, J.C. The degree of unwinding of the DNA helix by ethidium. I. Titration of twisted PM2 DNA molecules in alkaline cesium chloride density gradients. J. Mol. Biol. 89, 783–801 (1974).
[10] Pulleyblank, D.E. and Morgan, A.R. The sense of naturally occurring superhelices and the unwinding angle of intercalated ethidium. J. Mol. Biol. 91, 1–13 (1975).
[11] Shure, M. and Vinograd, J. The number of superhelical turns in native virion SV40 DNA and Minicol DNA determined by the band counting method. Cell 8, 215–226 (1976).
[12] Vosberg, H.-P., Grossman, L.I. and Vinograd, J. Isolation and partial characterisation of the relaxation protein from nuclei of cultured mouse and human cells. Eur. J. Biochem. 55, 79–93 (1975).
[13] Vosberg, H.-P. and Vinograd, J. Purification and demonstration of the enzymatic character of the nicking-closing protein from mouse L Cells. Biochem. Biophys. Res. Commun. 68, 456–464 (1976).
[14] Laemmli, U.K. Cleavage of structural proteins during the assembly of the head of bacteriophage T4. Nature 227, 680–685 (1970).

Chapter 10

TOWARDS AN IN VIVO ASSAY FOR THE ANALYSIS OF GENE CONTROL AND FUNCTION

J.B. GURDON* and D.D. Brown**

* MRC Laboratory of Molecular Biology, Hills Road, Cambridge CB2 2QH, England, and
** Department of Embryology, Carnegie Institution of Washington, 115 West University Parkway, Baltimore, Maryland 21210, U.S.A.

Abstract

Fertilized *Xenopus* eggs injected with macromolecules can serve as an *in vivo* assay system for analyzing gene control and function. This paper describes a preliminary investigation of the extent to which the genes contained in purified DNA can be used for this purpose. Experiments have been carried out with purified ribosomal DNA and 5S DNA of *X. mulleri*, and with mouse satellite DNA, injected into fertilized eggs of *X. laevis*.

The development of fertilized eggs becomes very abnormal if more than 4 ng of DNA is injected into each egg. Some of the injected DNA, previously labelled with ^{125}I, appears to become localized in nuclei and on chromosomes. Ten to fifty percent of the injected DNA persists up to the swimming tadpole stage, when introduced into fertilized eggs 4 days previously. Transcripts of mouse satellite DNA were not detected, even by sensitive methods. Transcripts of ribosomal DNA and 5S DNA were clearly detected during cleavage and late blastula stages, when the low level of endogenous synthesis of such transcripts allowed their detection by hybridization.

Abbreviations used: N. + F., stages of normal development (Nieuwkoop and Faber, 1956) - ref. 12. X.m., X.l., *Xenopus mülleri*, *Xenopus laevis*. rDNA, DNA complementary to 28S and 18S RNA and interspersed spacer sequences. 5SDNA, DNA complementary to 5S RNA and interspersed spacer sequences.

The Molecular Biology of the Mammalian Gene Apparatus:
edited by P. Ts'o 1977, Elsevier North-Holland Biomedical Press

Introduction

The controlled expression of genes in animal cells is probably brought about by regulatory macromolecules associated with chromosomes. An assay for the activity of such molecules is almost essential for their identification and for the investigation of their mode of action.

Cell-free systems for RNA synthesis include those based on whole nuclei, chromatin, pure DNA or reconstituted chromatin. In these cases, molecules whose regulatory function is to be tested are added to the system, and subsequent changes in the pattern of RNA synthesis are examined. The usefulness of in vitro transcription systems is at present limited by two considerations. First, the majority of the labelled RNA obtained from nuclei or chromatin results from the completion of chains already initiated in living cells before the cell-free system has been prepared; second, the number of transcripts synthesized by each gene in preparations of pure DNA or reconstituted chromatin is very small. For these reasons, in vitro transcription systems do not, at present, provide an entirely satisfactory test of the activity of macromolecules postulated to regulate gene transcription.

The design of a living-cell assay system

Some of the disadvantages of cell-free systems of transcription would be eliminated if it were possible to experimentally subject living cells, the genes of which continue to be transcribed over a long period of time, to the presence or absence of the regulatory molecules to be tested. Molecules larger than about 1000 daltons in molecular weight do not readily penetrate living cells, but can be injected into cells using appropriate equipment. Several possible approaches are available to determine the effects of the injected molecules on the cells. The most straightforward approach is to measure the synthesis of immediate gene-directed products which are the RNAs for which the genes code. A more indirect approach involves the recognition of the proteins which are coded for by the immediate gene products (mRNAs), or the measurement of those enzyme products for which known genes have coded. Least direct is to seek a morphological effect, such as a change in the differentiation of injected cells and their mitotic progeny. The more indirect the criteria for recognising gene activity, the greater the chance that the introduced molecules may operate at some post-transcriptional level of gene expression. It is therefore desirable to study, in an in vivo assay system, the activity of those genes whose immediate products can be conveniently measured. The genes for ribosomal RNA and 5SRNA, used in the experiments described below, fall into this category.

DNA injection

We have assumed that regulatory macromolecules will associate with most, if not all, of the genes whose activity they regulate, after injection into test cells. If the injected molecules do not associate with *all* copies of a regulated gene in a cell, or if they associate with these genes in some, but not in all, cells, the effect of the introduced molecules would be very hard to assess. The ideal way of overcoming this difficulty would be to introduce known genes as DNA to the test cells, with or without regulatory molecules already complexed with the DNA. Further advantages of this experimental system would be that many copies of the genes could be introduced, and that the type of gene could be selected so that its products could be distinguished from those of genes already active in the test cells.

This last requirement could be satisfied most easily by inserting into a test cell genes which are not normally expressed in that cell-type. This procedure would run the risk that the cytoplasm of the test cells would regulate the activity of these introduced genes, not normally expressed in that cell-type, so that they would be switched off. This is what would be predicted from previous experience with nuclear transplantation and cell fusion experiments (reviews in refs. 2 and 3). The most satisfactory way to avoid a background of transcripts similar to those being sought is to use genes of a different strain or species, but nevertheless of the same type as those expressed in the test cells. This is why, in the experiments which are described, we have used ribosomal RNA and 5S RNA genes of *X. mülleri*, injected into eggs of *X. laevis*. During gastrulation and later stages [4,5], 28S, 18S RNA and 5S RNA are actively transcribed in *Xenopus laevis* embryos. The mature 28S, 18S and 5S RNA molecules of *X. laevis* and *X. mülleri* cannot be distinguished by hybridization methods. However, the initial transcripts of the two rDNAs, that is, the 40S rRNA precursors, can be distinguished by hybridization, due to sequence differences in the two species [6]. There is no detectable 5S RNA precursor. However, spacer regions in both 5S DNA and rDNA of *X. laevis* and *X. mülleri* can be easily distinguished [1].

Toxicity of injected DNA

Fertilized eggs injected with more than a limited amount of DNA show irregular or defective cleavage in the injected region, and usually die soon after the blastula stage. A series of tests has shown that the maximum amount of DNA that can be tolerated by fertilized eggs is 2–4 ng. An injection of this quantity, though reducing viability compared to control saline-injected eggs, permits about one third of the embryos to reach the neurula stage, and about 15% to reach the feeding tadpole stage (table 1 and ref. 7). Other tests, details

TABLE 1

Toxicity of DNA injected into fertilized eggs.

Injected material	Amount injected per embryo	No. total embryos injected	% of total injected embryos reaching*			
			Blastula (st. 8)	Neurula (st. 15)	Swimming (st. 40)	Feeding (st. 45)
Saline	none (control)	103	90%	80%	79%	72%
Polyoma (native) DNA.I**	0.2 ng	163	79%	29%	25%	18%
Polyoma (native) DNA.I	2 ng	129	47%	12%	7%	6%
Mouse satellite (native)	2.5 ng	212	82%	36%	18%	10%
X. mülleri rDNA	3.5 ng	369	76%	35%	27%	23%
Calf thymus (native)	4 ng	22	86%	50%	36%	18%
Calf thymus (denatured)	4 ng	28	71%	0%	0%	0%

* Stages from Nieuwkoop and Faber[12].
** Form I configuration (double-stranded supercoiled closed circle).

The analysis of gene control and function   115

of which are not given here, have shown that uncleaved fertilized eggs are much more sensitive to injection than 2-cell eggs. DNA injected at the 2-cell stage enters both blastomeres as long as it is injected before the cleavage furrow is complete at the vegetal pole. It appears, from one series of experiments, that single-stranded DNA is several times more toxic than native DNA.

The nuclear DNA content of a fertilized egg is 6 pg, and that of a late blastula is about 60 ng. If the type of DNA injected constitutes only 0.2% of the genome (as is the case for ribosomal DNA in *X. laevis*), 2 ng of injected DNA provides a 200,000 fold excess over endogenous sequences of the same type at the 2-cell stage, and a 20-fold excess at the blastula stage.

Intracellular location of injected DNA

We have attempted to follow the fate of injected DNA using various ^{125}I-labelled DNAs. We considered this preferable to the use of [^{3}H]thymidine labelled DNA, since breakdown of the injected DNA might release [^{3}H]TdR or [^{3}H]TMP which could be reincorporated into endogenous DNA during the frequent periods of DNA synthesis which take place during cleavage.

Autoradiographic examination has been made of cleaving embryos and blastulae which resulted from 2-cell embryos injected with native or denatured calf thymus or mouse satellite DNA. In all experiments a similar result was obtained. Nuclei and chromosomes in some cells were heavily labelled (fig. 1), even though an intensely labelled cytoplasmic 'pool' of DNA was usually found concentrated in the cytoplasm in one region or in a few blastomeres. This result means *either* that some of the injected DNA was concentrated in host cell nuclei and associated with chromosomes, *or* that ^{125}I had been excised from the injected DNA in a utilizable form (possible as ^{125}I-dCMP). We have established that free ^{125}I does not become incorporated into DNA. Different nuclei in the same embryo were labelled at very different levels, and this favours the interpretation that at least some of the labelled DNA in nuclei was derived directly from the injected DNA and not from re-incorporation of its breakdown products. However, these experiments do not prove this point. We therefore investigated the persistence of sequences in the injected DNA by molecular hybridization methods.

Persistence of injected DNA

We have followed the fate of injected DNA by allowing the injected 2-cell embryos to develop for various lengths of time, extracting total DNA, and then testing for the presence of the injected DNA by molecular hybridization. DNA was immobilized on nitrocellulose filters and hybridized with [^{3}H]cRNA which had been transcribed in vitro from purified DNA of the type injected. In some

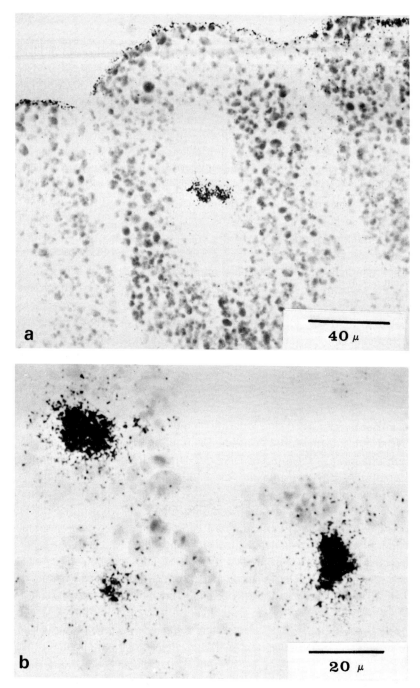

Fig. 1. Native calf-thymus DNA was iodinated (refs. 16, 17, and acknowledgements), and injected into fertilized eggs. The injected eggs were grown for a few hours, fixed, sectioned, and autoradiographed. a) Autoradiograph of a section of a mid-blastula showing a concentration of grains over chromosomes on a metaphase plate. b) Autoradiograph of three cells of a late blastula showing grains over interphase nuclei.

experiments an estimate of the efficiency of recovery of DNA from embryos was obtained by including in the hybridization mixture some ^{32}P-cRNA transcribed from total *X. laevis* DNA. Further details of the methods are described, or referred to, in the notes to table 2.

TABLE 2

Persistence of DNA injected into fertilized eggs.

1 Type of DNA injected	2 Stage of analysis [N. + F.]	3 Hours of incubation (at 19°C)	4 c.p.m.cRNA hybridized	5 Injected DNA µg injected	6 % recovered	7 Host DNA % recovered
X.m. rDNA	Unfert.eggs	5	775	105	29	—
X.m. rDNA	Blastula [8]	8	1796	115	56	150
X.m. rDNA	Neurula [14]	22	1624	105	38	200
X.m. rDNA	Tail-bud [24]	41	800	120	8	50
X.m. rDNA	Swimming [39]	84	1098	105	2	50
X.m. 5sDNA	Eggs and DNA mixed	0	223	60	6	—
X.m. 5sDNA	Unfert.eggs	0.3	5940	60	183	—
X.m. 5sDNA	Blastula [9]	11	3710	60	125	—
Mouse satellite	Eggs and DNA mixed	0	7540	83	14	—
Mouse satellite	Unfert.eggs	4	19,100	60	67	50
Mouse satellite	Blastula [6]	5	13,190	56	88	—
Mouse satellite	Hatching [34]	66	6020	60	14	—
Mouse satellite	Blastula [7]	6	43,000	75	200	—
Mouse satellite	Neurula [15]	22	41,600	75	200	250
Mouse satellite	Neurula [18]	30	24,900	80	93	150
Mouse satellite	Heart-beat [24]	41	19,500	77	88	100
Mouse satellite	Swimming [4]	96	6200	67	15	40

Column 1) The preparation and properties of the DNA samples are specified in the notes applicable to table 3, column 1. 2–3 ng of DNA was injected into each fertilized egg at the 2-cell stage, or into unfertilized eggs. In some cases (controls), DNA was mixed with unfertilized eggs, rather than injected, and frozen without incubation. Each sample contained 25–45 eggs.

Columns 2 + 3) The numbers in square brackets refer to the developmental stages of Nieuwkoop and Faber [12], which were reached after the injected embryos had been cultured at 19° for the numbers of hours indicated.

Column 4) The preparation of DNA included the following steps. Samples were homogenized in NaCl, Tris, EDTA, made 0.5% in SDS, extracted with phenol at 18°C, made 0.3 M in NaCl, and precipitated with ethanol at −20°C. DNA was denatured, immobilized on filters and hybridized with [^{3}H]cRNA, which had been prepared by in vitro transcription from a sample of the same DNA as was used for injection.

Colums 5 + 6) The varying total amount of DNA injected reflects different numbers of embryos in each sample. The recovery of injected DNA was estimated as follows. The total purified DNA from each sample was immobilized on nitrocellulose filters. Filters were prepared with known amounts of the same DNA as was used for injection. The known and unknown DNA-filters were hybridized at the same time with ^{3}H-labelled cRNA transcribed from DNA of the injected type. The known DNA-filters were used to construct a standard curve from the amounts of cRNA hybridized. Values are based on duplicate filters for each sample which varied sometimes by a factor of two. Losses of DNA during extraction were estimated by hybridizing the same filters with ^{32}P-cRNA transcribed from total *X. laevis* DNA.

Column 7) Since the numbers of cells were not accurately determined and the pre- or post-replicated state of chromosomal DNA not known (especially at early stages), the values for recovery of host DNA could be up to 100% inaccurate.

The persistence of injected *X. mülleri* ribosomal or 5S DNA, and mouse satellite DNA, is shown in table 2. The percentage of injected DNA recovered is shown in column 6. Because of the similarity of DNA containing ribosomal or 5S genes in *X. mülleri* and *laevis*, the accuracy of the results with mouse satellite DNA, which is quite dissimilar to *Xenopus* sequences, is greater than the accuracy of results with injected *Xenopus* DNA. Duplicate hybridizations, which were always carried out with each sample, gave values which differed by up to a factor of two, and therefore values in column 6 which are within a factor of two of each other are not necessarily significantly different. The values for recovery of injected DNA must be considered in the light of the decreasing efficiency of recovery of DNA from later stages (column 7).

The following conclusions can be drawn from these experiments.

1) Substantial amounts of injected DNA, in pieces at least large enough to be detectable by hybridization to cRNA, have been detected as long as 5 days after injection. During the four days that it takes to form a swimming tadpole, a 2-cell egg has undergone about 18 rounds of cell divisions to form nearly 5×10^5 cells.

2) From the blastula stage onwards, some of the injected DNA is broken down. After correcting for an approximately 25% recovery of endogenous DNA from later stages, it appears that 50–90% of the injected DNA is broken down during the four days needed for a swimming tadpole to be formed.

3) The results suggest, but do not prove, that the amount of injected DNA may increase between the 2-cell and blastula stages. It is known from other work that purified DNA is replicated after injection into unfertilized eggs [8,9]; when activated by penetration with a pipette, unfertilized eggs become metabolically similar to cleaving fertilized eggs.

Since substantial amounts of injected DNA persist in embryos which have reached the stage of ribosomal RNA synthesis (gastrulation onwards) we have questioned whether transcripts from the injected DNA can be detected.

Transcription of injected DNA

Eggs injected with DNA at the 2-cell stage were grown until a later stage, when they were given a second injection of [^{3}H]guanosine, and then frozen several hours after that. Labelled RNA was extracted from the frozen embryos and then hybridized, either with the same DNA as was used for injection, or with total bulk *X. laevis* DNA. In all cases, the DNA was immobilized on filters. Hybridization to purified *X. mülleri* rDNA or 5S DNA (table 3, column 4) shows how much [^{3}H]RNA complementary to this DNA was present. The amount of RNA which hybridized with total *X. laevis* DNA provided a measure of the amount of transcription from the endogenous host cell DNA at the same stages of development. In each case, some [^{32}P]RNA, which was homologous to the injected DNA, was added to the hybridization mixtures as an internal control for the efficiency with which [^{3}H]RNA synthesized by the embryo hybridized to the *X. mülleri* DNA. Knowing the efficiency with which pure transcripts (^{32}P-labelled RNA) hybridized to their homologous DNA (*X. mülleri* rDNA or 5S DNA), it was possible to estimate the proportion of [^{3}H]RNA which was composed of transcripts from the injected genes. Thus table 3, column 5, gives a corrected measure of the abundance of transcripts synthesized from the injected DNA.

The results in Table 3 permit the following conclusions:

1) Unequivocal evidence for the existence of transcripts of the injected DNA has been obtained in embryos injected with 5S DNA, and strong evidence was obtained for transcripts of ribosomal DNA; in both cases the evidence is clearest for RNA synthesized during cleavage, rather than at later stages.

2) Mouse satellite DNA is not transcribed to a detectable extent even though the assay ([^{3}H]cRNA) for discovering such transcripts is much more sensitive than that applicable to ribosomal and 5S DNA (see last section for reasons). Furthermore, relatively short labelling periods (of 1–2 h) were used in case any satellite DNA transcripts formed were unstable.

3) The relative rate of transcription from injected versus endogenous DNA decreases after the blastula stage. This is partly due to the greatly increasing rate of endogenous ribosomal and 5S RNA synthesis, but also is due to an apparently absolute decrease in the rate of transcription from injected DNA when gastrula and neurula stages are compared to cleavage stages. It is clear that the injected ribosomal and 5S genes are not, on the average, transcribed with the same efficiency as endogenous genes of comparable kinds. It is also clear that, on the average, the injected genes are not handled developmentally like endogenous genes of the same kind which show no detectable transcription until gastrulation [4,10]. It is possible that a small fraction of the injected DNA behaves like the equivalent type of endogenous genes, but that this effect is obscured because most of the injected DNA does not behave in this way.

TABLE 3

Transcription of DNA injected into fertilized eggs

1	2		3	4		5	
				Cpm RNA hybridized to:		Corrected rDNA transcripts	
Type of DNA injected	Duration of labelling Stages [N. + F] Hours		Total cpm in RNA and DNA	X. mülleri rDNA			as % of total
				^3H	^{32}P	cpm	^3H-nucleic acid
X.m. rDNA	Unfert. eggs [1]	5	5,100	7	98	7	—
X.m. rDNA	Cleavage [2–8]	10	182,000	224	100	224	0.1 %
X.m. rDNA	Cleavage [5–8]	5	86,000	100	117	86	0.1 %
None	Cleavage [5–8]	5	16,000	5	135	4	—
X.m.rDNA	Gastrula [8–12]	10	725,000	341	94	363	0.05%
X.m. rDNA	Neurula [14–16]	5	405,000	200	78	256	0.06%
None	Gastrula [8–12]	10	650,000	158	153	103	0.02%

						Corrected 5s DNA transcripts	
				X. mülleri 5s DNA			as % of total ^3H-nucleic acids
				^3N	^{32}P	cpm	
X.m. 5SDNA	Cleavage [2–8]	6	7,100	74	187	1360	19.2 %
None	Cleavage [2–8]	6	3,600	4	247	0	—
X.m. 5sDNA	Cleavage [2–9]	9	44,100	59	169	1160	2.6 %
None	Cleavage [2–9]	9	77,200	13	219	191	0.2 %
X.m. 5sDNA	Cleavage [7–9]	6	75,000	93	218	1530	2.0 %
None	Cleavage [7–9]	6	96,800	25	211	405	0.4 %
X.m. 5sDNA	Gastrula [9–14]	11	69,800	30	184	545	0.8 %
None	Gastrula [9–14]	11	182,900	63	238	630	0.3 %

				Cpm RNA hybridized to			Transcripts as % of total ^3H-nucleic acids
				mouse sat. DNA		Blank filters	
				1 μg	5 μg		
Mouse satellite	Cleavage [6–8]	1½	87,000	18	20	16	0.01%
None	Cleavage [6–8]	1½	162,500	31	37	20	0.01%
Mouse satellite	Cleavage [6–8]	2	134,000	48	52	53	0.01%
None	Cleavage [6–8]	2	101,600	16	17	16	—

Column 1) X.m. rDNA (ranging from 10^7–5×10^8 daltons) was prepared from ovaries of *X. mülleri* as described by Dawid et al.[13]. X.m. 5SDNA was approximately 10^7 daltons and was prepared from erythrocyte DNA of *X. mülleri* as described by Brown and Sugimoto [14]. Mouse satellite DNA was approximately 10^7 daltons and was prepared from mouse liver DNA by the method of Corneo et al.[15]. In all experiments listed in the table, between 3 and 4 ng of DNA were injected into each fertilized egg at the 2-cell stage, or into unfertilized eggs. Controls in which no DNA was injected ('none') received an injection of the same volume (c. 35 nl) of saline solution as was used for the DNA solution injected into experimental samples. For each test, between 25 and 45 eggs were injected.

Column 2) Embryos were labelled by a second injection of [^{3}H]guanosine at 5–10 mCi/ml. The numbers in square brackets refer to the normal stages of Nieuwkoop and Faber[12]. Label was injected at the first stage shown; embryos were incubated at 19°C for the number of hours indicated, and frozen at the second stage number shown in brackets.

Column 3) [^{3}H]RNA extraction included the following steps. Frozen samples were homogenized in 0.1 M Na acetate, pH 5.0, including 4 μg/ml PVS, made 0.5% in SDS, extracted with phenol at 18°C, made 0.3 M in NaCl, and ethanol precipitated at $-20°$C. The total cpm includes acid insoluble counts in RNA or DNA. When embryos were labelled during cleavage, a substantial proportion of the total cpm may have been in DNA; at later stages almost all acid insoluble label was in RNA.

Column 4) For X.m. rDNA experiments, [^{32}P]28S RNA, prepared from cultured cells, was added to the ^{3}H-labelled RNA extract to determine the efficiency of hybridization. In the case of 5S DNA injections, [^{32}P]cRNA transcribed from *X. mülleri* 5S DNA was used. Since the ^{32}P-labelled RNA was essentially a pure transcript, the efficiency with which it hybridizes to its gene on a filter can be used to estimate the total amount of ^{3}H-labelled transcripts which have the same sequence. In the case of experiments with satellite DNA, the supernatant was collected after hybridization to DNA and 5000 cpm of pure [^{3}H]cRNA made from mouse satellite DNA was added. New filters carrying 1 μg mouse satellite DNA were hybridized with this solution and bound 200–300 cpm; filters with 5 μg satellite DNA bound 350–950 cpm. Therefore, had any mouse satellite transcripts been present in the RNA extracted from injected embryos, they would have been readily detected.

Column 5) The relative amount of transcription from injected DNA is indicated, in column 5, by the ratio of [^{3}H]RNA c.p.m. and [^{32}P]RNA cpm bound to X.m. DNA (column 4). The values in column 5 were calculated as in the following example (taken from the top line of results for *X. mülleri* 5S DNA in table 3):

	cpm hybridized to 5S DNA
7100 ^{3}H-TCA-insoluble cpm	74
3400 ^{32}P-cRNA cpm added (for all samples in the table)	187

Calculation: All ^{32}P-cpm are pure 5S DNA transcripts. The fraction of these cpm which hybridize to pure 5S DNA (187 out of 3400 total cpm) shows the efficiency of hybridization (5.5% in this case). We assume that any ^{3}H-5S DNA transcripts will hybridize with the same efficiency. We can then calculate the total amount of ^{3}H-RNA (X) which are transcripts of 5S DNA, as follows:

$$\frac{187}{3400} = \frac{74}{X}. \therefore X = 1360 \text{ cpm}. \therefore \text{ out of 7100 total } ^3\text{H-cpm, 1360 are pure 5S DNA transcripts.}$$

5S DNA transcripts can also be expressed as a % of total [^{3}H]RNA: $\frac{1350}{7100} = 19\%$.

Discussion and conclusions

The experiments presented here constitute a preliminary test of the usefulness of injecting purified DNA into living cells in order to study transcriptional control, and to test the biological activity of chromosomal macromolecules.

Two aspects of the results described can be regarded as encouraging. One is the finding that a substantial amount of injected DNA persists throughout early development, that is, up to the swimming tadpole stage. Assuming that 50% of the injected DNA is present at that stage, a value based on mouse satellite DNA for which the assay was most accurate, these would be more copies of mouse satellite DNA sequences in a swimming tadpole than of any one of its own genes present as a single copy in each genome. Evidently there does not exist in embryonic cells an efficient mechanism for degrading foreign DNA. The other encouraging feature of the described experiments is the fact that at least some transcription certainly takes place from genes of the kind that are normally active in early development, and that no transcription was detected from mouse satellite DNA which is not transcribed in cells from which it is obtained. We presume from the results of other living-cell injection experiments (review in ref. 2), that the species origin (mouse, not frog) does not explain this difference.

A less satisfactory aspect of these experiments is that the injected genes are not handled with the same efficiency or in the same developmental stage specific way as the endogenous genes of a comparable kind.

We see the following prospects for the future application and development of the type of experiment described here. In its present form it has two potential applications. One is to test the biological role of molecules postulated to regulate transcriptional activity of purified genes. Since labelled transcripts are being formed more than 12 h. after DNA injection, at least some of these genes are presumably undergoing reinitiation of transcription. Potential regulatory molecules could be associated with purified DNA, injected, and the developing embryos tested for transcription from this DNA. A number of unknown factors could affect the value of this approach, one of which is the uncertainty whether molecules mixed or combined with DNA before injection will remain associated after injection. The second potential application of the type of experiment described concerns the differential handling of transcribed (28S, 18S, and 5S) DNA versus non-transcribed (satellite) DNA. Presumably, different egg components associate with these two classes of DNA so as to cause the transcription of one but not of the other. It might be possible to reisolate the two kinds of injected DNAs and compare the molecules associated with them. This approach would be made enormously easier if it were possible to arrange for a higher proportion of the injected ribosomal or 5S DNA molecules to be actively transcribed.

One of the longer-term objectives of the experiments described has been to explore the possibility of integrating injected DNA into host embryo genomes. The persistence of injected DNA through early development encourages further analysis of the kind carried out by Jaenisch and Mintz [11]. Several of the DNA-injected embryos prepared in the experiments described here have now become adult frogs, in which it is possible to look for persistent injected DNA by molecular hybridization.

Acknowledgements

We wish to thank Dr. Shigeki Mizuno very much for making available to us his DNA iodination method which is based on those described in references 16 and 17.

References

[1] D.D. Brown and K. Sugimoto, The structure and evolution of ribosomal and 5S DNA in *Xenopus laevis* and *Xenopus mülleri*. Cold Spring Hbr. Symp. Quart. Biol. 38, 501–505 (1974).
[2] J.B. Gurdon, The Control of Gene Expression in Animal Development, Oxford Univ. Press (1974).
[3] B. Ephrussi, Hybridization of Somatic Cells, Princeton Univ. Press (1972).
[4] D.D. Brown and Elizabeth Littna, R.N.A. synthesis during the development of *Xenopus laevis*, the South African clawed toad. J. Mol. Biol. 8, 669–687 (1964).
[5] L. Miller, Control of 5S RNA synthesis during early development of anucleolate and partial nucleolate mutants of *Xenopus laevis* J. Cell Biol. 59, 624–632 (1973).
[6] D.D. Brown, P.C. Wensink and E. Jordan, A comparison of the ribosomal DNAs of *Xenopus laevis* and *Xenopus mülleri*: the evolution of tandem genes, J. Mol. Biol. 63, 57–73 (1972).
[7] J.B. Gurdon, Molecular biology in a living cell. Nature 248, 772–776 (1974).
[8] J.B. Gurdon, M.L. Birnstiel and V.A. Speight, The replication of purified DNA introduced into living egg cytoplasm, Biochim. Biophys. Acta. 174, 614–628 (1969).
[9] R.A. Laskey and J.B. Gurdon, Induction of polyoma DNA synthesis by injection into frog-egg cytoplasm. Eur. J. Biochem. 37, 467–471 (1973).
[10] J.B. Gurdon, Nuclear transplantation and the control of gene activity in development. Proc. Roy. Soc. B, 176, 303–314 (1970).
[11] R. Jaenisch and B. Mintz, Simian virus 40 DNA sequences in DNA of healthy adult mice derived from preimplantation blastocysts injected with viral DNA. Proc. Nat. Acad. Sci. 71, 1250–1254 (1974).
[12] P.D. Nieuwkoop and J. Faber, eds., Normal Table of *Xenopus laevis* (Daudin) (North-Holland Pub. Co., Amsterdam) (1956).
[13] I.B. Dawid, D.D. Brown and R.H. Reeder, Composition and structure of chromosomal and amplified ribosomal DNAs of *Xenopus laevis*. J. Mol. Biol. 51, 341–360 (1970).
[14] D.D. Brown and K. Sugimoto, 5S DNAs of *Xenopus laevis* and *Xenopus mülleri*: evolution of a gene family. J. Mol. Biol. 78, 397–415 (1973).
[15] G. Corneo, E. Ginelli, C. Soave and G. Bernardi, Isolation and characterization of mouse and guinea pig satellite deoxyribonucleic acids. Biochemistry 7, 4373–4379 (1968).
[16] S.L. Commerford, Iodination of nucleic acids in vitro. Biochemistry 10, 1993–1999 (1971).
[17] M.S. Getz, L.C. Altenburg and G.F. Saunders, The use of RNA labelled in vitro with iodine–125 in molecular hybridization experiments. Biochim. Biophys. Acta 287, 485–494 (1972).

Chapter 11

TRANSCRIPTIONAL UNIT OF MOUSE MYELOMA*

RU CHIH C. HUANG

Department of Biology and the McCollum-Pratt Institute, The Johns Hopkins University, Baltimore, Md. 21218, U.S.A.

Mouse plasma tumor, or myeloma, synthesizes and usually secretes homogeneous immunoglobulin. The amount of immunoglobulin synthesized comprises up to 30% of the total protein made by the tumor cell. Great varieties of myelomas have been isolated for a range of immunoglobulin chain types [1]. Correlated with this large production of immunoglobulin, myeloma microsomes maintain a large percentage of specific immunoglobulin mRNA [2]. Among several light chain and heavy chain mRNAs now isolated, the structure of the kappa chain mRNA ($mRNA_k$) has been extensively studied [2,3]. This mRNA is approximately 1250 ($\pm$100) nucleotides in length, of which 640 nucleotides are responsible for the coding region, 321 nucleotides for the variable region of the kappa chain protein, and roughly the same number of nucleotides for its constant region. There is a substantial untranslated sequence in $mRNA_k$ situated both at the 5' end and also at the 3' end of the molecule, bridging the coding region and the poly A sequences.

For several years our laboratory has been engaged in studies related to the transcription and translation of $mRNA_k$ of different myelomas. Messenger RNA activity for kappa light chains of MOPC 41 was first established in a rabbit reticulocyte lysate system [4]. The in vitro protein product resembles that of the authentic kappa chain in all tryptic peptides except the NH_2 terminal sequence. Similar findings were reported for the cell-free synthesized kappa chain of myeloma MOPC 21. A precursor kappa chain isolated as an in vitro product was some fifteen amino acids longer than the secreted kappa chain isolated from the tumor [5]. We have extensively purified the $mRNA_k$. Reverse transcribed copy ($cDNA_k$) was made from $mRNA_k$ template. When [^{32}P]$cDNA_k$ was hybridized back with its template, it hybridized rapidly over about a 2 log range of CrT

---

* This work was supported by NIH grants 2R01-CA13953 and 8R01-AG00472.

The Molecular Biology of the Mammalian Genetic Apparatus:
edited by P. Ts'o © 1977, Elsevier North-Holland Biomedical Press

values and had a $CrT_{1/2} = 1.7 \times 10^{-3}$ [6]. From the CrT of known RNA standards and the molecular weight of $mRNA_k$, it was predicted that the purity of the $mRNA_k$ was higher than 90% [7]. We have used the $cDNA_k$ to titrate the gene coding for the kappa chain by way of a cDNA/DNA hybridization reaction [4]. It was found that in the haploid genome of the mouse, there are only a few copies, or perhaps as low as one copy of gene coding for the constant region of a kappa chain. Very recently Farace et al. have extended the gene titration studies to the entire sequence of the kappa chain of MOPC 41. They concluded that within a subgroup, there is only one or a very few genes coding for the entire kappa chain [8]. In other words, it is possible that in the genome of the MOPC 41 tumor cell, the entire $mRNA_k$ is transcribed from one copy of its genes.

Establishment of an in vitro transcriptional system

There are a few basic criteria which must be fulfilled for establishing an accurate in vitro transcriptional system. The system has to be active and stable to allow the RNA synthesis to proceed in vitro for a considerable length of time. The system should be able to initiate, elongate and terminate RNA chain properly. Also, the system must permit the correct strand of DNA to be copied. We have found that the isolated nuclei of MOPC 315 and 66.2 myeloma cell lines are very active in RNA synthesis and fulfil the requirements described above [6,9]. Nuclei which were isolated from these two cell lines synthesized RNA in vitro linearly for at least one hour at 25°C (fig. 1). Three known RNA polymerases (I, II, III) were found to be present in myeloma nuclei [10]. The size distribution of the in vitro RNA product resembles closely the profile of pulse labeled cellular RNA (fig. 2).

Transcription of 5S RNA and precursor tRNA in vitro was carefully analyzed. The size of the 5S synthesized in vitro is identical to that of cellular 5S as examined by acrylamide gel electrophoresis under denatured conditions (fig. 3). Precursor tRNA (4.5S) and matured tRNA (4S) were both observed [11]. The nuclei preparation apparently contains the enzyme responsible for the specific processing of the 4.5S to 4S tRNA species. Initiation of synthesis of the 5S RNA was carried out by [γ-^{32}P]GTP incorporation study in 66.2 myeloma nuclei. Guanosine tetraphosphate (^{32}pppGp) was identified after RNase T_1 digestion. Since cellular 5S RNA begins with guanosine triphosphate, we have concluded that in vitro 5S RNA is initiated correctly. In fact, we have estimated that at least 80% of the synthesized 5S RNA is initiated in vitro. DNA strand selection for the synthesis of 5S RNA is also preserved in vitro [11]. Synthesized 5S RNA was complementary only to the plus strand of 5S DNA, isolated according to the method of Brown et al. [12]. This experimental evidence indicated to us that

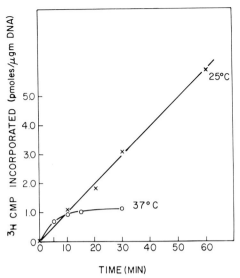

Fig. 1. Effect of temperature on RNA synthesis. Nuclei were incubated at either 37 or 25°C. At various times aliquots (2.5 × 10^5 nuclei) were removed and Cl$_3$CCOOH precipitable counts determined. 1 pmole = 5000 cpm.

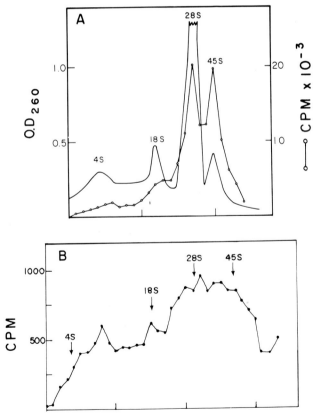

Fig. 2. Size of RNA synthesized by myeloma 66.2 tumor cells and nuclei. A) Nuclear RNA isolated after 10 min of pulse labeling with [^{3}H]uridine. B) Nuclear RNA isolated after 15 min of nuclear RNA synthesis in vitro with [^{3}H]GTP. ——— O.D.$_{260}$, ○——○ in vivo pulse label counts, ●——● in vitro made RNA.

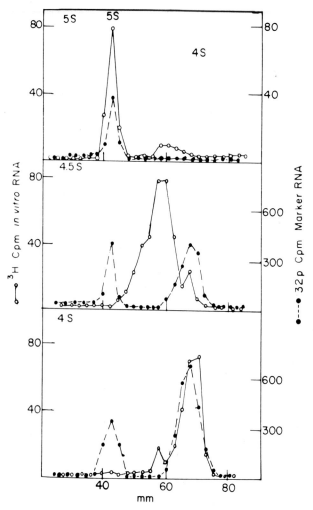

Fig. 3. Analysis of RNA by gel electrophoresis in 70% formamide. In vitro synthesized 4S, 4.5S and 5S RNAs were isolated and dissolved in 70% formamide and analyzed by electrophoresis in polyacrylamide gels polymerized in 70% formamide. Electrophoresis was for 16 h at 70 V at 27°C. ^{32}P-labeled 5S and 4S were added as markers. O——O ^{3}H in vitro product, ●——● ^{32}P RNA markers.

myeloma nuclei perhaps can be used as a model system for examining the factors essential for transcription of specific single copy genes.

Synthesis of kappa chain mRNA (mRNA$_k$) in vitro

Two myeloma tumor lines were chosen for this study, one serving as a control to the other. Cell line 66.2 synthesizes kappa light chain, and therefore also the kappa mRNA. The $mRNA_k$ comprises 0.025% of the total cellular RNA of this line. The control, MOPC 315, synthesizes and secretes IgA. Its light chain belongs to the class of lambda (λ_2). Both the amino acid sequences of the V-region and the C-region of a lambda light chain are considerably different from those of a kappa chain [13,14]. It was predicted that the two messengers, $mRNA_k$ and $mRNA_{\lambda_2}$, will have entirely different nucleotide sequences. For study of $mRNA_k$ synthesis in 66.2 cell nuclei, MOPC 315 nuclei will therefore be an ideal control.

It seems to us that one of the best ways to sort out the mRNA synthesized in vitro from the other RNA sequences is to use the isolated gene (DNA_k) as a probe for detecting the $mRNA_k$ sequence after a DNA_k/RNA hybridization reaction. We are currently developing a method to achieve this which will be discussed further in this paper. Alternatively, one may use a cDNA probe. In this assay, the stabilization of radioactive $cDNA_k$ against S_1 nuclease measures the mRNA content in a given preparation [15]. 66.2 cell nuclei contain a large proportion of endogenous $mRNA_k$ sequences. If not eliminated, the hybridization of the radioactive cDNA could therefore be due entirely to the endogenous mRNA, and not the in vitro product. Separation of the in vitro made RNA from that of in vivo sequences becomes a necessity if a cDNA probe is used.

Dale, Livingston and Ward have reported that a variety of enzymes are capable of utilizing 5-mercuridine triphosphate as a substrate in replacement of UTP provided that the mercury group is first blocked by a mercaptan [15]. We have used this method to 'tag' the in vitro made RNAs. Hg-RNAs were allowed to separate from cellular RNAs by binding the mercurated RNA to a sulfhydryl affinity column. An excellent separation was achieved (fig. 4).

Large scale nuclear RNA synthesis was possible using Hg-UTP as one of the substrates. Hg-RNA was isolated and allowed to hybridize to $cDNA_k$ in an $Hg-mRNA_k$ excess reaction [6]. CrT analyses were made for Hg-RNA synthesized both by nuclei of the 66.2 cell line and by nuclei of the MOPC 315 cell line (fig. 5). From the CrT analyses, it is estimated that the $mRNA_k$ concentration in the in vitro product of MOPC 315 cannot exceed 2% of that in 66.2 nuclear in vitro transcripts (table 1). This comparison was truly valid since the seeding control experiment (fig. 5) ruled out the possibility of endogenous $mRNA_k$ contamination in the in vitro transcripts of 66.2 cell nuclei. The template for $mRNA_k$ synthesis in MOPC 315 in vivo is also restricted. Essentially only a few, or no $mRNA_k$ sequences can be detected in the isolated cellular RNA of MOPC 315. The concentration of $mRNA_k$ in the in vitro transcripts of 66.2 nuclei was found comparable to those present in 66.2 cell nuclei (table 1).

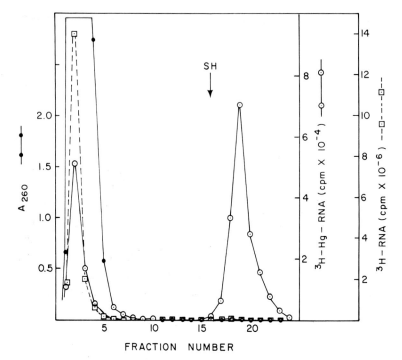

Fig. 4. Binding of Hg-RNA to a sulfhydryl affinity column. RNA was synthesized in isolated nuclei using either UTP (□) or Hg-UTP (○) and [^{3}H]GTP as label. The RNA was extracted and chromatographed on Sephadex G-50 in 0.1 M sodium acetate (pH 6) and 0.1% sodium dodecyl sulfate. The excluded material, containing all of the incorporated radioactivity, was put directly on a column of Bio-Rad Affi-Gel 401. After the bound RNA was washed with the same buffer, it was eluted by the addition of 0.1 M 2-mercaptoethanol to the buffer, indicated in the figure by 'SH'. (●) The nuclear RNA as measured by absorbance at 260 nm.

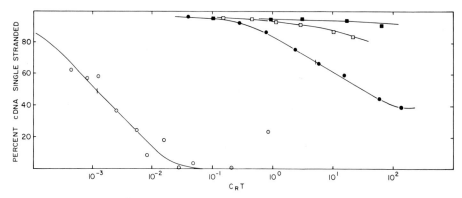

Fig. 5. Hybridization of [^{32}P]cDNA$_k$ with nuclear Hg-RNA synthesized in vitro. Nuclear Hg-RNA, synthesized in vitro, from either 66.2 (●) or MOPC 315 (□) was purified over Affi-Gel 401 and hybridized with cDNA. (■) A control hybrid of MOPC 315 Hg-RNA that had been seeded with unincubated 66.2 nuclei (ref. 6); (○) hybridization with purified mRNA$_k$.

TABLE 1

Kinetics of hybridization and mRNA$_k$ content of various RNA populations. In vitro nuclear Hg-RNA was synthesized using retained endogenous nuclear RNA polymerase activity.

Source of RNA hybridized	$C_rT_{1/2}$	Fractional mRNA$_k$ content
In vivo RNA:		
MOPC 41 mRNA$_k$	0.0017	1.0
MOPC 41 microsomes	1.6	110×10^{-5}
66.2 microsomes	3.7	46×10^{-5}
66.2 nuclei	2.6	65×10^{-5}
MOPC 315 microsomes	4000	0.04×10^{-5}
In vitro Hg-RNA:		
66.2 nuclei	5.2	32×10^{-5}
MOPC 315 nuclei	200	0.85×10^{-5}

A comparison between nuclear transcription and chromatin transcription by *E. coli* RNA polymerase

Since in vitro transcripts can be clearly isolated because of the Hg-UTP used in RNA synthesis, a direct comparison of the amount of mRNA$_k$ synthesized by nuclei vs. chromatin/*E. coli* polymerase became possible. Chromatin was isolated and purified to free the contaminating endogenous RNA polymerase [17]. *E. coli* RNA polymerase was used to transcribe the available DNA sequences in the chromatin. The concentration of mRNA$_k$ in the chromatin transcripts was found much lower than that found in the nuclei transcripts as judged by CrT analyses (table 2). For instance, chromatin transcripts were hybridized to cDNA$_k$ to give a CrT$_{1/2}$ of 120, while nuclei transcripts gave a value of 5.2. By this calculation, mRNA$_k$ concentration in the chromatin transcripts was about 25 times lower

TABLE 2

Comparisons between mRNA$_k$ content of in vitro transcripts of myeloma nuclei and chromatin/*E. coli* RNA polymerase.

Source of Hg-RNA hybridized	$C_rT_{1/2}$	Fractional mRNA$_k$ content
66.2 nuclei	5.2	32×10^{-5}
315 nuclei	200	0.85×10^{-5}
66.2 chromatin	120	1.4×10^{-5}
315 chromatin	1100	0.15×10^{-5}

than that in the nuclei transcripts. The reason for this difference is unknown at the present time. It may result from the recognition differences by *E. coli* RNA polymerase vs. myeloma RNA polymerase, or perhaps some labile control functional factor may be lost during the chromatin isolation, or the rate of processing the synthesized products is actually different in these two cases.

We have also compared the extent of mRNA$_k$ synthesis by the two cell line (66.2 vs. 315) chromatins. Methods for chromatin isolation are identical. *E. coli* RNA polymerase was used for transcription of both chromatins. We have found that the concentration of mRNA$_k$ in 66.2 chromatin transcripts is ten times higher than that in the MOPC 315 chromatin transcripts. In this case, 66.2 chromatin hybridizes to cDNA$_k$ to give a CrT$_{1/2}$ of 1.2×10^2 while those from MOPC 315 give a CrT$_{1/2}$ of 1.1×10^3. Thus, control of mRNA$_k$ synthesis is somewhat more stringent in MOPC 315 chromatin.

Some unsolved problems in analyzing transcription of the kappa chain gene

Many unanswered questions still exist: what is the size of the primary product for mRNA$_k$? What kinds of initiation nucleotides are there? How are the stages of initiation, elongation and termination of this specific mRNA$_k$ controlled? mRNA$_k$ can either be synthesized with pppAp or pppGp as its 5' terminal or be further modified as a capped structure [18]. The conventional way for detecting a RNA initiator is to cleave the specific product enzymatically and to look either for a highly charged nucleotide tetraphosphate or to look for a 'capped' structure such as m^{7}G$_{ppp}$X$_p$. This method, although applicable when a relatively large amount of specific product can be isolated [11], becomes extremely difficult to use when only limited transcript is available, since the identification is based mostly on the radioactivity of this one initiator nucleotide. We are currently developing a new method for this study using a modified nucleotide triphosphate as an initiator. The entire initiated RNA can be selectively separated out by means of affinity chromatography because the modified nucleotide triphosphate is used for initiation. In this case, the detection can be extended to the radioactivity of the whole RNA molecule. This method therefore will be several hundred times more sensitive than the conventional method currently available.

During the last decade we have learned a great deal about the specificities of histone/histone, histone/DNA and histone/non-histone/DNA interactions [19,20,21,22,23,24]. Several structural features of the reconstituted chromatin are found to be similar to those of native chromatin. For instance, at least certain functional specificities are preserved during the chromatin reconstitution. It seems, however, difficult to foresee how transcription of a single copy gene can easily be studied by gross reconstitution of total DNA and heterogenous protein propulations when the sequence of interest is only approximately $2 \times 10^{-3}\%$ of a total genome.

Isolation of specific genes thus becomes unavoidable if an understanding of the fine regulation of its transcription is desired. We have recently succeeded in isolating ribosomal RNA gene using mercurated rRNA to 'fish out' its gene in a modified R loop type experiment. In this study, 18S RNA and 28S RNA were mercurated in vitro according to the method of Dale et al. [25]. Degradation of RNA during mercuration is eliminated when 0.1% SDS was used during the entire reaction. Mercurated rRNA was allowed to hybridize with total cellular DNA (70% formamide, 0.5 M Na^+ at 52°C). The rDNA/mercurated rRNA hybrid was recovered from sulfhydryl agarose column after two recyclings. The DNA recovered represents approximately 0.3% of the input DNA, and can be stabilized by rRNA under standard DNA/RNA hybridization conditions [26].

Isolation of the genes coding for $mRNA_k$ by this procedure is now in progress. It is hoped that the specific kappa gene can be isolated and further replicated in a recombinant DNA environment [27]. We will then have a sufficient amount of DNA_k to further analyze the structure of a kappa chain gene and its transcriptional properties. The same method can, of course, be used to analyze other single copy genes provided that its specific gene product RNA can be isolated in a pure form.

References

[1] M. Potter, Immunoglobulin-producing tumors and myeloma proteins of mice. Physiol. Rev. 52, 631 (1972).
[2] J. Stavnezer, R.C.C. Huang, E. Stavnezer and J.M. Bishop, Isolation of messenger RNA for an immunoglobulin kappa chain and enumeration of the genes for the constant region of kappa chain in the mouse. J. Mol. Biol. 88, 43 (1974).
[3] C. Milstein, G.G. Brownlee, E.M. Cartwright, J.M. Jarvis and N.J. Proudfoot, Sequence analysis of immunoglobulin light chain messenger RNA. Nature 252, 354 (1974).
[4] J. Stavnezer and R.C.C. Huang, Synthesis of a mouse immunoglobulin light chain in a rabbit reticulocyte cell-free system. Nature New Biol. 230, 172 (1971).
[5] N.J. Cowan, T.M. Harrison, G.G. Brownlee and C. Milstein, The cell-free synthesis of immunoglobulin chains. Biochem. Soc. Trans. 1, 1247 (1973).
[6] M.M. Smith and R.C.C. Huang, Transcription in vitro of immunoglobulin kappa light chain genes in isolated mouse myeloma nuclei and chromatin. Proc. Nat. Acad. Sci. U.S.A. 73, 775 (1976).
[7] M.L. Birnstiel, B.H. Sells and I.F. Purdom, Kinetic complexities of RNA molecules. J. Mol. Biol. 63, 21 (1972).
[8] M.G. Farace, M.F. Aellen, P.A. Briand, C.H. Faust, P. Vassalli and B. Mach, No detectable reiteration of genes coding for mouse MOPC 41 immunoglobulin light chain mRNA. Proc. Nat. Acad. Sci. U.S.A. 73, 727 (1976).
[9] W.F. Marzluff, Jr., E.C. Murphy and R.C.C. Huang, Transcription of ribonucleic acid in isolated mouse myeloma nuclei. Biochemistry 12, 3440 (1973).
[10] L.B. Schwartz, V. Sklar, J.A. Jaelning, R. Weilnmann and R.G. Roeder, Isolation and partial characterization of the multiple forms of deoxyribonucleic acid-dependent RNA polymerase in the mouse myeloma, MOPC 315. J. Biol. Chem. 249, 5889 (1974).

[11] W.F. Marzluff, E.C. Murphy, Jr. and R.C.C. Huang, Transcription of the genes for 5S ribosomal RNA and transfer RNA in isolated mouse myeloma cell nuclei. Biochemistry 13, 3689 (1974).
[12] D.D. Brown, P. Wensink and E. Jordan, Purification and some characteristics of 5S DNA from *Xenopus laevis*. Proc. Nat. Acad. Sci. U.S.A. 68, 3175 (1971).
[13] P. Coffino and M.D. Scharff, Rate of somatic mutation in immunoglobulin production by mouse myeloma cells. Proc. Nat. Acad. Sci. U.S.A. 68, 219 (1971).
[14] E.P. Schulenberg, E.S. Simms, R.G. Lynch, R.A. Bradshaw and H.N. Eisen, Primary structure of the light chain from the myeloma protein produced by mouse plasmacytoma MOPC 315. Proc. Nat. Acad. Sci. U.S.A. 68, 2623 (1971).
[15] W.D. Sutton, A crude nuclease preparation suitable for use in DNA reassociation experiments. Biochim. Biophys. Acta 240, 522 (1971).
[16] R.M.K. Dale, D.C. Livingston and D.C. Ward, The synthesis and enzymatic polymerization of nucleotides containing mercury: Potential tools for nucleic acid sequencing and structural analysis. Proc. Nat. Acad. Sci. U.S.A. 70, 2238 (1973).
[17] R.C.C. Huang and P.C. Huang, Effect of protein bound RNA associated with chick embryo chromatin on template specificity of the chromatin. J. Mol. Biol. 39, 365 (1969).
[18] J.M. Adams and S. Cory, Modified nucleosides and bizarre 5'-termini in mouse myeloma mRNA. Nature 255, 28 (1975).
[19] R. Rubin and E.N. Moudrianakis, The F_3-F_{2a1} complex as a unit in the self-assembly of nucleoproteins. Biochemistry 14, 1718 (1975).
[20] R.D. Kornberg and J.O. Thomas, Chromatin structure: Oligomers of histones. Science 184, 865 (1974).
[21] L. Kleiman and R.C.C. Huang, Reconstitution of chromatin. The sequential binding of histone to DNA in the presence of salt and urea. J. Mol. Biol. 64, 1 (1972).
[22] J.A. D'Anna, Jr. and I. Isenberg, Fluorescence anisotropy and circular dichroism study of conformational changes in histone II_{b2}. Biochemistry 11, 4017 (1972).
[23] G. Felsenfeld, R. Axel, H. Cedar and B. Sollner-Webb, The specific template activity of chromatin. Ciba Foundation Symposium 28, 29 (1975).
[24] R.S. Gilmour and J. Paul, Role of non-histone components in determining organ specificity of rabbit chromatin. FEBS Letters 9, 242 (1970).
[25] R.M.K. Dale, E. Martin, D.C. Livingston and D.C. Ward, Direct covalent mercuration of nucleotides and polynucleotides. Biochemistry 14, 2447 (1975).
[26] R.C.C. Huang, unpublished results.
[27] Research involving recombinant DNA molecules. National Institutes of Health Guidelines, June 1976.

Chapter 12

TRANSCRIPTIONAL CONTROL IN HIGHER EUKARYOTES

J. PAUL, N. AFFARA, D. CONKIE, R.S. GILMOUR, P.R. HARRISON, A.J. MacGILLIVRAY, S. MALCOLM and J. WINDASS

Beatson Institute for Cancer Research, 132 Hill Street, Glasgow G3 6UD, Scotland

The problem of studying regulation in higher eukaryotes is quite simply one of complexity, the complexity of the genome under study, complexity of the regulatory elements and complexity of the formal problem. But, as a scientific activity, it is entirely justified by two considerations, its direct relevance to human developmental diseases, especially cancer, and the fact that investigations on mammalian tissues have already yielded an enormous amount of information about gene number, sequence complexity classes of nucleic acids, and the synthesis, processing and translation of messenger RNA.

To tackle a problem of this kind, two possible approaches can be used. One is to work with a relatively simple biological system. This is the approach, for example, of the fungal geneticist. It is clearly relevant to studies in higher eukaryotes but will be excluded from the present discussion which is concerned with more direct attacks. The second approach is to attempt to study a component of a complex system in isolation; this is what we have undertaken to do in our studies on erythropoiesis.

During maturation of erythroid cells, several functions are co-ordinately expressed. Although the genes are not linked, α and β globin chains are synthesised at the same rate; certain other proteins, such as spectrin, catalase and carbonic anhydrase accumulate and there is a great increase in the formation of haem. There is, therefore, a spectrum of clearly identifiable elements to examine. The problem is to acquire a detailed understanding of the control of synthesis of some of these elements. As one component of our work we have chosen to base our strategy on the preparation and use of highly specific and sensitive molecular probes (mainly cDNA) to enable us to follow the fate of transcripts from individual genes. We have particularly concentrated on the globin genes and their transcripts [1-5].

In the studies to be outlined, we have compared three situations, erythroid versus non-erythroid mammalian cells, mature versus immature erythroid cells

The Molecular Biology of the Mammalian Genetic Apparatus:
edited by P. Ts'o © 1977, Elsevier Nort-Holland Biomedical Press

and uninduced Friend erythroleukaemia cells versus the same cells induced to synthesise haemoglobin.

Erythroid versus non-erythroid cells

We can assign quite accurate numbers to some components of globin synthesis. For example, in the human there are probably two α genes and four or five genes of β type, comprising one or two β genes, two γ genes and one Δ gene. There are also genes for embryonic α-type chains; the number is not known but is probably small. In the mouse the total number of gene sequences of α and β-type is less than ten [6]. These numbers are the same in erythroid and non-erythroid tissue [4]. In studying the co-ordinate expression of all the α and β-type genes, one has therefore a fairly simple experimental system.

The number of globin messenger RNA molecules in a non-erythroid cell is less than ten and may be zero in some tissues (table 1), while in the fully differentiated reticulocyte, there are about 150,000. Moreover, in non-erythroid cells, haemoglobin is not detectable (but the methods used are not particularly sensitive), whereas a fully mature erythrocyte contains about 2.5×10^8 haemoglobin molecules. One therefore has a wide range of quantitative responses and these appear to implicate control at the transcriptional level. However, these observations do not yet prove this notion as we have inadequate information about turnover.

TABLE 1

Cell	Number of globin mRNA molecules in	
	Nucleus	Cytoplasm
Friend Ft uninduced	100	300
Friend Ft induced (5d)	1400	9000
L5178Y	17	5
LS	2	<1
Reticulocyte	—	150,000

Mature versus immature erythroid cells

Induction of normal erythroid maturation in the mouse can be taken as a reasonable model for most mammals (fig. 1). The first morphologically identifiable erythrocyte precursors are proerythroblasts and basophilic erythroblasts. These cells contain no measurable haemoglobin, which however accumulates at the next

Transcriptional control in higher eukaryotes

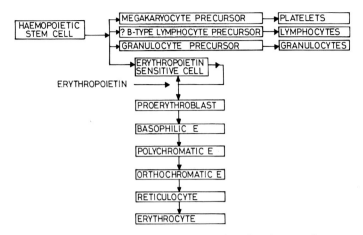

Fig. 1. Schematic representation of the maturation of erythrocytes from precursor cells in the mouse.

stage of maturation, the polychromatic erythroblast. The proerythroblast and basophilic erythroblast are, nevertheless, committed to form erythrocytes. They themselves arise from two kinds of precursors which have been demonstrated experimentally although not identified morphologically. One is a general haemopoietic stem cell or colony-forming cell (CFC) which can give rise to several types of blood cells; the other is the erythropoietin-sensitive cell (ESC) which can only replicate or, under the influence of erythropoietin, differentiate into a proerythroblast.

The CFC and ESC are not at present amenable to direct biochemical experimentation, but it is possible to determine whether accumulation of globin messenger RNA can be demonstrated in any of the identifiable erythroblasts by using globin cDNA to detect globin messenger RNA in individual cells by in situ hybridisation [7,8]. These studies revealed that, particularly in immature erythroid tissue, only a few proerythroblasts have diffuse distribution of globin messenger RNA but most contain very little (fig. 2). In marked contrast, basophilic erythroblasts have a high concentration of globin messenger. Moreover, in erythroid tissue which has been actively stimulated by erythropoietin by experimental means in mice or by the presence of erythropoietin in tissue culture, a much higher proportion of proerythroblasts exhibits globin messenger RNA in the cytoplasm [9]. These experiments show rather clearly that globin messenger RNA begins to accumulate in the proerythroblast. Terada and his colleagues [10] have made a similar observation that proerythroblasts from immature foetal liver have no translatable globin messenger RNA. Putting these findings together, they imply that one of the first events in the final stage of erythroid maturation is the accumulation of globin messenger in the proerythroblast. It should be noted that the expression of this messenger may, however, be controlled at a post-transcriptional level since haemoglobin is not detectable in basophilic erythroblasts.

Studies with Friend erythroleukaemia cells

To study these control mechanisms, investigations have been undertaken with Friend erythroleukaemia cells [11]. These cells resemble proerythroblasts in morphology. They do not respond to the hormone erythropoietin but, after treatment with dimethylsulphoxide or a number of other substances, they exhibit a range of erythroid characteristics, including the accumulation of mouse α and β globins [12,13,14] and globin messenger RNAs [15,16,17]. In preliminary studies, we measured the globin mRNA concentration in different cellular compartments during the course of induction and were able to detect two extreme patterns of behaviour [15,16]. In all Friend cells so far studied, the nuclear RNA contains a significant amount of globin messenger RNA. In one type of cell, a considerable increase in the concentration of globin messenger sequences in the nucleus occurs, together with a larger increase in the cytoplasm during induction of haemoglobin synthesis. In another line of cells, however, very little change in globin messenger RNA concentration occurs in the nucleus during induction; the increase in globin messenger RNA is confined entirely to the polysomes. These findings support the view that induction involves both transcriptional and post-transcriptional controls.

To extend our study of these phenomena, we prepared and studied mutants and hybrids formed between them. Some of the mutants were discovered in our stocks [18]; others were produced by mutagenisation and/or selection for non-inducibility by being grown continuously in a high concentration of DMSO [8]. Three non-inducible cells and their hybrids with inducible Friend cells have proved to be of particular interest.

i) The FtI^{d-} variant. In the inducible parent Friend cell (FtI^{d+}) (table 2), about 80% of the cells are haemoglobinised after 5 days of treatment with DMSO; this is accompanied by a 30-fold increase in cytoplasmic RNA and a 14-fold increase in nuclear globin RNA. By contrast, similar treatment of the FtI^{d-} variant results in haemoglobinisation of only about 0.5% of the cells and there is no significant increase in the globin mRNA content of cytoplasm or nucleus. Moreover, the basal levels of RNA in both compartments are lower in these cells than in the parent FtI^{d+} cells (although a significant amount of globin mRNA can be detected.

When a hybrid cell is prepared from cells with these two phenotypes, it also exhibits a very low level of haemoglobinisation. Treatment with DMSO promotes either no increase or a very modest increase in globin mRNA in both nucleus and cytoplasm, but it is of some interest that the basal level of globin mRNA is quite high in the cytoplasm of these cells. In this instance, therefore, the non-inducibility of haemoglobin synthesis seems to be due to a deficiency in globin gene expression at the nuclear level. Moreover, this defect behaves in a trans-dominant manner.

TABLE 2

Globin RNA levels in the inducible FtI^{d+} T parent, the non-inducible FtI^{d-}B variant and the corresponding hybrid. Values quoted are those of single experiments or the average of three or more (usually 4–6) independent experiments ± the standard deviation. DMSO (1.5%) treatment was for 5d.

Cell	% cells Hb + +DMSO	Cytoplasm		Nucleus	
		−DMSO	+DMSO	−DMSO	+DMSO
I^{d+} FtT	80	300 ± 150	9000 ± 3000	100,110	1400
I^{d-} FtB	~0.5	100 ± 40	150 ± 80	15,15	35 ± 10
I^{d+} × I^{d-}	~0.1	900 ± 300	2000 ± 1000	80 ± 40	90 ± 70

ii) *The FwI^{d-} variant (table 3)*. This variant resembles the FtI^{d-} variant in that only a small fraction of the cells becomes haemoglobinised on treatment with DMSO and there is no significant increase in the concentration of globin mRNA. However, when hybrids between this variant and normal Friend cells are prepared, they behave differently from hybrids with the FtI^{d-} variant in that they show an increased number of haemoglobinised cells and an increase in the amount of globin messenger RNA on treatment with DMSO. There is a considerable variation in the quantitative response among different clones. In this hybrid, there may be some evidence for a gene dosage effect but there is no evidence of trans-dominant repression, in contrast to the FtI^{d+} × FtI^{d-} hybrid.

TABLE 3

Globin mRNA levels in the FtI^{d+}B parent and the FwI^{d-}T variant and the corresponding hybrid.

Cell	% cells containing haemoglobin +DMSO	Molecules globin mRNA	
		−DMSO	+DMSO
I^{d+}FtB	80	170 ± 70	6000 ± 2000
I^{d-}FwT	~1	500 ± 250	500 ± 200
I^{d+} × I^{d-} uncloned	15–30	–	6000, 8000
clone 1/4	55	600, 800	20000 ± 8000
clone 1/5	6	600	1200, 1500

It was found that the FwI^{d-} variant, although not inducible by DMSO, can be induced with haemin and to a greater extent, with a combination of DMSO, and haemin (table 4). This would appear to implicate haem deficiency as the defect in this cell, but it is of particular interest that the accumulation of globin mRNA parallels the accumulation of haemoglobin. Hence, treatment with haem affects the accumulation of globin mRNA as well as the accumulation of globin, and one has to speculate whether this phenomenon results from feedback to the level of transcription of the globin gene or stabilisation of the messenger itself.

TABLE 4

Effect of DMSO and/or haemin treatment on accumulation of globin mRNA in Friend cell variants. Values (parts per million(ppm) of cytoplasmic RNA) are quoted for independent experiments, or the average ± the standard deviation of the results.

Inducible status	Other markers	Treatment of cells			
		Control	1.5% DMSO	0.1 mM Haemin	Haemin DMSO
I^{d+}	Ft	70 ± 30	800 ± 300	250 ± 100	800 ± 300
	FtB	7 ± 3	450 ± 150	55 ± 30	700, 1000
I^{d-}	LyT	< 2	< 2	< 2	< 4
	FtB	4 ± 2	5 ± 3	4, 8	13, 20
	FwT	20 ± 10	35 ± 10	50 ± 15	300 ± 80

iii) The LyI^{d-} variant (table 5). The Ly cell is a lymphoma cell line (L5178Y) and therefore not erythroid at all. Hence, it is not surprising that it accumulates few, if any, globin-specifying RNA sequences in the cytoplasm and nucleus after DMSO treatment. However, when a hybrid is formed by fusing this cell with an inducible Friend cell, there is a marked accumulation of globin mRNA in both cytoplasm and nucleus on treatment with DMSO, although no haemoglobin can be measured. Because of some resemblances between the behaviour of this hybrid line and the FwI^{d-} line, the effect of haem on haemoglobin induction was investigated; this increased both the rate of synthesis of globins and the accumulation of globin mRNA. It also increased the effect of DMSO on the globin mRNA levels.

These studies emphasise the fact that expression of globin genes is complex and that both transcriptional and post-transcriptional events can be involved, sometimes one being dominant, sometimes the other. One particular caution which must be remembered is that it cannot simply be assumed that the accumulation of a mRNA or a protein is necessarily due to transcriptional control. In all these studies, information about the stability of mRNA is lacking. Reliable data on RNA stability are very difficult to obtain and, therefore, we have found it desirable to try to tackle the problem directly, by studying transcription of the globin gene in chromatin.

TABLE 5

Globin mRNA levels in the FtI^{d+}B and LyI^{d-}T cells and the corresponding hybrid.

Cell		% cells Hb+	Molecules of globin RNA			
			Cytoplasm		Nucleus	
		+DMSO	−DMSO	+DMSO	−DMSO	+DMSO
I^{d+}	FtB	80	170 ± 70	6000 ± 2000	25, 35	1600
I^{d-}	LyT	Nil	< 20	< 20	< 5	< 10
I^{d+} × I^{d-}		Nil	900 ± 200	8000 ± 2500	160 ± 80	700 ± 200

Transcription from chromatin

Studies on transcription from chromatin go back many years to some of the original experiments undertaken in James Bonner's laboratory in 1963 [19] and have clearly demonstrated that chromatin is a less efficient template than DNA for RNA synthesis by a bacterial RNA polymerase [20,21,22,23]. Our own contribution to this work was the introduction of RNA/DNA hybridisation methods to study the qualitative nature of the transcript. Early studies from our own and James Bonner's laboratory (among others) demonstrated clear differences between the RNA transcribed from chromatin and that transcribed from DNA [21,24,25]. These hybridisations were conducted under what we now refer to as low C_0t conditions. Within the past two or three years, with the availability of cDNA as a probe, it has become possible to reexplore the problem, this time studying individual genes located in the unique component of the genome. When transcription of the globin gene from chromatin prepared from erythroid and non-erythroid tissue is studied, it is found that, whereas the globin gene is transcribed from the former with bacterial polymerase, it is not transcribed from the latter [26,27,28,29]. One of the great difficulties in interpreting these findings has been the presence of endogenous globin-coding RNA in chromatin [28], but several experiments have now been successfully undertaken which seem to eliminate this as a cause of confusion [26,28,29]. These experimental results, demonstrating accessibility of the globin gene in chromatin from erythroid tissue but not in chromatin from non-erythroid tissue, provide strong evidence for the existence of transcriptional controls mediated through structural modifications of chromatin.

In earlier work, evidence was obtained that the non-histone proteins of chromatin play an important role in determining the specificity of this structural

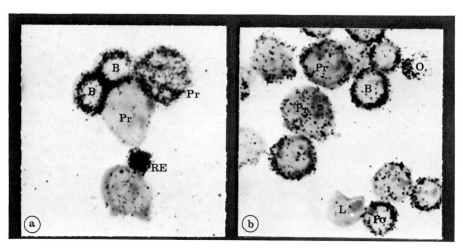

Fig. 2. In situ hybridisation of mouse globin cDNA to A) Liver from mouse embryo 13.5 days post-fertilization, B) Spleen from adult mice rendered anaemic by phenylhydrazine injection. Key: Pr, proerythroblast; B, basophilic erythroblast; Po, polychromatic erythroblast; O, orthochromatic erythroblast; RE, reticulocyte; L, lymphocyte.

modification [21]. These conclusions have now received further confirmation in relation to the globin gene. For example, it has been demonstrated that, although *E. coli* RNA polymerase cannot transcribe the globin gene in unmodified chromatin from brain tissue or from L5178Y cells, it can transcribe the globin gene from brain chromatin or LY5178 chromatin which has been reconstituted in the presence of non-histone proteins from an erythroid tissue such as mouse foetal liver [28,30]. These reconstituation experiments have proved difficult to do and are not highly reproducible. Part of the problem has been the presence of proteases in chromatin, which we have now been able to control by including the protease inhibitor, phenylmethyl sulfonylfluoride (PMSF), at all stages. Another problem has been adequate resolution against a background of endogenous RNA which contaminates the non-histone proteins; improved methods of preparation have now reduced this to very low levels. Finally, nucleases in reagents have been consistently troublesome. By devoting attention to minimising the effects of these factors, more reliable results have now been obtained and this has permitted us to undertake studies on individual non-histone protein fractions prepared by the technique of MacGillivray et al. [31]. (The fractions formerly called H1–H4 are now referred to as HAP1–HAP4 to avoid confusion with the new histone terminology). In these experiments (fig. 3), purified mouse DNA was combined with fraction HAP1 (together with histones) plus HAP2, HAP3 or HAP4 separately in the ratios DNA:histone:NHP = 1:1.5:1.5. Fraction HAP2, comprising the non-histone proteins with the lowest affinity for hydroxyapatite, is most active. This is still a very complex fraction but it

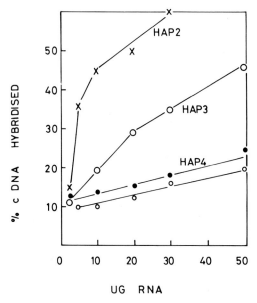

Fig. 3. Titration of mouse globin cDNA against RNA obtained by transcription with *E. coli* polymerase of nucleoprotein reconstituted from DNA, histones and fractions HAP2 (×——×), HAP3 (O——O) and HAP4 (●——●). Controls without added RNA polymerase (o —— o).

seems likely that it may now be possible, by use of reconstitution methods, to identify those classes of protein which contain regulatory elements.

Obviously, transcription from chromatin could also be used in an attempt to resolve the problem in the Friend cell of recognising true transcriptional control. In experiments to investigate transcriptional control in Friend cells, we were able to demonstrate marked differences in the ability of chromatin from uninduced and induced cells to act as templates for transcription of globin sequences [15,16].

Fractionation of chromatin

Some of our preconceived ideas about euchromatin and heterochromatin have led us, among others, to investigate the possibility of fractionating chromatin into active and inactive components. Because of the evidence that euchromatin is particularly rich in non-histone proteins as well as histones, it was postulated that the protein to DNA ratio should be higher in euchromatin than heterochromatin. Hence, several investigators have investigated the possibility of separating eu- and heterochromatin by buoyant density methods [32,33]. We were particularly interested in the possibility of using non-ionic aqueous buoyant density media, especially metrizamide. The first observations with this medium were essentially negative in that when chromatin was sheared to a few thousand base pairs by either mechanical methods or nuclease, no separation into buoyant density classes could be found [34,35]. Since most genes are at least 1,000 base pairs long, the transcriptional units may be considerably larger. Hence, this result implies that, provided active and inactive components are preserved in isolating chromatin, they must have approximately the same buoyant densities and, therefore, approximately the same protein to DNA ratios.

On shearing to fragments much smaller than gene size, we were, however, able to demonstrate heterogeneity in density. Very light nuclease shearing revealed a protein-poor component which appeared to be highly nuclease-sensitive since it very rapidly disappeared on continued digestion (fig. 4). It seems quite likely that it corresponds to the fraction described by Gottesfeld et al. [36]. On more extensive shearing either by sonication to about 360 base pairs or by nuclease hydrolysis to 200 base pairs or less, two other components appeared, one with more protein than the other. The protein-rich component represents about 25% of the DNA in the original chromatin, contains nearly all the non-histone proteins, and exhibits many of the characteristics associated with active chromatin. For example, it is associated with nascent RNA, has a somewhat higher polylysine-binding capacity and is a more efficient template for DNA-dependent RNA synthesis. Both fractions contain all the histones, although H1 is present in a lower ratio to the other histones in the particle of higher density.

Does this represent a fractionation into active and inactive components? The most convincing criterion for genome fractionation is the demonstration that those genes which are expressed in cells from which chromatin is derived are in the euchromatin fraction. The complexity of the DNA in the euchromatin fraction

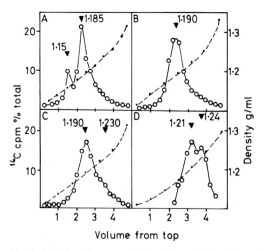

Fig. 4. Fractionation on metrizamide gradients of products obtained by digesting chromatin with micrococcal nuclease for varying times.

would also be expected to be less than in the heterochromatin fraction. We therefore studied the annealing characteristics of DNA from both heavy and light fractions prepared from chromatin which had been sheared either mechanically or by nuclease treatment. In no case did we find any difference between DNA fractions and the total DNA of the species. We also measured the globin gene concentration in the fractions derived from sonicated chromatin and observed no differences [35]. Finally, we studied the distribution of all those genes which are expressed as polyadenylated polysomal messenger RNA. In these experiments, total polyadenylated mRNA was transcribed with reverse transcriptase to provide a cDNA probe; this was then used as a tracer for the corresponding genes during re-annealing of DNA. No evidence for enrichment or depletion of these genes in the different chromatin fractions was observed. These findings demonstrate that the isolated fractions do not contain a specific component of the genome. By this criterion, we are clearly not fractionating chromatin into euchromatin and heterochromatin. What then does the fractionation mean? Since chromatin has to be sheared to a size corresponding to one or two nucleosomes before density fractions become apparent, it must be assumed that what we are demonstrating is two classes of nucleosomes, one associated with significant amounts of non-histone proteins. It is possible that proteins have moved in relation to each other during our preparation of chromatin and in these experiments this has not been excluded.

In the studies briefly outlined in this communication, we have obtained ample evidence that the control of globin synthesis is complex and involves post-transcriptional steps which may, in some circumstances, assume major importance. This has compelled us to ask again whether transcriptional controls of the globin gene actually exist. From studies on whole cells, the picture is by no means clear. Although the concentration of globin mRNA molecules is 50–100,000 times higher in mature reticulocytes than in non-erythroid cells, it has not been excluded that this is not entirely due to differences in globin mRNA stability, and definitive experiments to clarify the question are not easy to design. The strongest evidence for transcriptional controls would, indeed, seem to be the demonstration that chromatin from erythroid cells provides a more efficient template for transcription of the globin gene than does chromatin from non-erythroid cells. This may be the only way to obtain unequivocal information about transcriptional regulation in future work. However, the reproducibility of the technique is still not sufficiently high and efforts in this direction need to be intensified.

Acknowledgments

Original research referred to was supported by grants from the Medical Research Council and Cancer Research Campaign.

References

[1] D.L. Kacian and S. Spiegelman, In vitro synthesis of DNA components of human genes for globins. Nature New Biol. 235, 167 (1972).

[2] J. Ross, H. Aviv, E. Scolnick and P. Leder, In vitro synthesis of DNA complementary to purified rabbit globin mRNA. Proc. Nat. Acad. Sci. 69, 264 (1972).

[3] I. Verma, G.F. Temple, H. Fan and D. Baltimore, In vitro synthesis of DNA complementary to rabbit reticulocyte 10S RNA, Nature New Biol. 235, 163 (1972).

[4] P.R. Harrison, G.D. Birnie, A. Hell, S. Humphries, B.D. Young and J. Paul, Kinetic studies of gene frequency. I. Use of a DNA copy of reticulocyte 9S RNA to estimate globin gene dosage in mouse tissues, J. Mol. Biol. 84, 539 (1974).

[5] B.D. Young, P.R. Harrison, R.S. Gilmour, G.D. Birnie, A. Hell, S. Humphries and J. Paul, Kinetic studies of gene frequency. II. Complexity of globin complementary DNA and its hybridisation characteristics. J. Mol. Biol. 84, 555 (1974).

[6] B.D. Young and J. Paul, Absence from the mouse genome of DNA sequences related to the globin gene family, J. Mol. Biol. 96, 783 (1975).

[7] P.R. Harrison, D. Conkie, N.A. Affara and J. Paul, In situ localisation of globin messenger RNA formation. I. During mouse foetal liver development. J. Cell Biol. 6, 402 (1974).

[8] D. Conkie, N.A. Affara, P.R. Harrison, J. Paul and K. Jones, In situ localisation of globin messenger RNA formation. II. After treatment of Friend virus-transformed mouse cells with dimethylsulfoxide. J. Cell Biol. 63, 414 (1974).

[9] D. Conkie, L. Kleiman, P.R. Harrison and J. Paul, Increase in the accumulation of globin mRNA in immature erythroblasts in response to erythropoietin in vivo or in vitro. Exp. Cell Res. 93, 315 (1975).

[10] M. Terada, L. Cantor, S. Metafora, R.A. Rifkind, A. Band and P.A. Marks, Globin messenger RNA activity in erythroid precursor cells and the effect of erythropoietin. Proc. Nat. Acad. Sci. 69, 3575 (1972).

[11] C. Friend, W. Scher, J. Holland and T. Sato, Haemoglobin synthesis in murine virus-induced leukaemic cells in vitro: stimulation of erythroid differentiation by dimethylsulfoxide. Proc. Nat. Acad. Sci. 68, 378 (1971).

[12] Scher, W., J.G. Holland and C. Friend, Haemoglobin synthesis in murine virus-induced leukaemic cells in vitro. I. Partial purification and identification of haemoglobin. Blood 37, 428 (1971).

[13] W. Ostertag, H. Melderis, G. Steinheider, W. Kluge and S. Dube, Synthesis of mouse haemoglobin and globin mRNA in leukaemic cell cultures. Nature New Biol. 239, 231 (1972).

[14] S.H. Boyer, K.D. Wuu, A.W. Noyes, R. Young, W. Scher, C. Friend, H.D. Preisler and A. Bank, Haemoglobin synthesis in murine virus-induced leukaemic cells in vitro: structure and count of globin chains produced. Blood 40, 823 (1972).

[15] R.S. Gilmour, P.R. Harrison, J.D. Windass, N.A. Affara and J. Paul, Globin messenger RNA synthesis and processing during haemoglobin induction in Friend cells. I. Evidence for transcriptional control in clone M2. Cell Diff. 3, 9 (1974).

[16] P.R. Harrison, R.S. Gilmour, N.A. Affara, D. Conkie and J. Paul, Globin messenger RNA synthesis and processing during haemoglobin induction in Friend cells. II. Evidence for post-transcriptional control in clone 707. Cell Diff. 3, 23 (1974).

[17] J. Ross, Y. Ikawa and P. Leder, Globin messenger RNA induction during erythroid differentiation of cultured leukaemia cells. Proc. Nat. Acad. Sci. 69, 3620 (1972).

[18] J. Paul and I. Hickey, Haemoglobin synthesis in inducible uninducible and hybrid Friend cell clones. Exp. Cell Res. 87, 20 (1974).

[19] J. Bonner and R.C. Huang, Properties of chromosomal nucleohistones. J. Mol. Biol. 6, 169 (1963).

[20] J. Paul and R.S. Gilmour, Template activity of DNA is restricted in chromatin. J. Mol. Biol. 16, 242 (1966).
[21] J. Paul and R.S. Gilmour, Organ-specific restriction of transcription in mammalian chromatin. J. Mol. Biol. 34, 305 (1968).
[22] G.P. Georgiev, L.N. Ananieva and J.V. Kozlov, Stepwise removal of protein from deoxyribonucleoprotein complex and de-repression of the genome. J. Mol. Biol. 22, 365 (1966).
[23] K. Marushige and J. Bonner, Template properties of liver chromatin. J. Mol. Biol. 15, 160 (1966).
[24] R.C. Huang and A. Huang, Effect of protein-bound RNA associated with chick embryo chromatin on template specificity of the chromatin. J. Mol. Biol. 39, 365 (1968).
[25] I. Bekhor, G.M. Kung and J. Bonner, Sequence specific interactions of DNA and chromosomal proteins. J. Mol. Biol. 39, 351 (1969).
[26] R.S. Gilmour and J. Paul, Tissue-specific transcription of the globin gene in isolated chromatin. Proc. Nat. Acad. Sci. 70, 3440 (1973).
[27] R. Axel, H. Cedar and G. Felsenfeld, Synthesis of globin ribonucleic acid from duck-reticulocyte chromatin in vitro. Proc. Nat. Acad. Sci. 70, 2029 (1973).
[28] R.S. Gilmour, J.D. Windass, N.A. Affara and J. Paul, Control of Transcription of the Globin Gene. J. Cell Physiol. 85, 449 (1975).
[29] T. Barrett, D. Maryanka, P.H. Hamlyn and H.J. Gould, Non-histone proteins control gene expression in reconstituted chromatin. Proc. Nat. Acad. Sci. 71, 5157 (1974).
[30] J. Paul, R.S. Gilmour, N.A. Affara, G.D. Birnie, P.R. Harrison, A. Hell, S. Humphries, J.D. Windass and B.D. Young, The globin gene: structure and expression. Cold Spring Harbour Symp. on Quant. Biol. 28, 885 (1974).
[31] A.J. MacGillivray, A. Cameron, R.J. Krauze, D. Rickwood and J. Paul, The non-histone proteins of chromatin, their isolation and composition in a number of tissues. Biochim. Biophys. Acta. 277, 384 (1972).
[32] R. Hancock, Separation by equilibrium centrifugation in CsCl gradients of density-labelled and normal deoxyribonucleoprotein from chromatin. J. Mol. Biol. 48, 357 (1970).
[33] Y.V. Ilyin and G.P. Georgiev, Heterogeneity of deoxyribonucleoprotein particles as evidenced by ultracentrifugation in cesium chloride density gradient. J. Mol. Biol. 41, 299 (1969).
[34] G.D. Birnie, D. Rickwood and A. Hell, Buoyant densities and hydration of nucleic acids, proteins and nucleoprotein complexes in metrizamide. Biochim. Biophys. Acta. 331, 282 (1973).
[35] D. Rickwood, A. Hell, S. Malcolm, G.D. Birnie, A.J. MacGillivray and J. Paul, Fractionation of unfixed chromatin by buoyant-density centrifugation. Biochim. Biophys. Acta. 353, 353 (1974).
[36] J. Gottesfeld, W.T. Garrard, G. Bagi, R.F. Wilson and J. Bonner, Partial purification of the template-active fraction of chromatin: a preliminary report. Proc. Nat. Acad. Sci. 71, 2193 (1974).

Chapter 13

STRUCTURAL ANALYSIS OF NATIVE AND RECONSTITUTED CHROMATIN BY NUCLEASE DIGESTION

MICHAEL J. SAVAGE and JAMES BONNER

Division of Biology, California Institute of Technology, Pasadena, California 91125, U.S.A.

Summary

Native and reconstituted rat liver chromatin have been examined by digestion with staphylococcal nuclease and deoxyribonuclease II. Chromatin was reconstituted by salt-urea step dialyses, a reconstitution procedure developed by Bekhor et al. [5] and by Huang and Huang [23] and currently being used by many investigators. Chromatin thus reconstituted yields a staphylococcal nuclease limit digest pattern similar to that of native chromatin. However, the intermediate digestion products of reconstituted chromatin are markedly different than those of native chromatin. Native chromatin is sheared by the nuclease into subunits or multimers of the subunit. Multimers as high as the octomer were clearly resolved on 2.5% acrylamide gels. Digestion of reconstituted chromatin under identical conditions does not reveal multimers of the subunit above dimers. Limited digestion of native and reconstituted chromatin with DNase II reveals that although in native chromatin the nuclease sensitive regions correspond to the actively transcribed subset of the genome, the regions excised early in the digestion of reconstituted chromatin represent a random sample of total genomal DNA sequences. We conclude that the procedures which separate a subset of transcribed sequences from nontranscribed sequences in native chromatin do not yield a similar result with reconstituted chromatin due to the perturbed structure of the reconstituted material.

Introduction

The technique of chromatin reconstitution as originally developed in our lab [5] and by Huang and Huang [23] has been, and is currently being used to elucidate the regulatory roles of the various chromatin components. Several in-

The Molecular Biology of the Mammalian Genetic Apparatus:
edited by P. Ts'o © 1977, Elsevier/North-Holland Biomedical Press

vestigators have reported that chromatin reconstituted by a salt step gradient in the presence of urea [5,23] retains much of the transcriptional [4,5,12,15–17, 23,25,38,44] and structural [6,14,36,37,43] fidelity of native chromatin. Transcriptional fidelity is suggested by the experiments of Paul and Gilmour [38,39] and others [4,5,23,25,43] who have reported that transcription of native and reconstituted chromatin with exogenous *E. coli* RNA polymerase results in the production of tissue specific RNA sequences in approximately the same abundance. These results must be interpreted with caution in view of the presence of endogenous message sequences [4,17], and the difficulty in demonstrating that tissue specific RNA transcripts are produced de novo [17,38].

The structural similarity of native and reconstituted chromatin is suggested by chemical composition [9,32,34], binding of reporter molecules specific for the minor grove of DNA [43], and similarity of melting profiles [5,6,23]. However, recent evidence [3,9,11,13] indicates that chromatin proteins may be extensively degraded by chromatin bound protease during the dissociation and reconstitution of chromatin isolated from several tissues. Also Boseley et al. [7] using X-ray diffraction, have demonstrated that the presence of urea during reconstitution is detrimental to the production of higher order structure. We have analyzed native and reconstituted rat liver chromatin by digestion with exogenous endonucleases. Our findings are that even in the presence of protease inhibiting agents the reconstitution process does not yield material structurally similar to native chromatin. These results, as well as the ones mentioned above [3,7,9,11,13] cast doubt upon the structural and subsequent transcriptional fidelity of chromatin reconstitution.

Methods

Chromatin preparation and reconstitution

Chromatin was prepared from frozen rat livers as described by Douvas et al. [13] except that the grinding medium (saline EDTA) was made 1 mM in phenylmethyl sulfonyl fluoride from a 200 mM stock in isopropanol. For reconstitution experiments freshly prepared chromatin was dissociated in 5 M urea, 2 M NaCl, 10 mM Tris pH 7.6, and 0.1 mM EGTA, at a concentration of 0.5 mg DNA/ml. Reconstitution was accomplished by dialysis against 1.0 M, 0.8 M, 0.6 M, 0.4 M, 0.2 M, 0.1 M and finally 0.0 M NaCl, all in the presence of 5 M urea, 10 mM Tris pH 7.6 and 0.1 mM EGTA. Dialysis was for 4 h in each step. The urea was then removed by extensive dialysis against 10 mM Tris, 0.1 mM EGTA. Chromatin thus reconstituted was treated in an identical fashion with native chromatin.

Nuclease digestion

Staphylococcal nuclease (Worthington) digestions were carried out in the buffers described by Noll[33] and Axel[1] at a chromatin concentration of 25 A_{260}/ml. Staphylococcal nuclease was present at 50 units/ml and digestion was at 37°C. Reactions were terminated and DNA purified as described by Noll[33]. Polyacrylamide-agarose gel electrophoresis of the digestion products were analyzed on either 6% or 2.5% polyacrylamide-agarose gels as described by Loening[29].

DNase II (Worthington) digestion of native and reconstituted chromatin was exactly as described previously [18,20] except that the chromatin was dialyzed overnight vs. 200 volumes of 25 mM NaAc, 0.1 mM EGTA pH 6.6. DNA purification and renaturation analysis was done as described [18–20].

Results

Staphylococcal nuclease

It is now well established that the bulk of native chromatin exists in the form of repeating subunits [22], termed nucleosomes [35] or *v*-bodies [34]. Each subunit is thought to consist of about 200 base pairs of DNA held in a 6.5:1 packing ratio by eight histone molecules [21,26,46]. These subunits may be visualized by electron microscopy [14,35] or by analysis of the mild staphylococcal nuclease digestion products on sucrose gradients [14,27,31] or polyacrylamide gels [31,32,46]. Such analysis reveals that the chromatin subunit structure is a highly organized repeating pattern.

The internal structure of the chromatin subunit has also been examined by nuclease digestion [32,1]. These investigations suggest that subunit structure and organization arises from specific histone-histone and histone-DNA interactions in chromatin [41].

Reconstitution of the subunit structure has been demonstrated by electron microscopy [35] and nuclease digestion [1]. However, these reconstitution experiments were done in a different fashion than commonly employed by researchers interested in transcription of the reconstituted material [5,12,15,17,23,25,38,44]. Therefore we have examined rat liver chromatin, reconstituted by a salt-step gradient in the presence of urea, by digestion with staphylococcal nuclease and polyacrylamide gel electrophoresis of the resulting digestion products. The results of these experiments are shown in figs. 1 and 2.

We have analyzed the digestion kinetics of native and reconstituted chromatin in two different conditons. The first (fig. 1) is the buffer system described by Axel et al.[1]. Under these conditions the DNA of reconstituted chromatin

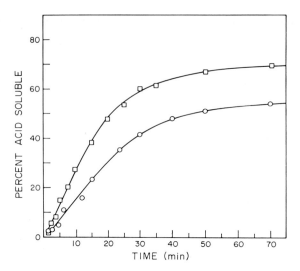

Fig. 1. Release of acid soluble products during staphylococcal nuclease digestions of native (○) and reconstituted (□) rat liver chromatin.

digests more rapidly and to a greater extent than does the DNA of native chromatin. The rate and extent of digestion of native rat liver chromatin is similar to results reported by others [2,27]. Native and reconstituted chromatin have also been digested by staphylococcal nuclease in the buffer described by Noll [33]. The kinetics of digestion (not shown) in this buffer are similar to those in fig. 1, reaching limit digest values of 52% and 67% acid soluble, for native and reconstituted chromatin, respectively. Thus the DNA of reconstituted chromatin is digested more rapidly and to a greater extent than the DNA of native chromatin in both buffers. The reason for the increased susceptibility of reconstituted chromosomal DNA is not clear. It may be due in part to aggregation of proteins in the reconstitution process, and thus to residual nuclease sensitive DNA [10].

The products of digestion in low [1] and high [33] ionic strength buffers were analyzed by polyacrylamide-agarose electrophoresis. The results are shown in fig. 2. These results are consistent with the observations of Griffith [21] and Simpson and Woodcock [42]. In the lowest ionic strength media [1] the v-bodies are extended, thus allowing multiple cleavage sites in the inter-bead spacers, leading to high backgrounds (fig. 2B). In the higher ionic strength buffer [33], inter-bead regions are condensed and the nuclease labile sites are better defined, thus generating a more discrete pattern of bands (fig. 2A). In the case of reconstituted chromatin, regardless of ionic conditions, no bands above the dimer appear (fig. 2A,B). This indicates that the spacing between v-bodies in reconstituted chromatin is not regular, an observation consistent with electron microscopic studies of Oudet et al. [35] who have analyzed chromatin reconstituted by salt-gradient dialysis. The disorderly arrangement of v-bodies may

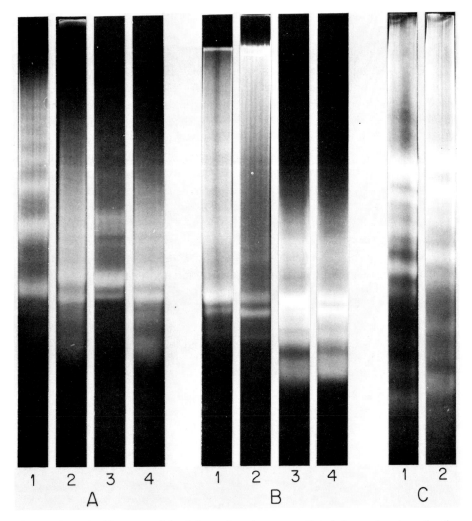

Fig. 2. Polyacrylamide-agarose slab gel electrophoresis patterns of staphylococcal nuclease digestion products. (A) Chromatin was digested to 10% PCA solubility of native (1) or reconstituted (2), and 30% PCA solubility of native (3) and reconstituted (4) in the buffer described by Noll [33] and analyzed on a 2.5% acrylamide gel. (B) Chromatin was digested to 5% PCA solubility of native (1) and reconstituted (2) or 20% PCA solubility of native (3) and reconstituted (4) in the buffer described by Axel [1] and analyzed on a 2.5% acrylamide gel. (C))hromatin was digested for 60 min in the buffer described by Axel [1] and analyzed on a 6% polyacrylamide gel, native chromatin, 54% PCA soluble (1) reconstituted chromatin, 67% PCA soluble (2).

be due to the order of binding of histones to DNA during dialysis in salt and urea. Since H_1 binds first, at ~0.4 M NaCl, 5 M urea [10,24], the association of the other histones into the subunit structure may not produce the regular inter-bead spacing seen in native chromatin, resulting in a random organization of the subunits along the DNA fiber.

Since the monomer and dimer bands are present in reconstituted material we decided to analyze the organization of the monomer v-body by extensive staphylococcal nuclease digestion. Axel et al. [1] have shown that this pattern is dependent upon all of the major histones, and may be reconstituted using the histones and homologous or heterologous DNAs. However, their reconstitution scheme differs from that commonly used by most investigators who have reported on the fidelity of chromatin reconstitution, in that the urea is removed at 0.6 M NaCl, an ionic strength at which it is known that histones have not yet associated with DNA [10,24]. Thus, in these experiments [1] most of the histones reassociate with DNA in the absence of urea. Chromatin reconstituted as described in Methods was digested and analyzed on 6% polyacrylamide gels (fig. 2C). It is clear that the digestion products of reconstituted chromatin reveal an intra-v-body organization similar to that in native chromatin, and it appears that the histone-histone and histone-DNA interactions required to generate the v-body conformation of native chromatin take place in the reconstitution scheme used herein. However, the spacing of the v-bodies along the DNA fiber appears to be randomized in the reconstitution process. Thus, although the v-body has the same internal structure, the inter v-body regions differ from those of native chromatin. This difference in v-body arrangement may be due to the fact that several classes of NHCP as well as H1 bind before and at the same ionic strength as do the histones in this dialysis scheme [10].

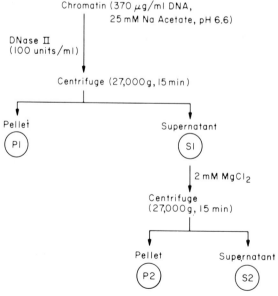

Fig. 3. DNase II fractionation scheme.

Deoxyribonuclease II

We have reported earlier on the fractionation of interphase chromatin by the endonuclease DNase II [18,30]. This fractionation scheme is based on the hypothesis that in physiological ionic strength the inactive regions of chromatin are condensed and insoluble while the active regions, which are extended in vivo [28] should remain soluble [30]. By limited digestion of chromatin with DNase II followed by centrifugation and selective precipitation of inactive chromatin in the presence of Mg^{++} ions, a fraction ($\sim 11\%$) of native chromatin remains in the supernatant (fig. 3). Extensive renaturation [20] analysis has demonstrated that the DNA of this chromatin fraction, termed S_2, represents a subset of the sequences in the rat genome, containing 12% of the middle repetitive and 11% of the single copy complexity of total rat DNA. Furthermore, the single copy component of the S_2 fraction is enriched in sequences coding for cellular RNA [18]. Thus it appears that transcriptionally active chromatin exists in a more extended, nuclease sensitive state, in ionic conditions simulating those found in vivo.

We have used this highly sequence specific fractionation scheme as a probe to study the structure of reconstituted rat liver chromatin. In these experiments chromatin was dissociated and reconstituted by a salt-step gradient in the presence of 5 M urea as described in the Methods section. After dialysis in low ionic strength buffer (10 mM Tris, 0.1 mM EGTA) the chromatin was resuspended and dialyzed overnight against 25 mM NaAc, 0.1 mM EGTA pH 6.6. This chromatin was then fractionated with DNase II as shown in fig. 3, and as described previously [18].

The digestion kinetics of native and reconstituted chromatin are compared in fig. 4. As can be seen in fig. 4b the plateau amount of chromatin in the S_2 fraction is higher with reconstituted chromatin (45%) than with native chromatin (22%) as substrate. The S_1 plateau is similar ($\sim 82\%$). That a greater portion of the chromatin is soluble in 2 mM Mg^{++} indicates that the protein components responsible for precipitation of the majority (75%) of the S_1 in native chromatin are arranged in a different fashion in reconstituted chromatin. This rearrangement results in precipitation of only 45% of the S_1 material. We have also analyzed the appearance of acid soluble material during the course of DNase II digestion (results not shown). These data imply that chromatin DNA is more accessible to DNase II in the reconstituted complex than in the native material, an idea consistent with the above finding, and the results obtained using staphylococcal nuclease.

In the DNase II digestion of native chromatin the rapidly liberated fraction corresponds to the template active regions [18]. We have therefore compared the DNA sequences liberated early during limited digestion of native and reconstituted chromatin. Native and reconstituted chromatins were digested as in

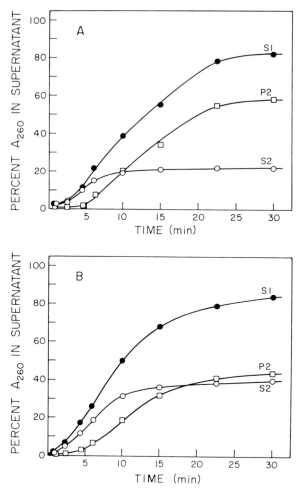

Fig. 4. Time course of digestion of native (A) and reconstituted (B) rat liver chromatin. Fractionation was carried out as described in fig. 3, and the percent in the supernatant determined by absorption at 260 nm in 0.9 N NaOH.

fig. 3 until ~11% of the chromatin DNA was found in the S_2 fraction. The digestion was terminated by raising the pH to 8.5 with 1 M Tris, pH 11 and the DNA was purified according to Gottesfeld et al.[19]. The DNA was further fractionated by chromatography on a Sephadex G-200 column in 50 mM Na_3PO_4, according to Britten et al.[8]. The excluded peak (>100 nucleotides in length) was pooled and used in reassociation experiments. The DNA fractions purified in this manner had an average single strand size of 310 and 340 nucleotides for the reconstituted and native S_2 preparations, respectively, as determined by alkaline sucrose density gradients. Whole rat DNA was purified and sheared as described [20] and had an average single strand length of 350 nucleotides.

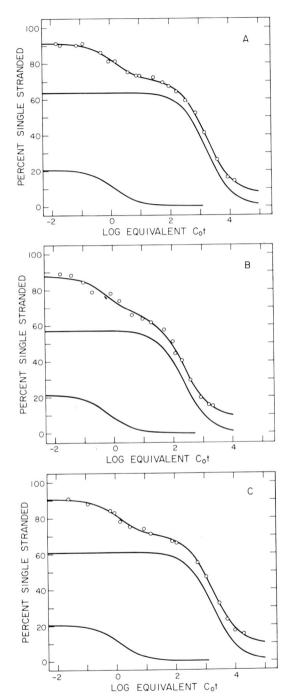

Fig. 5. Renaturation profiles of purified total (A), native S_2 (B), and reconstituted S_2 (C) DNAs. The solid lines represent a two component least-squares analysis of the data.

Fig. 5 shows the optical reassociation profiles of total rat liver DNA and S_2 DNAs prepared from native and reconstituted chromatin. By this analysis the S_2 fraction of native chromatin was found to have an observed C_0t 1/2 of 250 for the single copy sequences. This value is in good agreement with the data reported earlier [20]. In contrast, the C_0t 1/2 observed for the single copy component of reconstituted S_2 chromatin DNA is 1.74×10^3, a value which suggests that it is a random selection of all sequences available, since the C_0t 1/2 observed for single copy in the rat genome is 2.04×10^3. Thus, the sequences which are rapidly excised by the action of DNase II on reconstituted chromatin do not represent the same subset as those liberated when native chromatin is the substrate.

From these experiments it appears that the structural aspects which allow DNase II to recognize the actively transcribed regions of native chromatin are not present in the reconstituted material. Whether this means that the sequences available for transcription in reconstituted chromatin represent a random sample of the rat genome, or that the structural conformation is dramatically changed such that the transcriptionally active genes are no longer susceptible to rapid excision by DNase II is not known. We are attempting to answer this question by hybridization of sequence specific probes to RNA transcribed in vitro from native and reconstituted chromatins.

Discussion

In the years since 1969 [5,23] when the salt-urea step dialysis method of reconstitution was first used, a large number of reports have appeared utilizing this technique. Comparative tests of native and reconstituted chromatin suggest that the reconstituted material may have many of the same properties as native chromatin. Such tests include chemical composition [36,43], comparison of the in vitro transcription of native and reconstituted chromatin of repetitious [5,15,16,23] and single copy [4,12,38,44] sequences and limit digestion using Staphylococcal nuclease [1].

However, recent analysis of reconstitution has led some investigators to the conclusion that extensive proteolysis occurs throughout the reconstitution process [3,9,11]. A recent report by Chae and co-workers [10] indicates that during the reconstitution process the bulk of the non-histones reassociate with the chromatin complex during the last step of dialysis, a point at which most of the histones have already reassociated with the DNA. These studies [10] also reveal that many of the non-histones do not reassociate with the chromatin in the presence of 5 M urea, the concentration at which Stein et al. [43,44] remove the reconstituted chromatin complex from solution. Thus in their reconstituted chromatin at least some of the non-histones are missing. Also the X-ray diffraction pattern of chromatin reconstituted by a salt-step gradient in the presence of urea casts serious doubt on the structural fidelity of this reconstituted material [40]. More recently Bradbury and co-workers [7] using X-ray diffraction, have reported that chromatin reconstituted in the presence of urea gives anomalous ring patterns, possibly due to the prevention of specific protein-protein interactions.

We have analysed the structural organization of reconstituted rat liver chromatin with two nucleases which have differing, but highly specific modes of action on native chromatin. First, staphylococcal nuclease has been shown to provide a convenient test of the secondary structure of chromatin [33]. Secondly, we have used deoxyribonuclease II as a gentle, specific means of fractionating chromatin into template active and template inactive regions [18–20]. Using these probes we have established that chromatin reconstituted by a salt-step gradient in the presence of urea yields material that is structurally different from native chromatin.

Figs. 1 and 4 demonstrate that reconstituted rat liver chromatin has a greater portion of its DNA in nuclease sensitive structures than does native chromatin. Since, in our reconstitution experiments, greater than 95% of the chromosomal proteins were 'reconstituted' to the chromatin, we conclude that some of the proteins aggregate to the protein surface of the reconstituted nucleohistone [10] leaving more free DNA exposed to the nucleases. The presence of the protease inhibitors, DFP and PMSF, has obviated the problem of endogenous proteolysis during reconstitution. SDS-polyacrylamide gel electrophoresis of the native and reconstituted chromosomal proteins show that no appreciable protein degradation has occurred during reconstitution, in agreement with results reported by others [9–11,13].

As shown in fig. 3, the sequences of chromatin accessible to DNase II, which probably correspond to the open, accessible regions of chromatin [30,39], represent a select subset in native chromatin and a random population of total genomic DNA in reconstituted chromatin. These results indicate that the open sequences available for transcription in reconstituted chromatin are a random selection of the rat genome, therefore the reconstitution process is not sequence specific for the population of sequences transcribed in native interphase chromatin.

We cannot eliminate the possibility that those regions sensitive to DNase II in reconstituted chromatin do not represent the regions which are transcribed by exogenous *E. coli* RNA polymerase. We do not know whether the polymerase molecule transcribes [39] the more open, nuclease labile regions. While little is known about the way in which RNA polymerases recognize the regions of chromatin which are to be transcribed, it is probable that the recognition process involves specific structural features in the open chromatin regions. Indeed, R.C. Huang and her colleagues have recently [45] demonstrated the very labile nature of in vitro transcription systems, and emphasize that the native structures present in gently isolated chromatin may be essential for transcriptional fidelity, using exogenous or endogenous polymerase.

We conclude that for a reconstitution scheme to be useful in elucidating the elements and mechanisms which control eukaryotic gene expression, both structural and transcriptional fidelity must be demonstrated. We are currently examining alternative reconstitution methods in hopes of attaining this goal.

References

[1] R. Axel, W. Melchior, Jr., B. Sollner-Webb and G. Felsenfeld, Specific sites of interaction between histones and DNA in chromatin. Proc. Nat. Acad. Sci. 71, 4104 (1974).
[2] R. Axel, Cleavage of DNA in nuclei and chromatin with staphylococcal nuclease. Biochemistry 14, 2921 (1975).
[3] N.R. Ballal, D.A. Goldberg and H. Busch, Dissociation and reconstitution of chromatin without appreciable degradation of the proteins. Biochem. Biophys. Res. Comm. 4, 972 (1975).
[4] T. Barrett, D. Maryanka, P.H. Hamlyn and H.J. Gould, Non-histone proteins control gene expression in reconstituted chromatin. Proc. Nat. Acad. Sci. 71, 5057 (1974).
[5] I. Bekhor, G.M. Kung and J. Bonner, Sequence-specific interaction of DNA and chromosomal protein. J. Mol. Biol. 39, 351 (1969).
[6] I. Bekhor, Physical studies on the effect of chromosomal RNA on reconstituted nucleohistones. Arch. Biochem. Biophys. 155, 39 (1973).
[7] P.G. Boseley, M. Bradbury, G.S. Butler-Browne, B.G. Carpenter and R.M. Stephens, Physical studies of chromatin. Eur. J. Biochem. 62, 21 (1976).
[8] R.J. Britten, D.E. Graham and B.R. Neufeld, Analysis of repeating DNA sequences by reassociation. Methods Enzymol. 39, 363 (1974).
[9] D.B. Carter and C.-B. Chae, Chromatin-bound protease: Degradation of chromosomal proteins under chromatin dissociation conditions. Biochemistry 15 (1976).
[10] C.-B. Chae, Reconstitution of chromatin: Mode of reassociation of chromosomal proteins. Biochemistry 14, 900 (1975).
[11] C.-B. Chae, R.A. Gadski, D.B. Carter and P.H. Efird, Integrity of proteins in reconstituted chromatin. Biochem. Biophys. Res. Comm. 67, 1459 (1975).
[12] J.-F. Chiu, Y.-H. Tsai, K. Sakuma and L.S. Hnilica, DNA-binding chromosomal non-histone proteins: Isolation, characterization, and tissue specificity. J. Biol. Chem. 250, 9431 (1976).
[13] A.S. Douvas, C.A. Harrington and J. Bonner, Major non-histone proteins in rat liver chromatin: Preliminary identification of myosin, actin, tubulin, and tropomyosin. Proc. Nat. Acad. 72, 3902 (1975).
[14] J.T. Finch, M. Noll and R.D. Kornberg, Electron microscopy of defined lengths of chromatin. Proc. Nat. Acad. Sci. 72, 3320 (1975).
[15] R.S. Gilmour and J. Paul, RNA transcribed from reconstituted nucleoprotein is similar to natural RNA. J. Mol. Biol. 40, 137 (1969).
[16] R.S. Gilmour and J. Paul, Role of non-histone components in determining organ specificity of rabbit chromatins. FEBS Letters 9, 242 (1970).
[17] R.S. Gilmour and J. Paul, The in vitro transcription of the globin gene in chromatin. Florida (1975).
[18] J.M. Gottesfeld, W.T. Garrard, G. Bagi, R.F. Wilson and J. Bonner, Partial purification of the template-active fraction of chromatin: A preliminary report. Proc. Nat. Acad. Sci. 71, 2193 (1974).

[19] J.M. Gottesfeld, J. Bonner, G.K. Radda and I.O. Walker, Biophysical studies on the mechanism of quinacrine staining of chromosomes. Biochemistry 13, 2937 (1974).

[20] J.M. Gottesfeld, G. Bagi, B. Berg and J. Bonner, Sequence composition of the template-active fraction of rat liver chromatin. Biochemistry 15, 2472 (1976).

[21] J.D. Griffith, Visualization of prokaryotic DNA in a regularly condensed chromatin-like fiber. Proc. Nat. Acad. Sci. 73, 563 (1976).

[22] D.R. Hewish and L.A. Burgoyne, Chromatin sub-structure. The digestion of chromatin DNA at regularly spaced sites by a nuclease deoxyribonuclease. Biochem. Biophys. Res. Comm. 52, 504 (1973).

[23] R.C.C. Huang and P.C. Huang, Effect of protein-bound RNA associated with chick embryo chromatin on template-specificity of the chromatin. J. Mol. Biol. 39, 365 (1969).

[24] L. Kleiman and R.-C.C. Huang, Reconstitution of chromatin. The sequential binding of histones to DNA in the presence of salt and urea. J. Mol. Biol. 64, 1 (1972).

[25] L.J. Kleinsmith, J. Stein and G. Stein, Dephosphorylation of non-histone proteins specifically alters the pattern of gene transcription in Reconstituted chromatin. Proc. Nat. Acad. Sci. 73, 1174 (1976).

[26] R.D. Kornberg, Chromatin structure: A repeating unit o histones and DNA. Science 184, 868 (1974).

[27] E. Lacey and R. Axel, Analysis of DNA of isolated chromatin subunits. Proc. Nat. Acad. Sci. 72, 3978 (1975).

[28] V.C. Littau, V.G. Allfrey, J.H. Frenester and A.E. Mirsky, Active and inactive regions of nuclear chromatin as revealed by electron microscope autoradiography. Proc. Nat. Acad. Sci. 52, 93 (1964).

[29] A.E. Loening, The fractionation of high-molecular-weight ribonucleic acid by polyacrylamide-gel electrophoresis. Biochem. J. 102, 251 (1967).

[30] K. Marushige and J. Bonner, Fractionation of liver chromatin. Proc. Nat. Acad. Sci. 68, 2941 (1971).

[31] M. Noll, Subunit structure of chromatin. Nature 251, 249 (1974).

[32] M. Noll, Internal structure of the chromatin subunit. Nucleic Acids Res. 1, 1573 (1974).

[33] M. Noll, J.O. Thomas and R.D. Kornberg, Preparation of native chromatin and damage caused by shearing. Science 187, 1203 (1975).

[34] A.L. Olins and D.E. Olins, Spheroid chromatin units (ν-bodies). Science 183, 330 (1974).

[35] P. Oudet, M. Gross-Bellard and D. Chambon, Electron microscopic and biochemical evidence that chromatin structure is a repeating unit. Cell 4, 281 (1975).

[36] J. Paul and I.R. More, Properties of reconstituted chromatin and nucleohistone complexes. Nature New Biol 239, 134 (1972).

[37] J. Paul and I.A.R. More, Ultrastructural and biochemical characteristics of reconstituted chromatin and synthetic nucleohistones. Exp. Cell Res. 82, 399 (1973).

[38] J. Paul, R.S. Gilmour, N. Affara, G. Birnie, P. Harrison, A. Hell, S. Humphries, J. Windass and B. Young, The globin gene: structure and expression. Cold Spring Harbor Symp. Quant. Biol. 38, 885 (1973).

[39] J. Paul and R.S. Gilmour, in: The Structure and Function of Chromatin, *In vitro* transcription of the globin gene in chromatin (Ciba Foundation Symposium, 28) Elsevier. Excerpta Medica., North Holland (1975).

[40] B. Richards and J. Pardon, The molecular structure of nucleohistone (DNH). Exp. Cell Res. 62, 184 (1970).

[41] R.L. Rubin and E.N. Moudrianakis, The F3-F2 al complex as a unit in the self-assembly of nucleoproteins. Biochemistry 14, 1718 (1975).

[42] R.T. Simpson and J.R. Whitlock, Jr., Chemical evidence that chromatin DNA exists as 160 base pair beads interspersed with 40 base pair bridges. Nucleic Acids Res. 3, 117 (1976).

[43] G.S. Stein, R.J. Mans, E.J. Gabbay, J.L. Stein, J. Davis and P.D. Adawadkar, Evidence for the fidelity of chromatin reconstitution. Biochemistry 14, 1859 (1975).

[44] G. Stein, W. Park, C. Thrall, R. Mans and J. Stein, Regulation of cell cycle stage-specific transcription of histone genes from chromatin by non-histone chromosomal proteins. Nature 257, 764 (1975).

[45] M.M. Smith and R.-C. Huang, Transcription in vitro of immunoglobulin kappa light chain genes in isolated mouse myeloma nuclei and chromatin. Proc. Nat. Acad. Sci. 73, 775 (1976).

[46] B. Sollner-Webb and G. Felsenfeld, A comparison of the digestion of nuclei and chromatin by staphylococcal nuclease. Biochemistry 14, 2915 (1975).

[47] J.O. Thomas and R.D. Kornberg, An octamer of histones in chromatin and free in solution. Proc. Nat. Acad. Sci. 72, 2626 (1975).

Chapter 14

NEW APPROACHES TO CELL GENETICS COTRANSFER OF LINKED GENETIC MARKERS BY CHROMOSOME MEDIATED GENE TRANSFER

FRANK H. RUDDLE* and O. WESLEY McBRIDE**

*Yale University, Kline Biology Tower, New Haven, Conn. 06520, U.S.A. and **Laboratory of Biochemistry, National Cancer Institute, Department of Health, Education and Welfare, NIH, Bethesda, Maryland 20014, U.S.A.

Somatic cell genetics has progressed rapidly during the past half decade. The application of cell hybridization methodologies has contributed to the development of human gene mapping in an impressive way. The Third Human Gene Mapping Conference at Baltimore [1] cites 120 loci which have been assigned to specific chromosomes [1]. All the human chromosomes have now been assigned markers, with the single exception of chromosome 22 which is among the smallest in the set. These developments are all the more encouraging when one considers that the rate of acquisition of comparable genetic data should accelerate as new techniques are introduced, and as the number of laboratories participating in this work increases.

In this article, we shall first briefly review the standard procedures of gene mapping which are based primarily on the segregation of chromosomes from interspecific cell hybrids. Secondly, we shall discuss a new approach to genetic analysis which makes use of isolated metaphase chromosomes as the vehicles for gene transfer between cells. In this section, we shall present hitherto unpublished data which support the occurrence of the cotransfer of genes in the context of chromosome transfer systems.

Somatic cell genetic analysis depends on the transfer and segregation of genetic elements. In the standard cell hybridization systems, cells of different specific origin are fused together. In such hybrids, one of the input parental chromosome sets is partially eliminated. The chromosomes are segregated largely on a random basis. Thus, one can generate a series of hybrids which possess different subsets of chromosomes derived from one of the parental types. In the primate/rodent combinations which we shall discuss here, it is the primate (human) chromosomes which are lost.

The Molecular Biology of the Mammalian Genetic Apparatus:
edited by P. Ts'o © 1977, Elsevier North-Holland Biomedical Press

Three types of genetic tests can be applied to a series of somatic cell hybrid populations. These have been referred to as (a) the synteny test, (b) the assignment test, and (c) the regional assignment test. All three depend on correlating attributes relating to the segregating chromosomes among the individual hybrid clones. In the synteny test, one tests for correlation between phenotypic markers of genes within the segregating genome. The positive correlation between any two markers indicates the presence on a single chromosome of genes which encode or regulate their expression. Neither the distance relationships between the genes nor the identity of the specific chromosome involved are provided directly by this test. A positive synteny test result can provide such information indirectly if the position of one marker has already been defined. The assignment test extends the correlation to specific chromosomes in the segregating set. The presence and absence of the chromosomes is determined directly by cytological analysis. In the assignment test, correlations are made between single phenotypes or syntenic sets of phenotypes and individual chromosomes. This procedure permits the assignment of genes to specific chromosomes, but does not provide data on the location of genes within chromosomes. Such regional localization information can be obtained by means of an extension of the assignment test. This method makes use of chromosome translocations or deletions which reduce the chromosomes to smaller segregating units. Thus one might start with a situation where a particular marker (A) had been assigned to chromosome 2. If one could make a hybrid series in which the long arm of chromosome 2 segregated independently as a consequence of a translocation, or if one arm was absent as a consequence of deletion, then the A marker could be assigned to one arm or the other. By using different translocation or deletion breakpoints in a series of independent experiments, one can order loci on chromosomes and restrict the loci to smaller and smaller regions. These genetic tests have been discussed in a recent review (Ruddle and Creagan, 1975), and the reader is directed to it for more detailed accounts of specific cases.

A new approach to cell genetic analysis has been introduced by the findings of McBride and Ozer [10]. These investigators demonstrated that genetic information could be transferred to recipient cells by exposing such cells to isolated metaphase chromosomes. McBride and Ozer performed their first experiment in the following manner. Chinese hamster cells possessing hypoxanthine guanine phosphoribosyltransferase (HPRT) wild type enzyme activity were used as a source of donor metaphase chromosomes. The metaphase chromosomes were prepared by a simple modification of the method originally devised by Maio and Schildkraut [8]. The liberated metaphase chromosomes were then separated from intact cells, nuclei, and smaller debris by centrifugation and unit gravity sedimentation. Mouse (A9) cells which are HPRT deficient were used as recipient cells. The donor hamster chromosomes were mixed with intact disaggregated recipient mouse cells in suspension at a high concentration ($\sim 10^7$ cells per ml) in multiples of one donor chromosome set to one recipient cell. The mixture was gently mixed for a period of 2–3 h at 37°C, and the cells were then inoculated into dishes at standard plating densities ($\sim 10^4$ cells per cm^2). Poly-L-ornithine was added to the incubation mixture to foster the uptake of chromosomes. Using these general procedures, a high proportion of recipient cells phagocytized donor chromosomes. Ten to fifty percent of recipient cells ingested at least one donor chromosome. In order to select for transformed cells, the treated cell population was maintained on a selective medium (HAT) [7]. Under these conditions only cells expressing HPRT enzyme activity can survive. In McBride and Ozer's original report a frequency of transformation of 1×10^{-7} was recorded. The same general frequency has been reported in subsequent confirmatory experiments performed in a similar manner with comparable materials. The transfer of the donor gene to the recipient was demonstrated by showing that the HPRT enzyme was of the Chinese hamster type in terms of its chromatographic and electrophoretic properties. McBride and Ozer also performed experiments which showed that the transferred HPRT phenotype was retained intact in some but not all transformants in the absence of HAT selection pressure.

Subsequent reports [17,19] have completely confirmed the original findings of McBride and Ozer. One of these reports [17] made use of human cells as donor material with mouse as recipient. HPRT deficiency of the recipient cells was used for the purpose of selection. In this system, one can obtain some estimate of the size of the genome (transgenome) transferred into the transformed cells (transgenote) by making use of regional assignment data. *Hprt* has been assigned to the human X chromosome. The locus resides in the mid-region of the X long arm, and it is flanked by the phosphoglycerate kinase locus (*Pgk*) which is more proximal to the centromere, and by *Gpd* (glucose-6-phosphate dehydrogenase) which is the most distal of the three loci from the centromere. The experiments of Willecke and Ruddle yielded three transformants of independent origin. None of the transformed cells expressed any human autosomally assigned genes. In regard to the X chromosome, only the *Hprt* locus was expressed. It can be estimated on the basis of chromosome length and DNA content data that the distance between *Pgk* and *Gpd* is approximately 1.0% of the haploid human genome. Therefore, since neither human PGK or G6PD enzymes were observed it could be inferred that the transgenome in these few instances could be no larger than 1% of the human genome. This estimate assumes that the *Pgk* and *Gpd* genes are physically absent, and not inactivated. Similar results and conclusions were reported by Burch and McBride [3].

It would be useful to extend this analysis of transgenome size to selectable genes which are closely linked to appropriate indicator genes. One system exists which is amenable to study in terms of mammalian gene transfer systems as they are currently defined. This system involves the thymidine kinase (*Tk*) locus which has been shown to be tightly linked to the galactokinase (*Gk*) locus in man.

Somatic cell genetic studies of the conventional type first demonstrated this linkage relationship. Miller et al. [13] and Ruddle and Chen [15] using cell hybrids showed that *Tk* could be assigned to chromosome 17. Boone et al [2] showed subsequently using a clone in which the long arm of 17 was translocated to a mouse chromosome that *Tk* could be assigned to the long arm of 17 (17q). McDougall et al. [11] making use of a series of adenovirus 12 induced breaks in 17q showed that *Tk* could be assigned to a very small region of 17q (17q21-17q22). Elsevier et al. [5] using material carrying the same breakpoint deletions was able to demonstrate that *Gk* could be mapped into the same small interval as *Tk*. This interval measures approximately 0.2% of the total haploid genome. Thus it would be useful to determine whether *Tk* and *Gk* might be cotransferred in chromosome mediated gene transfer experiments. It is interesting to note that *Tk* and *Gk* are linked not only in *Homo* but also in *Pan* [4], *Mus* [6], and *Rattus* [12].

The first experiments on the cotransfer of *Tk* and *Gk* were performed by McBride and Ruddle, and a preliminary report was made in 1975 by Burch and McBride [3]. These experiments employed Chinese hamster chromosomes as the donor material and mouse A9 cells as the recipient. A detailed account of these experiments will be reported elsewhere, but the basic findings will be described here. Metaphase chromosomes [9] were isolated from Chinese hamster fibroblasts and incubated with thymidine kinase deficient mouse B_{82} cells [7]. Colonies of thymidine kinase positive cells appeared on independent plates with a frequency of $4-5 \times 10^{-7}$ in a HAT selective system. The species of origin of the cytoplasmic form of thymidine kinase in cell-free extracts of these independent clones was determined by assessment of thermostability of the enzyme after dialysis of the extracts against buffer (pH 6.8) containing 2 mM EDTA. The thymidine kinase from Chinese hamster cells exhibited markedly greater thermostability than the mouse (L929 or A9) species of this enzyme as determined by the time course of heat inactivation at either 55° or 65°. The thymidine kinase in extracts derived from 12 independent clones resulting from chromosome mediated gene transfer had the same thermostability as that observed for the chromosomal donor hamster species, indicating that all of these clones resulted from gene transfer rather than reversion. In contrast, the thymidine kinase in two authentic revertants of the B_{82} recipient cell line was considerably more thermolabile than even the enzyme from the L and A9 cells.

The species of galactokinase (fig. 1) present in ten of these same clones was determined by starch gel electrophoresis [14] and the chromosome donor species was detected in two of these clones. Mass cultures of these two clonal cell lines were also each subjected to back selection in 0.1 mM BuDR. Analysis of extracts of these two resultant thymidine kinase deficient lines demonstrated the persistence of the mouse species of galactokinase, whereas the Chinese hamster species was no longer present. This provides convincing evidence for the cotransfer of the TK and galactokinase genes and for the persistence of this linkage in the

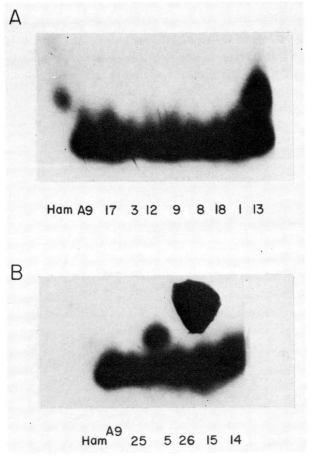

Fig. 1. Starch gel electrophoresis of galactokinase activity [10]. Extracts of individual clones resulting from gene transfer are indicated numerically; 25 and 26 are extracts of B82 revertants. Ham and A9 are extracts of Chinese hamster and mouse cells respectively. Numbered channels refer to extracts of transformants. Transformants 5 and 13 (cotransformants) possess both Ham and A9 galactokinase activity. The large blotch in fig. 1b is an artifact.

genome of the recipient cells. All clones were also examined for ten other non-linked genetic markers and only the mouse recipient species were found. These ten clones exhibited a wide range of stabilities of the TK + phenotype when cultured in non-selective medium, ranging from over 50% loss in one month to less than 1% loss in 4 months. Marked stability of the TK + phenotype was found in one of the lines containing the chromosomal donor species of both TK and galactokinase, while the other similar line demonstrated about 25% loss of the TK + phenotype after 4 months ($\sim$120 cell generations) culture in non-selective medium.

These results suggest that TK and galactokinase are closely linked in Chinese hamster cells. Close linkage for these two genes is already established in human cells and both loci have been localized to the q21-22 region of chromosome 17 [5,11]. It is estimated that this region comprises no more than 10% of the long arm of chromosome 17 and thus about 0.2% of the human genome (i.e. about 7×10^6 nucleotide base pairs). Of course these two genes could be represented in a region of only a few thousand base pairs if they are directly adjacent to one another. These results point out the inherent limitations in determining the size of the transgenome by this procedure. The minimum size could be slightly less than 10^4 base pairs and the maximum could be slightly less than 10^8 base pairs (based on the X-chromosome results) but this presents an uncertainty of about 4 orders of magnitude. However, the cotransfer of these two genetic markers by CMGT indicates that this method may have considerable practical importance for the ordering of closely linked genes whose linkage cannot be readily dissociated by other methods. It may be assumed that the frequency of cotransfer of two linked genes should be inversely proportional to their map distance. However, at least one selectable marker is required for this type of analysis of any pair of genes at the present time.

Recently, an experiment has been conducted which provides direct information on the *Tk/Gk* linkage relationship in the human genome. Willecke et al. [18] have used human HPRT + WI-L2a cells as a chromosome source, and the TK deficient mouse L cells, B82 and LM(TK −), as recipient cell populations. The validity of the transformants was determined by the expression of the human TK phenotype as defined by its electrophoretic properties. Twenty three human enzyme markers, corresponding to 15 different human chromosomes, were absent. The only human marker expressed other than human TK was GK. GK was expressed in 2 out of 8 transformants, a frequency comparable to that found in the Chinese hamster/mouse system. This result strongly suggests that linkage relationships between *Gk* and *Tk* in Chinese hamster and man are similar. The conservation of the *Gk/Tk* linkage relationship in species as evolutionarily divergent, as primates and rodents suggests, possibly, a biologically significant association between these two genes, and other as yet unidentified genes in this region.

References

[1] Baltimore Conference (1975): Third International Workshop on Human Gene Mapping. D. Bergsma, ed. Birth Defects. Orig. Art. Ser. XII, 1976. The National Foundation – March of Dimes.
[2] C. Boone, T.R. Chen, and F.H. Ruddle, Assignment of three human genes to chromosomes (LDH-A to 11, TK to 17, and IDH to 20) and evidence for translocation between human and mouse chromosomes in somatic cell hybrids. Proc. Nat. Acad. Sci. U.S.A. 69, 510–514 (1972).

[3] J.W. Burch and O.W. McBride, Human gene expression in rodent cells after uptake of isolated metaphase chromosomes. Proc. Nat. Acad. Sci. U.S.A. 72, 1797–1801 (1975).
[4] S. Chen, J.K. MacDougall, R.P. Creagan, V. Lewis and F.H. Ruddle, Genetic homology between man and the chimpanzee: syntenic relationships of genes and adenovirus-12 induced gaps using chimpanzee-mouse somatic cell hybrids. Somatic Cell Genetics, in press. (1976).
[5] S.M. Elsevier, R.S. Kucherlapati, E.A. Nichols, R.P. Creagan R.E. Giles, F.H. Ruddle, K. Willecke and J.K. McDougall, Assignment of the gene for galactokinase to human chromosome 17 and its regional localization to band q21-22. Nature 251, 633–635 (1974).
[6] C. Kozak and F.H. Ruddle, unpublished results.
[7] J.W. Littlefield, The use of drug-resistant markers to study the hybridization of mouse fibroblasts. Exp. Cell Res. 41, 190–196 (1969).
[8] J.J. Maio and C.L. Schildkraut, Isolated mammalian metaphase chromosomes. II. Fractionated chromosomes of mouse and Chinese hamster cells. J. Mol. Biol. 40, 203–216 (1969).
[9] O.W. McBride, J.W. Burch and F.H. Ruddle, Cotransfer of closely linked mammalian genes by chromosome mediated gene transfer. In preparation, for Somatic Cell Genetics.
[10] O.W. McBride and H.L. Ozer, Transfer of genetic information by purified metaphase chromosomes. Proc. Natl. Acad. Sci. U.S.A. 70, 1258–1262 (1973).
[11] J.K. McDougall, R. Kucherlapati and F.H. Ruddle, Localization and induction of the human thymidine kinase gene by adenovirus 12. Nature New Biol. 245, 172–175 (1973).
[12] J.K. McDougall and F.H. Ruddle, unpublished results.
[13] O.J. Miller, P.W. Allderdice, D.A. Miller, W.R. Breg and B.R. Migeon, Human thymidine kinase gene locus: assignment to chromosome 17 in a hybrid of man and mouse cells. Science 173, 244–245 (1971).
[14] E.A. Nichols, S.M. Elsevier and F.H. Ruddle, A new electrophoretic technique for mouse, human and Chinese hamster galactokinase. Cytogenet. Cell Genet. 13, 275–278 (1974).
[15] F.H. Ruddle and T.R. Chen, Utilization of centric heterochromatin for chromosome identification in somatic cell hybrids, in: Perspectives in Cytogenetics. S.W. Wright and B.F. Crandall, eds., Charles Thomas, Colorado Springs (1971).
[16] F.H. Ruddle and R.P. Creagan, Parasexual approaches to the genetics of man; in Annual Review of Genetics. H.L. Roman, A. Campbell and L.M. Sandler, eds., vol. 9, 407–497 (1975).
[17] K. Willecke and F.H. Ruddle, Transfer of the human gene for hypoxanthine-guanine phosphoribosyltransferase via isolated human metaphase chromosomes into mouse L-cells. Proc. Natl. Acad. Sci. U.S.A. 72, 1792–1796 (1975).
[18] K. Willecke, R. Lange, A. Krüger and T. Reber, Cotransfer of two linked human genes into cultured mouse cells. Proc. Natl. Acad. Sci. U.S.A. 73, 1274–1278 (1976).
[19] G.J. Wullems, J. van der Horst and D. Bootsma, Somatic Cell Genet. 1, 137–152 (1975).

Chapter 15

STUDIES IN MAMMALIAN CELL REGULATION: CELL SURFACE ANTIGENS AND THE ACTION OF CYCLIC AMP*

THEODORE T. PUCK

Eleanor Roosevelt Institute for Cancer Research, Department of Biophysics and Genetics, University of Colorado Medical Center, Denver, Colorado 80262, U.S.A.

Introduction

E. coli has been the equivalent in biology of the hydrogen atom in modern physics. It has furnished the simplest model for the operation of the living cell. However, the principal function of *E. coli* is reproduction. Each of the daughter cells produced in this fashion is a completely independent organism which can grow and fulfill its genetic potentialities in complete isolation. The mammalian cell also has the power to reproduce itself and in appropriate circumstances can also grow as an independent organism. However, in addition to reproduction, mammalian cells in the body are capable of recognizing and reacting with each other and the complex components of their fluid medium so that from an original single cell, the fertilized egg, communities of cells differentiated into discrete tissues and organs are produced. These are coordinated in an extremely complex fashion to produce a multi-cellular organism capable of integrated behavior. Cells of the different tissues retain a common genome. However, in each of the different states of cellular differentiation, different subsets of the genetic complement are active while the rest is inactive. Presumably, the inactive genes are of two kinds – those which are shut off in fairly permanent fashion and those which are readily turned on or off in accordance with the needs of the cell in particular environmental situations. The principal problem of differentiation then would appear to lie in understanding the control of the genome activity in cells of the different tissues in different stages of development. Presumably, the characteristic

---

* This investigation was supported by a grant from the Louis B. Mayer Foundation and by Grant No. CA15793-01A1, awarded by the National Cancer Institute, DHEW (contribution No. 225).

The Molecular Biology of the Mammalian Genetic Apparatus:
edited by P. Ts'o © 1977, Elsevier North-Holland Biomedical Press

morphological structures of different tissue cells exist for the purpose of maintaining cells of each type in a form which permits expression of the particular genes whose function is required by the cell's specialized role in the body.

It would appear that several different kinds of structural elements are needed to maintain a specific state of differentiation. Sensing elements must exist, presumably on the cell surface, which monitor the presence of other cells so that only the needed cell-cell associations will be set up and appropriate cell interactions can occur. Sensors also must monitor the molecular content of the bathing fluids for every cell. This information so collected from the environment must be transmitted to the nucleus where appropriate mechanisms for activation or inactivation of selected genome regions can occur. These considerations imply the existence of a transmission system by which the information is conducted from the cell surface to the nucleus. Acting on the information received, the genome alters its biosynthetic responses to adapt to the needs for an altered cell response.

Important progress has been achieved in delineating the mammalian cell genome. Many laboratories are now engaged in identification of specific gene products, fixing their chromosomal location and mapping of specific genetic loci on each chromosome. The rate of progress is sufficiently rapid so that despite the fact that the mammalian cell contains approximately 500 times as much DNA as *E. coli*, the human genetic map is gradually being filled in. All but two of the human chromosomes now have at least one marker gene and some have as many as 20. When one recalls that approximately two decades ago, even the total number of the human chromosomes was unknown and not a single human gene had been assigned to any specific autosome, the rate of progress can be appreciated.

One should expect that each differentiated state of the mammalian cell should be distinguished by four characteristic parametric sets: an arrangement of cell surface structures which collect information from neighboring cells and from the fluid medium; an appropriate set of information transmission mechanisms which may involve molecular, mechanical, or other kinds of informational conduction to carry the necessary signals to the nucleus and perhaps to other cytoplasmic structures as well; a genomic situation in which all but one particular set of genes is sequestered from the reactive state, leaving this set in a condition of continuing activity or capable of modulational control in accordance with signals from the surrounding environment; and other specialized cytoplasmic structures characteristic of the given state of differentiation. Presumably, then, the normal differentiation state of a cell involves a harmonious relationship between these four sets of structures. The characteristic morphologies of the cells of different tissues reflect the different cell structures needed to achieve the specialized functions and the regulatory processes demanded by their states of differentiation.

Thus, a characteristic set of cell surface antigens should exist for each specific state of differentiation. Indeed, the existence of specialized membrane receptors for hormones and other metabolites are already recognized for many different cell types. Moscona, in his historic studies [1], demonstrated specific membrane structures which bring about recognition between the cells of different embryonic tissues. One might conceive of a simple system in which each differentiation process imposes on its constituent cells one or a few unique surface elements. However, a much more flexible system would result if a certain basic set of cell surface elements exists such that the cells of each particular differentiation state are characterized by a particular combination of these elements. If this latter picture is valid, a certain code of differentiation should exist in which cells of a given type will exhibit a particular set of cell membrane structures.

Biochemical genetic studies of tissue-specific cell membrane structures

A variety of approaches already exists for study ot the special molecular constituents which are inserted into the membranous structure of mammalian somatic cells. Ideally, one would like a system which exhibits the following characteristics: a) It should permit identification of surface structures on the cells of all or at least many different kinds of tissues; b) it should also provide identification of the genetic locus or loci controlling expression of each element; c) it should make possible complementation analysis and other kinds of genetic examination of such loci; d) it should make possible ready identification of genetic and physiological mechanisms regulating the activity of each such component; and e) ideally, this method of approach should be applicable to the human tissue-specific cell surface substances, in such a way as to make possible exploration of human disease situations.

The system adopted in our laboratory employs the following strategy:

1) The Chinese hamster ovary cell (CHO) culture developed in this laboratory furnishes the basic approach. This cell is remarkably stable and versatile, exhibiting a stable, hypo-diploid karyotype with 20 chromosomes; a rapid generation time of approximately 11 h; and a plating efficiency of virtually 100%, either for cells plated on plastic surfaces or embedded in agar suspension. Its nutritional requirements for growth have been identified [2].

2) Specific auxotrophic mutants can readily be prepared from this cell by mutagenesis and isolated by means of the BudR-near visible light procedure previously described [2]. Both normal and temperature-sensitive mutant stocks can be prepared and grown in stable fashions. At present, approximately 30 such mutants have been prepared and characterized. Identification of the site of the biochemical block has already been accomplished for most of these mutants and is being investigated in the others.

3) Such mutants can be hybridized with human cells in a medium which lacks the specific nutrilite required by the CHO auxotroph employed. The medium also

fails to promote reproduction of the human parental cell involved. Under these circumstances, each developed colony represents a hybrid cell containing chromosomes of both species.

4) During the first two weeks of growth in selective medium, these hybrids tend to lose human chromosomes. However, the particular human chromosome which is furnishing the gene which is deficient in the parental CHO auxotroph cannot be lost or else the cell will die. After several weeks of growth, such hybrids tend to become stabilized with respect to their chromosome components so that further loss of human chromosomes occurs rarely, if at all. It thus becomes possible to isolate hybrids containing one or a small number of human chromosomes and to grow these as stable, clonal stocks.

5) The human chromosomes carried in such hybrids can be identified by cytogenetic and isozymic analysis. If the hybrid contains only a single human chromosome and if no previous rearrangement has occurred, any human gene product which is displayed is due to a determinant on the included human chromosome. These hybrids have been useful in locating the human chromosomes responsible for approximately 10 human genes.

6) It was demonstrated by Oda and Puck [3] that a high degree of specificity exists in the immunologic reactivity of cell surface antigens of man as compared to those of the Chinese hamster. Thus, in the presence of complement, antisera prepared in the rabbit against human cells kill human cells in dilutions as low as 0.01% without affecting Chinese hamster cells at all, even in concentrations 100 times greater. Antisera against the Chinese hamster cell reveal a similar specificity. The basis of these lethal actions was shown to lie in the cell surface antigens.

7) Consequently, it became a simple matter to test these hybrids for the expression of human cell surface antigens contained on their human chromosomal complements. Antisera against cells from various human tissues were prepared in rabbits and sheep. The antisera were collected and tested against a variety of hybrids of the Chinese hamster cell containing single or multiple human chromosomes. When a given hybrid fails to be killed by a specific antiserum against a human cell, it can be concluded that the chromosome in question does not furnish to the hybrid cell surface lethal antigens capable of reacting with the specific antibodies present in the test antiserum. For example, hybrids containing human chromosome number 12 fail to be killed under the standard conditions by an antiserum prepared against the human fibroblast or the HeLa cell. Therefore, it may be concluded that human chromosome 12 fails to donate to its hybrid cell membrane any antigenic activity common to those on the human fibroblast or HeLa membranes. However, when such killing action is observed it can be concluded that the human chromosomes present in the hybrid do yield cell surface antigens capable of immunological reactions with the specific human antibodies present in the antiserum employed. The presence of human chromosome num-

ber 11 in such hybrids causes them to become highly susceptible to the killing action of antisera produced against the HeLa cell, the human fibroblast, and other human cells. Human-CHO hybrids containing human chromosome number 11 were named A_L cells.

By exhaustive adsorption of such a lethal antiserum with an A_L culture, it was possible to remove all of the antibody activity against A_L antigens. The resulting antiserum was then tested against other hybrids. The next one found to be killed was called the B_L hybrid. The antiserum was again adsorbed with this hybrid and the resulting preparation tested with other hybrids so that in similar fashion, C_L, D_L, and E_L hybrids each with a characteristic surface antigenic activity could be identified [4,5,6,7].

8) The next problem in this approach involves resolution of each of the A_L, B_L, C_L ... families of antigens into their individual antigenic components. We have begun with the A_L antigens carried on human chromosome number 11.

Resolution of the A_L family of antigens into its individual constituents would be relatively simple if the proposal set forth in the Introduction were actually true. That is, if different combinations of a common group of cell surface antigens are distributed on the cells of different tissues, it should be possible to resolve composite antibody activities present in antisera made against particular cell types by selective adsorption with cells from different human tissues. For example, if some but not all of the A_L antigens were shared on cells of one tissue while others were present on other tissue cells, a scheme of adsorption of antisera against one tissue by cells of another could be devised which would yield antisera with a smaller number of specific antibody activities. Such a differential distribution of antigen activities on different cell types was confirmed. The A_L combination of antigens defined as those contributed to the hybrid by chromosome 11 has been resolved so far into three different activities which have been named a_1, a_2, and a_3. The following example illustrates this action [5,6,7].

An antiserum prepared in the horse against human lymphoblasts was demonstrated to be highly lethal for all hybrids containing the A_L character. Similarly, an antiserum prepared in the rabbit against human red cells is also highly lethal for A_L hybrids. However, exhaustive adsorption of the anti-lymphoblast serum with human red cells fails to remove more than a small part of the killing activity for A_L hybrids from the anti-lymphoblast serum. Obviously, then, the red cell shares some, but not all, of the lethal antigens common to the A_L hybrid and the human lymphoblast.

By straightforward extension of this kind of reasoning, the existence of three separate antigenic activities was demonstrated of which a_1 and a_3 are present on the surface of the human red cell, while a_2 is present on human lymphoblasts and other tissues but not on the red cell. By exposure of hybrid cells to mutagenic agents and selection of the survivors in the presence of particular antisera, mutants were isolated with different phenotypic expressions with respect to a_1, a_2, and a_3. All four of the forms, $a_1^+a_2^+$, $a_1^+a_2^-$, $a_1^-a_2^+$, and $a_1^-a_2^-$ have been prepared. Of the eight possible phenotypes which are to be expected, involving all three antigens, six have already been prepared [8].

9) In continuing experimental study of these tissue-specific cell surface antigens, it has been possible to demonstrate that a_1 is immunologically identical with glycophorin, a glycoprotein which has been identified in the membrane of the human red cell [9]; preliminary studies suggested that a_1 is present also in cells derived from the adult human brain but virtually absent from cells of the adult human kidney; a_1 and a_3 have been mapped regionally and have been shown, along with the gene for lactate dehydrogenase A, to be situated on the short arm of human chromosome 11, while a_2 is carried on the long arm of this chromosome [10]; and finally, it has been demonstrated that the a_1 phenotype is made up of several different complementation groups, mutation in any one of which results in a_1^- behavior [11]. Since all of these different genes are present on the human chromosome number 11, the possibility of the existence of common genetic regulatory mechanisms is an intriguing possibility currently under investigation.

10) In the course of these investigations it has been possible to adapt the fluorescent antibody technique for the examination of these cell surface antigens. Thus, biopsies are now being taken from various human adult and embryonic tissues and examined for the existence of these specific antigens. It is hoped in this way the presence of these antigens can be used as developmental markers for the process of tissue differentiation and to explore disease situations which may involve alterations of the cell surface antigens. Presumably, the presence of an abnormal distribution of antigens on the cell of a particular tissue represents an error of development. Therefore, these approaches may provide means for investigation and understanding of a variety of developmental diseases in man.

11) These studies are continuing in an attempt to extend identification of these cell surface antigens and their genetic determinants to as wide a range of hybrids as possible in order to obtain a wide variety of chromosomally determined loci.

The action of cyclic AMP and various hormones

In earlier studies we demonstrated that increase in the concentration of cellular cyclic AMP in a transformed cell causes the cell to revert morphologically and metabolically to a normal fibroblast-like form. In place of the compact, pleomorphic morphology in which the cell surface is studded with rapidly oscillating knobs or blebs, the cell elongates, the knobs disappear with a consequent tranquilization of the surface membrane, and the cells tend to associate together in parallel with their long axis [12,13,14]. The growth behavior of the cell also changes. In the native state single CHO cells grow into colonies with virtually 100% plating efficiency, regardless of whether growth occurs on plastic surfaces or in agar suspension. The addition of cyclic AMP does not alter the capacity of the cells to grow on surfaces but eliminates completely their ability to grow in suspension (table 1). These effects have now been shown in a variety of laboratories to apply to a reasonably large number of transformed cells and the phenomenon has been named 'Reverse Transformation'.

TABLE 1

Effect of DBcAMP on single cell growth on surfaces and in suspension. Basal medium = F12 + 5% fetal calf serum macromolecular fraction throughout. Number of colonies is indicated as a result of plating 200 cells.

	Growth on surface	Growth in agar suspension
No addition	189	193
DBcAMP 5×10^{-4}M + Testololactone 10 μg/ml	124	0

Agents like testosterone, testololactone, and prostaglandin were found to synergize Reverse Transformation. On the other hand, either colcemid, which disrupts microtubular structure or cytochalasin B, which disorganizes microfilaments, were found to prevent or reverse the characteristic action of cyclic AMP derivatives and its synergists. Consequently, it was postulated that the Reverse Transformation phenomenon involves organization of cellular microtubules and microfibrils into a single system in which the cells were induced to assume a characteristic morphology and to display metabolic behavior consistent with their particular state of differentiation. Disruption of this organized system, which helps maintain normal habitus and metabolic control, was postulated to result in pathologic cell behavior which could include malignancy.

The role of the microtubules in the Reverse Transformation phenomenon was confirmed in electron microscopic studies carried out in collaboration with Dr. Keith Porter [15]. In the case of the CHO-K1 cell, raising the cellular level of cyclic AMP causes the microtubular structures to be converted from a fairly sparse, random arrangement to a dense set of parallel structures which extend throughout the cell in parallel to its long axis so that a typical fibroblast-like morphology results. Experiments have demonstrated that protein synthesis is not required for Reverse Transformation [16]. Therefore, it may be concluded that the pre-existing tubulin is organized into the highly ordered arrangement of the cell in its Reverse Transformed state.

Cells other than the CHO cell can assume different patterns of microtubular organization under the influence of cyclic AMP. It has been demonstrated that neuroblastoma cells in culture extend dendrites when treated with cyclic AMP derivatives so as to assume a typical differentiated morphology [17]. We have shown that when CHO cells are hybridized with cells from a normal brain, the resulting multiplying cells which form display a typical undifferentiated appearance characteristic of many tissue culture cells. However, when dibutyryl cyclic AMP plus testololactone is added to such a culture, each of the cells extends long dendritic processes which connect together to form a network. This process is also completely reversible [18]. Because of these experiments, it was necessary to assume that each cell is already committed to some form of microtubular organization. Under the influence of cyclic AMP, the specific aggregation of tubulin into the pre-determined pattern takes place. The detailed chemistry of these events remains to be determined.

Experiments were carried out to determine whether the changes induced by the action of cyclic AMP will involve the behavior of the cell surface antigens which we have studied. Experiments have demonstrated that the addition of dibutyryl cyclic AMP plus testololactone prevents the cell killing that would otherwise occur by the addition of specific antibody and complement. A typical experiment is shown in table 2. The effect was shown to be completely reversible, normal killing again resulting when the cyclic AMP derivative was removed from the medium.

It is possible to construct a highly preliminary model in an attempt to unify some of these actions. The tentative nature of this proposal is emphasized.

Some chains of informational transmission from the cell surface to the nucleus have already been identified, as in the sequence of events by which the steroid hormones cause specific cystosol proteins to change their configurations so as to bind to sites in the chromatin [19]. The microtubular and microfibrillar systems which become organized under the aegis of cyclic AMP action may be essential

TABLE 2

Effect of DBcAMP on killing of a_1^+, a_3^+ hybrid by 0.02% of rabbit anti (a_1^+, a_3^+) serum in 0.75% complement, which was present in all plates.

10^{-3} M DBcAMP + 10 µg/ml of testololactone	Plating efficiency (%)	
	Antiserum absent	Antiserum present
absent	100	1.0
present	100	94.0

for other kinds of informational transfer processes. Since these structures maintain the overall cell morphological integrity, it is hardly conceivable that their collapse, such as occurs in the transformed cell, would fail to interfere with a variety of communication and regulational functions. However, in addition to such general action, a more specific mode of regulatory activity involving microtubules and microfibrils may exist.

The microtubular system appears to traverse the interior of the cell over distances ranging from the exterior membrane to the nucleus. It is conceivable that specific attachment of the microtubules to regions in the chromatin may occur. Under appropriate stimulation, the system so formed may act to shield or to expose particular regions of the genome, so as to allow a relatively coarse form of genetic regulation. The proposal that microtubules organized by cyclic AMP are attached to and affect activity of chromosomal regions may seem less radical a hypothesis when it is recalled that during mitosis the microtubules of the spindle become attached to the centromeric regions of each chromosome. The current proposal simply extends this function to one which also operates in mammalian cells during interphase. This conception also affords a possible explanation for the regions of reiterated DNA that are found distributed over the mammalian chromosomes. Like those characteristic of the centromeric regions, some of these may mark the sites of microtubular attachment. These considerations could also explain why so many malignant cells, in which reproductive control has been lost, have assumed the transformed habitus in which the microtubular system is disorganized. The well-known role of cyclic AMP in causing changes in the synthesis of specific proteins could also fit this picture (although it is quite possible that enzyme induction by cAMP may proceed through completely different pathways). Finally, there has been reported a description of the presence of regions of tubulin concentrated in the nucleus of interphase cells in culture, as observed by means of immunofluorescence procedures [20].

The studies described here illustrate the recent growth in power of genetic, biochemical, immunological and ultrastructural modes of examination of the mammalian cell. It is a great privilege to be able to participate in a symposium honoring the conceptual and experimental milestones contributed by James Bonner to Cell Biology, and those of Jerome Vinograd, whose brilliant investigations on the behavior of the genetic substance have been so tragically terminated. It is gratifying to note how the California Institute of Technology continues decade after decade to illuminate brilliantly the dark regions of biology and to furnish a nurturing home for scientific creativity.

References

[1] Moscona, A.A., Surface specification of embryonic cells: lectin receptors, cell recognition, and specific cell ligands. In The Cell Surface in Development, A.A. Moscona (John Wiley & Sons, New York) pp. 67–100 (1974).

[2] Puck, Theodore T., The Mammalian Cell As A Microorganism, (Holden-Day, Inc., San Francisco) pp. 1–219 (1972).

[3] Oda, M. and T.T. Puck, The interaction of mammalian cells with antibodies. J. Exp. Med. 113, 599–610 (1961).

[4] Puck, T.T., P. Wuthier, C. Jones and F.T. Kao. Genetics of somatic mammalian cells. XIII. Lethal antigens as genetic markers for study of human linkage groups. Proc. Nat. Acad. Sci. U.S.A. 68, 3102–3106 (1971).

[5] Wuthier, Paul, Carol Jones and Theodore T. Puck, Surface antigens of mammalian cells as genetic markers, II. J. Exp. Med. 138, 229–244 (1973).

[6] Jones, Carol, Paul Wuthier and Theodore T. Puck. Genetics of somatic cell surface antigens. III. Further analysis of the A_L marker. Somatic Cell Genet. 1, 235–246 (1975).

[7] Kao, Fa-Ten, Carol Jones and Theodore T. Puck. Genetics of somatic mammalian cells: Genetic, immunologic and biochemical analysis with Chinese hamster cell hybrids containing selected human chromosomes. Proc. Nat. Acad. Sci. U.S.A. 1, 193–197 (1976).

[8] Jones, Carol and Theodore T. Puck, Manuscript in preparation.

[9] Moore, Emma E., Carol Jones and Theodore T. Puck, Cell surface antigens IV: Immunological correspondence between glycophorin and the a_1 human cell surface antigen. Cytogenet. Cell Genet., in press (1976).

[10] Kao, Fa-Ten et al., Manuscript in preparation.

[11] Jones, Carol and Emma E. Moore, manuscript in preparation.

[12] Hsie, A.W. and T.T. Puck, Morphological transformation of Chinese hamster cells by dibutyryl adenosine cyclic 3′:5′-monophosphate and testosterone. Proc. Nat. Acad. U.S.A. 68, 358–361 (1971); and Johnson, G.S., R.M. Friedman and I. Pastan. Restoration of several morphological characteristics of normal fibroblasts in sarcoma cells treated with adenosine 3′:5′-monophosphate and its derivatives. Proc. Nat. Acad. Sci. U.S.A. 68, 425–429 (1971).

[13] Hsie, Abraham W., Carol Jones and Theodore T. Puck. Mammalian cell transformation in vitro. II. Further changes in differentiation state accompanying the conversion of Chinese hamster cells to fibroblastic form by dibutyryl adenosine cyclic 3′:5′-monophosphate and testosterone. Proc. Nat. Acad. Sci. U.S.A. 68, 1648–1652 (1971).

[14] Puck, Theodore T., Charles A. Waldren and Abraham W. Hsie, Membrane dynamics in the action of dibutyryl adenosine cyclic 3′:5′-monophosphate and testosterone on mammalian cells. Proc. Nat. Acad. Sci. U.S.A. 69, 1943–1947 (1972).

[15] Porter, Keith, T.T. Puck, A.W. Hsie and Donna Kelley. An electron microscope study of the effects of cyclic AMP on Chinese hamster ovary cells. Cell 2, 145 (1974).

[16] Patterson, David and Charles A. Waldren, The effect of inhibitors of RNA and protein synthesis on dibutyryl cyclic AMP mediated morphological transformation on Chinese hamster ovary cells in vitro. Biochem. Biophys. Res. Commun. 50, 566–573 (1973).

[17] Prasad, K.N. and A.W. Hsie. Morphological differentiation of mouse neuroblastoma cells in vitro by dibutyryl adenosine 3′:5′-monophosphate. Nature New Biol. 233, 141–142 (1971); and Furmanski, P., D.J. Silverman and M. Lubin, Expression of differentiated functions in mouse neuroblastoma mediated by dibutyryl 3′:5′-monophosphate. Nature 233, 413–415 (1971).

[18] Puck, Theodore T. and Fa-Ten Kao, studies in progress.

[19] Jensen, E.V. et al. A two-step mechanism for the interaction of estradiol with rat uterus. Proc. Nat. Acad. Sci. U.S.A. 59, 632 (1968).

[20] Brinkley, B.R., G.M. Fuller and D.P. Highfield, Cytoplasmic microtubules in normal and transformed cells in culture: Analysis of tubulin antibody fluorescence. Proc. Nat. Acad. Sci. U.S.A. 72, 4981–4985 (1975).

Chapter 16

THYMIDINE KINASE: SOME RECENT DEVELOPMENTS

JOHN W. LITTLEFIELD

Johns Hopkins University, School of Medicine, Baltimore, Maryland, U.S.A.

Cultured animal cells deficient in thymidine kinase (EC 2.7.1.21) leak pyrimidine deoxyribonucleosides [1], and this kinase has been regarded as a simple 'salvage' enzyme. However, it is now recognized that thymidine kinase occurs in animal cells in more than one form, representing different locations in the cell and perhaps more than one genetic locus. There is continued uncertainty as to the role of the kinase in the control of DNA synthesis, the relation of the kinase to the uptake of thymidine across the cell membrane, and the nature of the kinase introduced into cells by viruses. It is useful to review our understanding of these processes in the light of present knowledge of the molecular biology of this enzyme.

Purification and properties

Okazaki and collaborators in 1964 and 1967 [2–5] showed that thymidine kinase highly purified (1200-fold) from *E. coli* was a typical allosteric protein, activated by dCDP (or dADP) and inhibited by dTTP. All these deoxynucleotides appeared to cause 'dimerization', presumably to slightly different conformations. The affinity for thymidine was increased by dCDP and decreased by dTTP. In contrast to the dimer (molecular weight of 89,000–91,000), the monomer (molecular weight of 42,000) was quite temperature-sensitive, even at moderate temperatures and especially in the absence of substrate. Inactivation by heat was a rapid and reversible process. In fact, the monomer behaves like a common temperature-sensitive mutant protein, with maximal enzyme activity at 30°C and considerable loss of activity above this temperature.

The *E. coli* enzyme was also isolated by other investigators using preparative disc gel electrophoresis [6] or affinity chromatography with agarose containing an analogue of thymidine [7]. These studies confirmed the molecular weight of the

The Molecular Biology of the Mammalian Genetic Apparatus:
edited by P. Ts'o © 1977, Elsevier North-Holland Biomedical Press

monomer. When the enzyme was bound to agarose, it retained some activity; this complex could be a model for the mitochondrial membrane-bound kinase to be described below.

Subsequently, the kinase was purified about 1500-fold from the Yoshida rat sarcoma [8]. This preparation showed two peaks of activity on DEAE chromatography, both of which were inhibited by sodium dodecyl sulfate or by dTTP. The molecular weight of Peak I was approximately 70,000, and that of Peak II approximately 120,000. It seemed likely that Peak I was a dimer and Peak II a tetramer; a monomer might have been missed due to low activity or instability. It was interesting that both Peaks I and II were found in tumor tissues, but only Peak I in embryonic or regenerating liver [9].

More recently the kinase in regenerating liver has been purified 2600-fold [10]. This preparation still contained several protein bands. The molecular weight was about 69,000, which is comparable to the *E. coli* dimer, to Peak I from the Yoshida tumor, and to a less pure regenerating liver preparation previously examined in detail [11]. Less pure preparations from other mammalian sources have also been studied. Such preparations differ somewhat from the *E. coli* enzyme in regard to behavior with inhibitors and activators [12], but nonetheless these agents probably cause conformational changes in the mammalian kinase analogous to the allosteric behavior of the *E. coli* enzyme.

Electrophoretic forms of the kinase

Different electrophoretic forms of the kinase occur in extracts from different animal tissues. Thus fetal liver, tumors, and fibroblasts transformed with SV40 virus have been shown to contain kinase which migrates more slowly than that in extracts of adult tissues [13,14]. It is also more heat-sensitive, and differs in regard to pH optimum and interactions with nucleotides. Initially it was suggested that this fetal form was one of several fetal proteins expressed together in tumors. But it now seems likely that the predominant form of the kinase simply reflects the activity of cells in regard to growth [15]. Cells approaching confluency in culture show a shift from the slow-moving to the fast-moving form of the enzyme, while the converse occurs upon re-initiation of growth. Moreover, the electrophoretic patterns often contain more than these one or two peaks, including additional activity at the origin, or moving even more rapidly than the fast-moving peak [15].

While these investigations were in progress, an electrophoretic form of thymidine kinase characteristic of mitochondria was noted in several laboratories [16–20]. This was first indicated by the continued incorporation of radioactive thymidine into mitochondrial DNA by cultured cells known to be markedly deficient in the kinase. The amount of mitochondrial kinase varied from a few percent in actively growing cells to 50–80% in adult tissues [17,19]. The mitochondrial form probably is the same as the fast-moving electrophoretic form described above, and the slow-moving form probably is present in the cytosol fraction.

Recently Kit and collaborators have shown that the mitochondria of HeLa cells contain more than electrophoretic form of the kinase [20]. They have suggested that one form may represent an intermediate between cytosol and mitochondrial kinases, possibly involving modification 'through proteolytic cleavage, through covalent modification of the protein chain (e.g., phosphorylation, adenylation, or sialidation), or through deamidation or acetylation, respectively, of polypeptide glutamine-asparagine and lysine residues' [20]. Subfractionation of mitochondria indicates that most of the kinase is present in the inner membrane and matrix, rather than near the surface of the organelle [21]. The kinase remains membrane-bound when mitochondria are opened by osmotic lysis [17,19].

Relation between cytosol and mitochondrial kinases during growth

A marked increase in total kinase occurs during the logarithmic growth phase of the culture cycle of animal cells, reflecting active DNA synthesis and cell division in many individual cells. During the mitotic cycle of each cell, total kinase activity increases in the late G_1 and early S phases, and plateaus or decreases thereafter [22–26]. Both these increases have been regarded as being due to the production of new enzyme with rapid turnover, and this may be the case. But the electrophoretic investigations described above raise the possibility that the kinase may shuttle between mitochondria and cytosol. Indeed mitochondrial kinase activity does not increase during active cell division, but does increase when growth slows [15,16]. Since a marked increase in total kinase activity occurs during active cell division, the conversion of mitochondrial enzyme to cytosol enzyme would have to be accompanied by activation, and vice versa. In fact the mitochondrial kinase activity is markedly increased by solubilizing the enzyme [17,19] or by treatment with phospholipase C of *C. welchii* [27], and to some extent just by opening mitochondria [17,19]. This increase in activity might relate to the dimerization and activation of the *E. coli* enzyme by dCDP or dADP mentioned above.

An important question is whether mitochondrial kinase is coded by a nuclear or mitochondrial gene. Kit and collaborators used ethidium bromide and chloramphenicol to inhibit both nucleic acid and protein synthesis in the mitochondria of HeLa cells [28]. Mitochondrial kinase continued to be formed as the cells grew slowly for two to three generations. Moreover, some mitochondrial kinase appeared to leak into the cytosol. These experiments suggested that the mitochondrial kinase is coded by a nuclear gene, synthesized in the cytosol, and transferred into the mitochondria. Further evidence for such a sequence of events comes from somatic cell hybrids between mouse cells deficient in cytosol kinase and human or monkey cells [27].

The next important question is whether cytosol and mitochondrial kinases are coded by the same or different nuclear genes. The fact that variant cells, deficient in cytosol kinase but retaining mitochondrial kinase, occur commonly in established animal cell cultures [18–20,26] immediately suggests that two different genes are involved: if both cytosol and mitochondrial kinases were coded by the same nuclear gene, with the latter derived from the former, why would a cell contain mitochondrial kinase but no cytosol kinase? The experiments with somatic cell hybrids mentioned above [28] also suggest that cytosol and mitochondrial kinases represent two different genes, and that these two genes are located on different chromosomes.

However, we are still uncertain as to the reason for the deficiency of cytosol kinase in such common variant cells. Epigenetic rather than genetic mechanisms may be responsible [29]. Kinase may be particularly unstable in the cytosol of these cells, for example. Or perhaps all kinase may be formed in the cytosol, transported into mitochondria, and converted back to cytosol kinase during active growth, with a block in this final conversion explaining the defect in these cells. Such a block would involve an enzyme which normally 'processed' the kinase. 'Processing' enzymes, modifying proteins which move from one organelle or cell fraction to another, have been described for β-glucuronidase [30] and other proteins. Furthermore, the reason for the loss of cytosol kinase in common variant cells can be expected to be heterogenous, especially following chemical mutagenesis [29]. A detailed analysis of these common kinase-deficient variants is badly needed [31].

Uptake of thymidine

The uptake of thymidine through the cell membrane appears to be a mediated process involving Michaelis–Menten kinetics at low thymidine concentrations. Passive diffusion is added at high thymidine concentrations [32]. The kinase is believed to be in excess intracellularly [33], so that uptake may be the rate-limiting process in the utilization of thymidine for DNA synthesis.

The uptake of thymidine increases during DNA synthesis in animal cells [33–37], as does cytosol kinase activity. Is there any relation between cytosol kinase and the uptake of thymidine? Here opinions are conflicting [32]. Kinase has not been found in cell membranes [33]; but if loosely bound, it could easily be released into the cytosol. Indeed, Schuster and Hare have proposed that the kinase is located on the inner side of the cell membrane, perhaps in equilibrium with cytosol kinase, and that it functions by group translocation uptake [38]. Plagemann and collaborators have concluded that uptake and phosphorylation of thymidine are different processes, although they agree that some kinase may be located on the inner surface of the cell membrane [33].

Additional evidence for involvement of the kinase in the uptake of thymidine derives from experiments on variant animal cells in culture. Breslow and Goldsby reported in 1969 that a variant cell frequently occurred in cultures of Chinese hamster cells which was partially resistant to the thymidine analogue 5-bromodeoxyuridine, due to markedly decreased uptake of pyrimidine deoxyribonucleosides [39]. These variants had also lost about 50% of the total kinase activity of the parental culture. Subsequently, similar variants resistant to low levels of 5-bromodeoxyuridine due to markedly decreased uptake were isolated from haploid frog lines [40,41]. Here also the total kinase activity was decreased. Very recently Freed and Hames [42] have found that one such uptake-deficient line has lost a heat-sensitive form of thymidine kinase, which constitutes about one-third of the parental activity. This form may be involved in thymidine uptake [42].

The possible relation between this new thermolabile form of kinase and that in the cytosol is as intriguing as the possible relation between cytosol and mitochondrial forms. Recent studies with cytochalasin, a drug which interacts with cell membranes, suggest that 'elements involved in thymidine transport enter the cell surface membrane as the cells move from G_1 to S' [37]. This element might be cytosol kinase.

It is interesting that, as mentioned above, considerable uptake of thymidine occurs in variant cells markedly deficient in cytosol kinase, again especially during DNA synthesis [36]. In these cells most of the intracellular thymidine does not become phosphorylated. Perhaps in these cells the cytosol kinase is absent or masked, and only the thermolabile form is present.

Herpes virus – induced kinase

Many virus infections cause increased thymidine kinase activity. Large viruses, such as herpes and vaccinia, appear to contain the genetic information for a thymidine kinase of their own [43], while small viruses, such as polyoma and SV40, utilize a host cell kinase [44,45]. One large virus, adenovirus 12, appears to lack the information for its own kinase, but may somehow cause an increase in host cell kinase by injuring the long arm of chomosome number 17, on which the gene for the kinase is located in human cells [46].

The herpes virus-induced kinase is found in the cytosol, but has greater electrophoretic mobility and kinetic differences from the usual cytosol enzyme

[43]. It can also be distinguished from mitochondrial kinase, at least when mouse cells are infected. However, when human cells are used, it is difficult to distinguish the virus-induced enzyme from the mitochondrial form in regard to electrophoretic mobility and certain other properties [43]. As noted above, mitochondria contain an intermediate electrophoretic form, and also mitochondrial enzyme can leak into the cytosol under certain circumstances [20]. Thus, while the evidence so far, including studies with herpes virus mutants which produce temperature-sensitive kinases [47], supports the conclusion that herpes induces its own enzyme [43], it remains possible that the virus-induced kinase is a new or modified cellular enzyme.

Presently there are conflicting reports as to whether the genetic information for thymidine kinase can be transferred with mammalian chromosomes [48,49], but permanent transfer seems possible with herpes virus. In 1971 Munyon and collaborators described the isolation of kinase-positive cells from cultures of kinase-deficient mouse cells subjected to nonlytic infection with ultraviolet light – inactivated herpes virus [50]. This new kinase has an electrophoretic mobility and heat stability which is intermediate between the fast and slow cellular forms, and is similar to that of the enzyme in cells lytically infected with herpes virus [51]. This suggests again that the enzyme is coded by a viral gene. However, the kinase activity introduced by inactivated herpes virus behaves like mitochondrial kinase during the mitotic and culture cycles: its activity fails to increase during the S phase of mitotic cycle and remains low during active cell division, increasing thereafter [51]. Thus, in this situation also, it is possible that the virus has caused the appearance of a modified host cell kinase.

Furthermore, the subsequent behavior of these kinase-positive cells is most interesting [52,53]. Such cells are isolated as clones by means of selective culture medium, which requires the presence of the kinase for survival. In these clones the frequency of cells which lose the kinase is remarkably high, and among these latter kinase-deficient subclones regain of kinase is equally frequent. If the kinase-positive cells are allowed to grow at length in non-selective medium, the number of cells which contain kinase decreases exponentially. However, the ability to form kinase is retained, since in all these cells kinase activity can be regained at a frequency which, while low, is much higher than in the original kinase-deficient parental line. Furthermore, clones which have twice regained kinase in this way proceed to lose it once more at the same exponential rate, characteristic for each clone. Incidentally, the regain of kinase occurs spontaneously, regardless of the presence of the selective medium, and is not enhanced by treatment of the cells with mutagens or by cell fusion.

It has been suggested that the kinase-positive cells have acquired a viral kinase gene which resides intracellularly in a form not under the same sort of control as cell cytosol kinase, and somehow subject to reversible loss or inactivation [53]. Yet, as discussed above, the enzyme introduced by the virus may in fact be of cellular rather than viral origin. Further, as also discussed above, cell kinase may shuttle between cytosol and cell organelles, with changes in activity and thermostability. Until we know the reason for the loss of cytosol kinase in the original kinase-deficient parental cell line, we will be uncertain as to the effects of the virus.

For example, in some kinase-deficient variant cell cultures, rare spontaneous 'revertants' occur, the frequency of which is increased with mutagens [31] or certain viruses [29]. Perhaps herpes virus also is acting like a mutagen to stimulate reversion. In this case reversion might represent the acquisition of instability in an epigenetic process regulating the amount of cytosol kinase by affecting its rate of degradation or its relation to mitochondrial or cell membrane kinase forms.

Summary

There are now several uncertainties concerning the number of cellular forms of thymidine kinase, their relations to each other, how they may be modified, and the mechanisms regulating these processes. These uncertainties prevent us from increasing our understanding of intriguing biological phenomena in which the kinase appears to participate; for example, in the control of DNA synthesis, in the uptake of thymidine, and in lytic and non-lytic infections with herpes and other viruses. We need to characterize the structure, rates of synthesis and degradation, and intracullular locations and movements of the different cellular kinases, and to determine conclusively the origin of the viral kinases. For this, purification methods from small quantities of cells or cell organelles will be necessary, perhaps using affinity chromatography [7] or kinase-specific antibodies. Although much more complex than originally conceived, the thymidine kinase system provides an unusual opportunity to further our knowledge of genetic and epigenetic control mechanisms in animal cells.

Acknowledgement

I would like to acknowledge the support of U.S.P.H.S. research grant No. CA-16754 while this review was being written.

References

[1] T. Chan, M. Meuth and H. Green, Pyrimidine excretion by cultured fibroblasts: effect of mutational deficiency in pyrimidine salvage enzymes. J. Cell Physiol. 83, 263–266 (1974).

[2] R. Okazaki and A. Kornberg, Deoxythymidine kinase of *Escherichia coli*. I. Purification and some properties of the enzyme. J. Biol. Chem. 239, 269–274 (1964).

[3] R. Okazaki and A. Kornberg, Deoxythymidine kinase of *Escherichia coli*. II. Kinetics and feedback control. J. Biol. Chem. 239, 275–284 (1964).

[4] N. Iwatsuki and R. Okazaki, Mechanism of regulation of deoxythymidine kinase of *Escherichia coli*. I. Effect of regulatory deoxynucleotides on the state of aggregation of the enzyme. J. Mol. Biol. 29, 139–154 (1967).

[5]. N. Iwatsuki and R. Okazaki, Mechanism of regulation of deoxythymidine kinase of *Escherichia coli*. II. Effect of temperature on the enzyme activity and kinetics. J. Mol. Biol. 29, 155–165 (1967).

[6] P. Voytek, P.K. Chang and W.H. Prusoff, Purification of deoxythymidine kinase by preparative disc gel electrophoresis and the effects of various halogenated nucleoside triphosphates on its enzymatic activity. J. Biol. Chem. 246, 1432–1438 (1971).

[7] W. Rohde and A.G. Lezius, The purification of thymidine kinase from *Escherichia coli* by affinity chromatography. Hoppe-Seylver's Z. Physiol. Chem. 352, 1507–1516 (1971).

[8] T. Hashimoto, T. Arima, H. Okuda and S. Fujii, Purification and properties of deoxythymidine kinases from the Yoshida sarcoma. Cancer Res. 32, 67–73 (1972).

[9] H. Okuda, T. Arima, T. Hashimoto and S. Fujii, Multiple forms of deoxythymidine kinase in various tissues. Cancer Res. 32, 791–794 (1972).

[10] D.E. Kizer and L. Holman, Purification and properties of thymidine kinase from regenerating rat liver. Biochim. Biophys. Acta 350, 193–200 (1974).

[11] E. Bresnick, K.D. Mainigi, R. Buccino and S.S. Burleson, Studies on deoxythymidine kinase of regenerating rat liver and *Escherichia coli*. Cancer Res. 30, 2502–2506 (1970).

[12] Y.-C. Cheng and W.H. Prusoff, Mouse ascites sarcoma 180 deoxythymidine kinase. General properties and inhibition studies. Biochemistry 13, 1179–1185 (1974).

[13] A.T. Taylor, M.A. Stafford and O.W. Jones, Properties of thymidine kinase partially purified from human fetal and adult tissue. J. Biol. Chem. 247, 1930–1935 (1972).

[14] D.L. Bull, A.T. Taylor, D.M. Austin and O.W. Jones, Stimulation of fetal thymidine kinase in cultured human fibroblasts transformed by SV40 virus. Virology 57, 279–284 (1974).

[15] R. Adler and B.R. McAuslan, Expression of thymidine kinase variants is a function of the replicative state of cells. Cell 2, 113–117 (1974).

[16] S.J. Adelstein, C. Baldwin and H.I. Kohn, Thymidine kinase in mouse liver: variations in soluble and mitochondrial-associated activity that are dependent on age, regeneration, starvation, and treatment with actinomycin D and puromycin. Dev. Biol. 26, 537–546 (1971).

[17] H. Masui and L.D. Garren, On the mechanism of action of adrenocorticotropic hormone. J. Biol. Chem. 246, 5407–5413 (1971).

[18] B. Attardi and G. Attardi, Persistence of thymidine kinase activity in mitochondria of a thymidine kinase-deficient derivative of mouse L cells. Proc. Nat. Acad. Sci. U.S.A. 69, 2874–2878 (1972).

[19] A.J. Berk and D.A. Clayton, A genetically distinct thymidine kinase in mammalian mitochondria. J. Biol. Chem. 218, 2722–2729 (1973).

[20] S. Kit, W.-C. Leung and L.A. Kaplan, Distinctive molecular forms of thymidine kinase in mitochondria of normal and bromodeoxyuridine-resistant HeLa cells. Eur. J. Biochem. 39, 43–48 (1973).

[21] S. Kit and W.-C. Leung, Submitochondrial localization and characteristics of thymidine kinase molecular forms in parental and kinase-deficient HeLa cells. Biochem. Genet. 11, 231–247 (1974).

[22] E. Stubblefield and G.C. Mueller, Thymidine kinase activity in synchronized HeLa cell cultures. Biochem. Biophys. Res. Commun. 20, 535–538 (1965).

[23] E. Stubblefield and S. Murphee, Synchronized mammalian cell cultures. II. Thymidine kinase activity in colcemid-synchronized fibroblasts. Exp. Cell. Res. 48, 652–656 (1967).

[24] J.W. Littlefield, The periodic synthesis of thymidine kinase in mouse fibroblasts, Biochim. Biophys. Acta 114, 398–403 (1966).

[25] H.R. Zielke and J.W. Littlefield, Different timing of increases in activities of four X-chromosome linked enzymes during the cell cycle of synchronized human lymphoblasts. Exp. Cell. Res. (in press).
[26] J.W. Littlefield, Studies on thymidine kinase in cultured mouse fibroblasts. Biochim. Biophys. Acta 95, 14–22 (1965).
[27] S. Kit and W.-C. Leung, Genetic control of mitochondrial thymidine kinase in human-mouse and monkey-mouse somatic cell hybrids. J. Cell Biol. 61, 35–44 (1974).
[28] J.S. Salser and M.E. Balis, Distribution and regulation of deoxythymidine kinase activity in differentiating cells of mammalian intestines. Cancer Res. 33, 1889–1897 (1973).
[29] J.W. Littlefield, Variation, Senescence, and Neoplasia in Cultured Somatic Cells (Harvard University Press, Cambridge) 1975 (in press).
[30] K. Paigen, R.T. Swank, S. Tomino and R.E. Ganschow, The molecular genetics of mammalian glucuronidase. J. Cell. Physiol. 85, 379–392 (1975).
[31] D.J. Roufa, B.N. Sadow and C.T. Caskey, Derivation of TK − clones from revertant TK + mammalian cells. Genetics 75, 515–530 (1973).
[32] J. Hochstadt, The role of the membrane in the utilization of nucleic acid precursors, CRC Critical Reviews in Biochemistry 2, 259–310 (1974).
[33] P.G.W. Plagemann and D.P. Richey, Transport of nucleosides, nucleic acid bases, choline and glucose by animal cells in culture. Biochim. Biophys. Acta 344, 263–305 (1974).
[34] B.R. Fridlender, E. Medrano and J. Mordoh, Synthesis of DNA in human lymphocytes: possible control mechanism. Proc. Nat. Acad. Sci. U.S.A. 71, 1128–1132 (1974).
[35] B.A. Roller, K. Hirai and V. Defendi, Effect of cAMP on nucleoside metabolism. II. Cell cycle dependence of thymidine transport. J. Cell. Physiol. 84, 333–342 (1974).
[36] D.D. Cunningham and R. A. Remo, Specific increase in pyrimidine deoxynucleoside transport at the time of deoxyribonucleic acid synthesis in 3T3 mouse cells. J. Biol. Chem. 248, 6282–6288 (1973).
[37] L.P. Everhart, Jr. and R.W. Rubin, Cyclic changes in the cell surface. I. Change in thymidine transport and its inhibition by cytochalasin B in Chinese hamster ovary cells. J. Cell Biol. 60, 434–441 (1974).
[38] G.S. Schuster and J.D. Hare, The role of phosphorylation in the uptake of thymidine mammalian cells. In Vitro 6, 427–436 (1971).
[39] R.E. Breslow and R.A. Goldsby, Isolation and characterization of thymidine transport mutants of Chinese hamster cells. Exp. Cell Res. 55, 339–346 (1969).
[40] J.J. Freed and L. Mezger-Freed, Origin of thymidine kinase-deficient (TK −) haploid frog cells via an intermediate thymidine transport deficient (TT −) phenotype. J. Cell. Physiol. 82, 199–212 (1973).
[41] L. Mezger-Freed, Thymidine kinase deficiency in cultured haploid cells: gene mutation or gene regulation? J. Cell. Physiol. (in press).
[42] J.J. Freed and I.M. Hames, Loss of a thermolabile thymidine kinase activity in bromodeoxyuridine-resistant (transport deficient, kinase positive) haploid cultured frog cells. Exp. Cell Res. (in press).
[43] S. Kit, W.-C. Leung, D. Trkula and G. Jorgensen, Gel electrophoresis and isoelectric focusing of mitochondrial and viral-induced thymidine kinases. Int. J. Cancer 13, 203–218 (1974).
[44] J.W. Littlefield and C. Basilico, Infection of thymidine kinase-deficient BHK cells with polyoma virus. Nature 211, 250–252 (1966).
[45] C. Basilico, Y. Matsuya and H. Green, On the origin of the thymidine kinase induced by polyoma virus in productively infected cells. J. Virol. 3, 140–145 (1969).
[46] J.K. McDougall, R. Kucherlapati and F.H. Ruddle, Localization and induction of the human thymidine kinase gene by adenovirus 12. Nature New Biol. 245, 172–175 (1973).
[47] G.M. Aron, P.A. Schaffer, R.J. Courtney, M. Benyesh-Melnick and S. Kit, Thymidine kinase

activity of herpes simplex virus temperature-sensitive mutants. Intervirology 1, 96–109 (1973).
[48] Y. Boyd and H. Harris, Correction of genetic defects in mammalian cells by the input of small amounts of foreign genetic material. J. Cell Sci. 13, 841–861 (1973).
[49] S. Kit, W.-C. Leung, G. Jorgensen, D. Trkula and D.R. Dubbs, Acquisition of chick cytosol thymidine kinase activity by thymidine kinase-deficient mouse fibroblast cells after fusion with chick erythrocytes. J. Cell Biol. 63, 505–514 (1974).
[50] W. Munyon, E. Kraiselburd, D. Davis and J.P. Mann, Transfer of thymidine kinase to thymidine kinase-less L cells by infection with ultraviolet-irradiated herpes simplex virus. J. Virol. 7, 813–820 (1971).
[51] S.-S. Lin and W. Munyon, Expression of the viral thymidine kinase gene in herpes simplex virus-transformed L cells. J. Virol. 14, 1199–1208 (1974).
[52] R.L. Davidson, S.J. Adelstein and M.N. Oxman, Herpes simplex virus as a source of thymidine kinase for thymidine kinase-deficient mouse cells: suppression and reactivation of the viral enzyme. Proc. Nat. Acad. Sci. U.S.A. 70, 1912–1916 (1973).
[53] E.R. Kaufman and R.L. Davidson, Control of the expression of a herpes simplex virus thymidine kinase gene incorporated into thymidine kinase-deficient mouse cells. Somatic Cell Genet. 1, 153–164 (1975).

Chapter 17

IMAGE ANALYSIS, FLOW FLUOROMETRY AND FLOW SORTING OF MAMMALIAN CHROMOSOMES*

MORTIMER L. MENDELSOHN, ANTHONY V. CARRANO,
JOE W. GRAY, BRIAN H. MAYALL and MARVIN A. VAN DILLA

Biomedical and Environmental Research Division, University of California, Lawrence Livermore Laboratory, Livermore, California 94550, U.S.A.

Introduction

As the major packaging units of genetic material, chromosomes have broad relevance to basic and applied genetics, to clinical medicine and to public health. The traditional morphological approach to the mammalian chromosome is a subjective, non-quantitative and essentially non-molecular biological approach. The thrust of the Livermore effort in chromosomology is to emphasize the physical and chemical characterization of the chromosome by developing and using appropriate instruments to quantitate the DNA in individual mammalian chromosomes, to process chromosomes rapidly and precisely and to purify chromosomes for subsequent chemical and biological applications. Progress in these areas has been rapid and exciting in the past few years, and this chapter will attempt to summarize that part of the work that bears on the molecular biology of the mammalian genetic apparatus.

We have approached the measurement and manipulation of chromosomes with two very different techniques. The first is absorption cytophotometry coupled to image analysis, a method which measures relative DNA content and centromere index of preidentified chromosomes within the well flattened metaphase cell on a microscope slide. The second is flow fluorometry and sorting, a technique which requires chromosomes in suspension, measures relative DNA content and sorts or separates chromosomes within specified ranges of DNA content. The

---

* This work was performed under the auspices of the U.S. Energy Research and Development Administration (Contract# W-7405-ENG-48) and was supported in part by USPHS Grant GM20291.

The Molecular Biology of the Mammalian Genetic Apparatus:
edited by P. Ts'o © 1977, Elsevier North-Holland Biomedical Press

image-analytic approach uses traditional preparations and can be directly coupled to traditional analysis; as presently practiced it is slow but immediately applicable to human cytogenetics. The flow fluorometric approach uses non-conventional preparations and a non-conventional chromosomal rather than cellular framework for the analysis; it is extremely rapid, unique in its ability to purify chromosomes, and potentially but not immediately applicable to clinical cytogenetics.

Absorption cytophotometry and image analysis

The absorbance of light traversing a minute region in the object plane of a microscope can be measured by methods analogous to those of conventional spectrophotometry. For a microscopic object isolated within a defined area, such absorbance measurements can be summed point by point throughout the area to give the relative chromophore content of the object [1]. Irregularly shaped, non-overlapping chromosomes lying helter-skelter within a metaphase cell can be measured by the same principle, provided the boundaries of the chromosomes can be used to define ad hoc photometric areas within which the absorbance measurements are summed for the enclosed chromosome. This task is best accomplished by an image-analytical process [2]. The absorbance measurements are made in a regular array over the entire metaphase cell. Once digitized, they constitute a high fidelity digital image which is ideally suited for image processing and optical weighing of component parts.

In the Livermore system, the digital image is generated by CYDAC, a flying-spot cathode-ray tube scanner built around a conventional microscope. The image typically consists of 192 by 192 absorbance measurements digitized to 255 levels. The measurements are taken at 6 kHz with a sample and line spacing of 0.25 μm and a 1% uncertainty in the unmodulated light intensity. The digital image is near the limit of optical resolution and when properly reconstructed has the appearance of a good photomicrograph (fig. 1).

The image processing is done by a small general-purpose digital computer coupled to a human operator through an interactive picture display system. To begin, the chromosomes are thresholded at mid-grayness and displayed as black blobs on a white background. The operator edits this image to separate touching chromosomes, to reunite fragments, to eliminate artifacts, and to name or number the chromosomes. The computer then extends the boundary of each chromosome by 1 μm in order to include all relevant grayness and sums the absorbance measurements within the new boundary. The centromere is located by its relatively low absorbance and a DNA-based centromeric index is measured by taking the ratio of absorbances of the larger arm to the total.

With human chromosomes our usual practice is to preidentify all chromosomes by quinacrine banding; the chromosomes then are restained with gallocyanin-

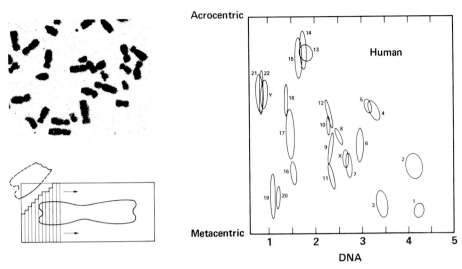

Fig. 1. Image-analysis of human chromosomes. The chromosomal images in the upper left were scanned directly through the microscope and digitized by CYDAC. Such digital data has the fidelity and resolution to outline chromosomes, to cumulate their absorbance (thus measuring their relative DNA content), and to find their centromeres. The strips shown in the lower left are oriented perpendicularly to the chromosome axis; their cumulated absorbance yields a DNA profile which has a pronounced local minimum at the centromere. The graph on the right plots relative DNA content against the centromeric index (large arm to total DNA content) for the quinacrine-identified chromosomes from 6 metaphase cells of a normal male. The numbers identify the chromosome type and locate the means, while the ellipses describe the 50% tolerance regions of the 6 to 12 chromosomes of each type. Over 90% of the chromosomes can be correctly classified into 20 groups using such data.

chrome alum, a broad-spectrum absorbing chromophore with good specificity and stoichiometry for DNA. A typical result for 6 cells from one normal male is shown in fig. 1. The two parameters, DNA content and centromeric index, give 20 clusters corresponding to 18 chromosome types, plus the D group (13, 14 and 15) and the F group (21, 22, Y). The complete analysis of 10 metaphase cells from one individual requires one week with the present system.

We are still relatively inexperienced with the genetic applications of this technique, but have been able to make several surprising and interesting observations [3]. Perhaps the most fundamental is the finding that every normal individual so far studied has two or more chromosomes which deviate significantly ($P < 0.01$) from the population mean. The deviants range widely over the karyotype and are of two roughly equal types: 1) homolog pairs in which the two members of the pair differ from each other, and 2) homolog pairs whose mean differs from the mean derived from other normal individuals. Most of the deviant chromosomes have normal banding patterns and are not detectably deviant by any other cytogenetic criterion. For several individuals, lymphocytes taken five years apart have given essentially identical results. In two family studies we have clear-cut evidence that almost all of the deviant chromosomes found in the lymphocytes of a child can also be found in the lymphocytes of one or the other parent; thus the bulk of chromosomes with deviant DNA content are directly inherited, but clearly some deviants arise de novo, presumably from mechanisms such as meiotic crossing over. These studies are continuing, as are efforts to clarify which type of DNA is involved and what significance the deviants have in human genetics. Our working assumption is that the variations are due to heterochromatic, non-informational DNA and that somehow homolog differences as large as 10% are well tolerated in meiosis. The deviants are potentially useful for human linkage studies but will complicate routine karyotyping by machine.

Our image-analytic studies have included one type of abnormal chromosome, the Philadelphia chromosome of chronic myelogenous leukemia. Table 1 shows the salient data for a typical case [4]. In the leukemic cells of this individual, the abnormal 22, the Philadelphia chromosome, has lost 0.43 units of DNA, while one chromosome 9 has gained the corresponding amount. (Note that we define 100 units as the autosomal metaphase DNA content. The DNA content of any one chromosome type, estimated from 10 cells, has a standard deviation of the mean of 0.03 unit or three-thousandths of the autosomal genome or 3 million base pairs or slightly less than one *E. coli* genome.) This and similar results in three other cases confirm that this unique cancer-specific aberration is in fact a translocation with no measurable loss of DNA.

As a final example of image-analytic chromosomal measurements and as an introduction to flow-fluorometric measurements, fig. 2 shows the distribution of chromosome DNA content of a Chinese hamster cell line. The mean of each chromosome type has been convoluted with a Gaussian distribution. Displayed in this manner, each peak represents chromosomes of common DNA content and

TABLE 1

Image-analytic DNA measurements in chronic myelogenous leukemia.

Chromosome 22		Chromosome 9	
Normal	Philadelphia	Abnormal	Normal
0.90	0.47	2.78	2.36
	0.43		

100 = autosomal metaphase DNA content = 10^{-11} gm.

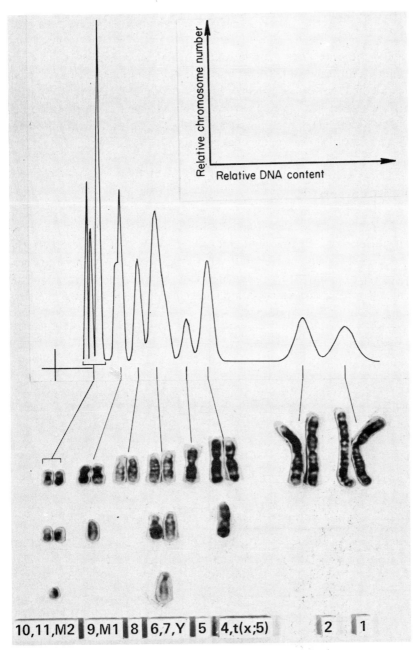

Fig. 2. A simulated DNA distribution based on image-analysis of Chinese hamster chromosomes. Preidentified chromosomes in metaphase cells of the M3-1 line were scanned by CYDAC and analyzed to obtain the mean relative DNA content of each chromosome type. This information plus the number of chromosomes of each type in the karyotype plus a constant coefficient of variation of DNA content normally distributed about each mean was used to compute a simulated DNA distribution for an unbiased population of Chinese hamster chromosomes. Representative banded chromosomes are shown in relation to their karyotype assignment and their location in the distribution.

the area under the peak gives the relative frequency per cell of the constituent chromosomes. With preidentified chromosomes the components of each peak are known and the DNA distribution becomes a well-defined signature or partial karyotype. The upper panel in fig. 3 shows an almost identical distribution obtained by flow fluorometry of approximately 100,000 suspended chromosomes measured at a rate of 1,000 chromosomes per second.

Flow fluorometry and sorting

In a flow fluorometer, fluorochromed objects are carried singly by laminar flow through a beam of exciting light and produce a burst of fluorescence. The fluorescence is detected and converted to an electrical pulse whose height is proportional to the total fluorescence. The height of the pulse is analyzed and the resulting digital value is accumulated in the memory of a multichannel analyzer.

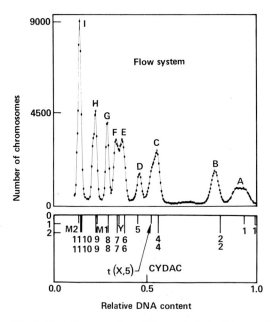

Fig. 3. Flow-fluorometric and image-analytic data on chromosomes from Chinese hamster M3-1 line cells. The flow-based DNA distribution in the upper panel is strikingly similar to the simulated image-analytic-based distribution in fig. 2. The assignment, number per cell and image analytic mean relative DNA content for the chromosomes are shown in the lower panel, and again the correspondence to the flow data is striking. Based on CYDAC measurement, the homologous number 1 chromosomes differ by 6% in their DNA content. The corresponding first peak of the flow distribution is broadened and in later experiments is clearly notched. For the nine peaks in this flow-fluorometric DNA distribution the modes reliably indicate relative DNA content and the areas indicate relative chromosome number. Reproduced from [11].

The objects must be monodispersed, held in suspension, and delivered along a precise trajectory at a constant velocity. A schematic of the Livermore bicolor flow fluorometer is shown in fig. 4. We use an argon ion laser as the exciting source and the beam is shaped by cylindrical lenses to provide uniform illumination across the range of trajectories. Objects are measured at rates up to 1,000 per second [5].

A flow sorter uses the same principle of measurement but adds a mechanism to separate objects whose fluorescence falls within a predetermined range of values [6,7,8]. In our sorter, shown in fig. 5, objects are carried in a laminar stream which emerges from a nozzle and crosses the laser beam while still cylindrical. Under the influence of a supersonic oscillator, the stream breaks up into uniform droplets. When the fluorescence from an object falls within the desired range, a momentary electrical charge is applied to the fluid column some 200 μsec later at the instant when the droplet containing the object is breaking away. All droplets traverse an electrical field which separates the positively charged, the negatively charged and the uncharged droplets into three containers. In practice, 3 droplets are charged to cover the uncertainty of object transit from the laser beam to droplet break-off. One in 50 droplets contains an object, and the throughput is 1,000 objects per second.

To prepare chromosomes using suspension cultures, actively dividing cells are blocked with Colcemid, treated with hypotonic (0.075 m) KCl for 30 min, and sheared by passing twice through a 22 gauge needle. For monolayer cultures,

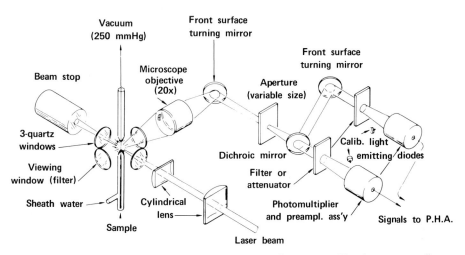

Fig. 4. Schematic diagram of the Livermore bicolor flow fluorometer. The chromosomes flow at a constant velocity of 10 m/s in a stream of about 10 μm diameter. As each chromosome crosses the laser beam a fluorescent pulse is emitted, converted to an electrical pulse, and processed by a multichannel pulse-height analyzer. The bicolor system has the potential to measure two fluorochromes simultaneously. Reproduced from [5].

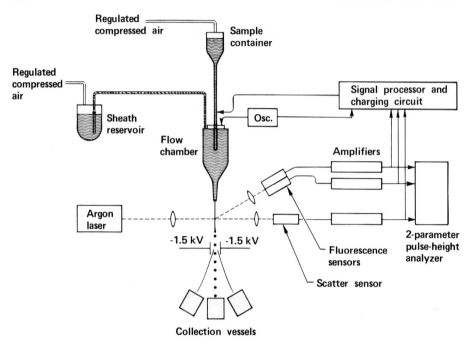

Fig. 5. Schematic diagram of the Livermore flow sorter. In this device the signal produced by the fluorescence of a single chromosome is compared to one or two ranges of preset values of relative DNA content. When the signal falls within range the fluid column is given a brief positive or negative charge at the moment the droplet containing the chromosome is breaking away from the column. The droplet and the enclosed chromosomes are thus deflected into the appropriate container. As with the flow fluorometer, 1,000 chromosomes can be examined per second.

Colcemid may be eliminated by shaking to harvest the mitotic cells. Typically the crude suspensions containing 10^7 chromosomes are stained in 10^{-4} M ethidium bromide and are used directly for fluorometry [9,10].

The chromosomal DNA distribution for the M3-1 line of Chinese hamster cells, as shown in fig. 3, can be described by a series of Gaussian distributions and a background continuum which increases exponentially toward lower DNA contents [11]. The expected 14 types of chromosomes in these cells yield nine peaks whose modes and areas correspond to within 4 and 9% respectively of those obtained by CYDAC (fig. 3, lower panel). Coefficients of variation of peaks due to single chromosome types range from 2.2. to 3.3%.

The chromosomal content of each peak of the DNA distribution can be determined directly by sorting, banding the separated chromosomes and identifying the chromosomes morphologically. Table 2 summarizes one such experiment done with Chinese hamster cells. The results agree with the assignments based on CYDAC measurements. More importantly, they establish that single chromosome types can be collected with a purity of about 90%. Contamination is presumably due largely to the underlying background continuum and since it

TABLE 2

Dominant fraction of Chinese hamster M3-1 chromosomes found after flow sorting

		\multicolumn{8}{c}{Chromosome}							
		1	2	4/X	5	6/7/Y	8	9/M1	10/11/M2
Peak*	A	0.90							
	B		0.89						
	C			0.88					
	D				0.91				
	E-F					0.95			
	G						0.91		
	H							0.94	
	I								0.99

* See fig. 3 for peak notation.

generally involves smaller chromosomes, we believe a likely mechanism is aggregation prior to fluorometry.

Flow fluorometry of an abnormal karyotype of a subline of M3-1 is shown in fig. 6 [5]. The cells have a stable, balanced translocation in which a single number 1 and a single number 4 chromosome exchange DNA and end up as two new chromosomes with intermediate DNA contents. The DNA distributions reflect this by 1) a reduction in area of peaks A and C equivalent to the loss of one chromosome each; 2) an increase in the area of peak B equivalent to the gain of one chromosome; 3) appearance of a new peak with area corresponding to one chromosome and DNA content increased above peak C by the identical amount that peak B is below peak A.

Reconstruction experiments involving known proportions of the two lines of M3-1 cells indicate that this translocation is detectable by flow fluorometry if present in 5% of the cells. Thus, the method is suitable for detecting aberrations of DNA content present uniformly in all or a sizable subset of cells. Heterogeneous or random aberrations will produce diffuse signals and will appear in the DNA distribution as an increment to the background continuum [12].

Fig. 7 shows the DNA distribution of the chromosomes of the Indian muntjac [10]. The male muntjac has seven chromosomes with two pairs, hence five chromosome types. However the measured distribution has seven peaks, an anomaly caused by the tendency of the X+3 chromosome to fragment on shearing. This chromosome has a long heterochromatic neck region lying between the X and the 3 components. Fragmentation occurs at the junctions of neck to arm, yielding objects having various combinations of the DNA contents. This includes isolated heterochromatic neck regions which sort with the heterochro-

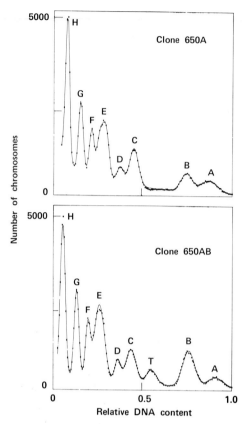

Fig. 6. Flow-fluorometric DNA distribution of chromosomes of control (clone 650A) and translocated (clone 650AB) M3-1 lines of Chinese hamster cells. The translocation between one number 1 and one number 4 chromosome changes peaks A, B and C and introduces a new peak, T. Peaks A and C have lost one chromosome each; peaks B and T have gained one chromosome each; and the two translocated chromosomes are equally and oppositely displaced in DNA content. Reproduced from [5].

matic Y chromosome and provide a unique opportunity to study this specialized DNA by biochemical methods.

Human chromosomes from cultured skin cells give the DNA distributions shown in the upper panel of fig. 8 [5]. Seven broad peaks are detected by flow fluorometry. The middle panel of fig. 8 shows a distribution computed from CYDAC means convoluted with Gaussians having a coefficient of variation of 4%. The two distributions (fluorescence and absorbance) are nearly identical, confirming the validity of the flow measurements, allowing us to identify the constituents of each peak, and estimating the overall coefficient of variation of flow measurement as 4%. The lower panel shows a computed distribution based again on CYDAC data but with a coefficient of variation of 2%. At this level of

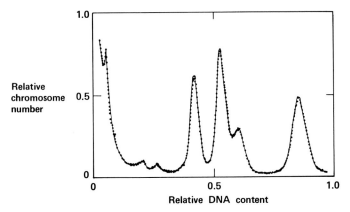

Fig. 7. Flow-fluorometric DNA distribution of chromosomes of a male Indian muntjac cell line. The seven peaks in this distribution are due to the five chromosome types of this remarkable deer plus three distinct fragments of the large, heterochromatically necked, X + 3 chromosome. Sorting on these peaks gives single chromosomes of up to 97% purity.

resolution the simulated distribution shows 18 peaks (out of the 24 normal chromosome types in the human karyotype). We are currently studying the sources of variation of chromosomal flow measurements in the hopes of achieving coefficients of variation of 2% on human material. Improved resolution may be obtained by flow devices which have narrower flow trajectories, more stable flow velocities, more intense light sources and better control of electronic stability. Improvements may also result from better methods of chromosome preparation and stabler, more precise methods of fluorochroming chromosomal DNA.

Flow processing of chromosomes has many important implications to clinical cytogenetics and genetic research. In clinical applications it could become a new way to perform standard karyotypes in search of trisomies, sex chromosome anomalies, inherited chromosomal aberrations and mosaicism. For such use in humans it will require improved performance, the development of routine methods for producing bulk chromosomes, and the reorientation of cytogeneticists away from conventional morphological analysis. We are hoping as well that with the potential addition of centromere detection the flow methods will also be able to resolve heterogeneous aberrations [12]; this could have a major impact on population screening, evaluation of clastogens and perhaps even cancer diagnosis.

The molecular biological potential of flow processing is centered around the ability to sort chromosomes. Selected outlying chromosomes from judiciously chosen species can now be sorted with 90 to 97% purity, and as described above, there are even situations where chromosome parts can be dealt with effectively. For the bulk of chromosomes, the key to effective sorting is to improve the resolution of DNA measurements or to in some way mark particular chromosomes. One approach to marking is to use the bicolor capability to sense two

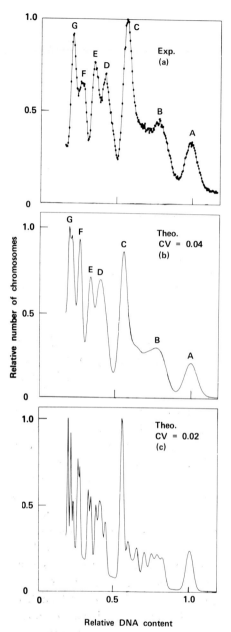

Fig. 8. Flow-fluorometric and image-analytic DNA distributions of human chromosomes. The upper panel is a flow-fluorometric distribution of ethidium bromide stained chromosomes from a foreskin cell line. It shows seven peaks. The middle panel is an image-analytic distribution simulated with a coefficient of variation of 4% and mean values taken from the lymphocytes of several normal males using gallocyanin-chrome alum stained chromosomes. Apart from minor differences in several peaks, the two distributions are remarkably similar. The lower panel is based on the same data as the middle panel but the distribution is simulated at a coefficient of variation of 2%. It is only at this level of resolution that the full complexity of the human karyotype becomes apparent. Reproduced from [5].

simultaneous DNA fluorochromes which differ in chromosomal affinity because of differences in AT to GC ratios or heterochromatin. Differences in staining affinity of DNA fluorochromes have been found [13] but have yet to be used in this way. Another approach is to bootstrap the purification. An example of this would be the preparation of fluorescent antibody to a particular chromosome using a partially purified sort, an aberrant cell line in which an ambiguous chromosome is displaced, or heterokaryons with a reduced human complement. Once such antibodies are available they could be used to sort subsequent material, and incidentally, could be very valuable for automated detection of translocations.

Purified chromosomes can be used for biochemical analysis, template activity, pharmacologic studies of reactivity, antigenicity and biological transduction. Critical studies on linkage, chromosome structure, mutagenicity, gene control, evolution and genetic engineering become possible in a variety of species, including man. However, the feasibility of such studies hinges on a difficult logistical problem. The average Chinese hamster metaphase chromosome contains about 5×10^{-13} g of DNA. At an overall flow rate of 1,000 chromosomes per second, a pair of homologs is sorted every 2 out of 23 chromosomes, or 87 chromosomes per second. Thus the yield for an average pair would be 4.5×10^{-11} g/sec. At this rate it would take 367 min to sort 1 μg of DNA and 17 years to sort 1 mg. The logistics can be improved by selecting larger chromosomes, by choosing species with fewer chromosomes, by sorting two chromosome types in parallel (using opposite charges), by new instruments which have higher throughput or more sorting channels, and by preprocessing to concentrate the desired chromosomes using bulk techniques such as centrifugation. Alternatively, the analytical procedures could be improved to reduce the required quantity of chromosomes to micrograms or less. We are currently making sorting runs over an eight-hour period and are collecting 2.5×10^6 Chinese hamster chromosomes, enough to make several applications feasible.

Conclusion

The techniques we have just described, as well as the many other advances described in these volumes should make the mammalian chromosome increasingly available for critical study by the methods of molecular biology. One can anticipate rapid and important advances in our understanding of chromosome structure and our application of this to improving the human condition, all very much in the spirit and tradition of James Bonner.

References

[1] B.H. Mayall and M.L. Mendelsohn, Deoxyribonucleic acid of stained human leukocytes. II. The mechanical scanner of CYDAC, the theory of scanning photometry and the magnitude of residual errors. J. Histochem. Cytochem. 18, 383 (1970).

[2] M.L. Mendelsohn and B.H. Mayall, Chromosome identification by image analysis and quantitative cytochemistry, in: Human Chromosome Methodology (J.J. Yunis, ed) Academic Press, Inc., New York, 1974, p. 311.

[3] B.H. Mayall, A.V. Carrano, D.H. Moore II, L.K. Ashworth, D.E. Bennett, E. Bogart, J.L. Littlepage, J.L. Minkler, D.L. Piluso and M.L. Mendelsohn, Cytophotometric analysis of human chromosomes, in: Automation of Cytogenetics, Conf. 751158, Asilomar Workshop, Nov. 30-Dec. 2, 1975 (M.L. Mendelsohn, ed.), Energy Research and Development Administration Rept. (Available from National Technical Information Service, U.S. Dept. of Commerce, 5285 Port Royal Road, Springfiels, Virginia 22151) p. 135 (1976).

[4] B.H. Mayall, A.V. Carrano and J.D. Rowley, DNA cytophotometry of chromosomes in a case of chronic myelogenous leukemia. Clin. Chem. 20, 1080 (1974).

[5] J.W. Gray, A.V. Carrano, D.H. Moore II, L.L. Steinmetz, J. Minkler, B.H. Mayall, M.L. Mendelsohn and M.A. Van Dilla, High-speed quantitative karyotyping by flow microfluorometry. Clin. Chem. 21, 1258 (1975).

[6] M.J. Fulwyler, Electronic separation of biological cells by volume. Science 150, 910 (1965).

[7] H.R. Hulett, W.A. Bonner, J. Barrett and L.A. Herzenberg, Cell sorting: Automated separation of mammalian cells as a function of intracellular fluorescence. Science 166, 747 (1969).

[8] M.A. Van Dilla, L.L. Steinmetz, D.T. Davis, R.N. Calvert and J.W. Gray, High-speed cell analysis and sorting with flow systems: Biological applications and new approaches. IEEE Trans. Nucl. Sci. NS-21, 714 (1974).

[9] W. Wray and E. Stubblefield, A new method for the rapid isolation of chromosomes, mitotic apparatus, or nuclei from mammalian fibroblasts at near neutral pH. Exp. Cell Res. 59, 469 (1970).

[10] A.V. Carrano, J.W. Gray, D.H. Moore II, J.L. Minkler, B.H. Mayall, M.A. Van Dilla and M.L. Mendelsohn, Purification of the chromosomes of the Indian muntjac by flow sorting. J. Histochem. Cytochem. 24, 348 (1976).

[11] J.W. Gray, A.V. Carrano, L.L. Steinmetz, M.A. Van Dilla, D.H. Moore II, B.H. Mayall and M.L. Mendelsohn. Chromosome measurement and sorting by flow systems. Proc. Nat. Acad. Sci. 72, 1231 (1975).

[12] M.A. Van Dilla, A.V. Carrano and J.W. Gray, Flow karyotyping: current status and potential development, in: Automation of Cytogenetics, Conf. 751158, Nov. 30-Dec. 2, 1975 (M.L. Mendelsohn, ed.), Energy Research and Development Administration Rept. (Available from National Technical Information Service, U.S. Dept. of Commerce, 5285 Port Royal Road, Springfield, Virginia 22151) p. 145 (1976).

[13] M.A. Van Dilla, J.W. Gray, A.V. Carrano, J.L. Minkler and L.L. Steinmetz, New directions for flow cytometry: chromosome analysis, in: Proc. of Symposium on Pulse Cytophotometry, Sept. 17–20, 1975, Munster, Germany, 1976. (In press).

Chapter 18

CHROMOSOMAL ALTERATIONS IN CARCINOGEN TRANSFORMED MAMMALIAN CELLS

J.A. DIPAOLO

Cytogenetics and Cytology Section, Biology Branch, Carcinogenesis Program, Division of Cancer Cause and Prevention, National Cancer Institute, Bethesda, Md. 20014, U.S.A.

General considerations

Cancer represents a series of diseases that may be caused by a variety of environmental agents belonging to diverse classes. It differs from the usual diseases such as those caused by bacteria in that malignancy results in the alteration of cells that belong to the body; hence malignant cells are the direct descendants of one's own somatic cells. Thus, it is logical and appealing to conclude that alterations in the function of chromosomes or chromatin result in the neoplastic properties of malignant cells. The concept that malignant cells are self-perpetuating and part of an evolving clone with a genetically unbalanced complement of chromosomes (aneuploidy) dates from the time of Boveri [1]. This conclusion was given further impetus by the studies of Makino [2], who studied malignant cells of an azo dye Yoshida ascites tumor and determined that it contained a new metacentric chromosome. This apparently was the beginning of the concept of stem lines and the later observation concerning numerical deviation of chromosome numbers.

Subsequently, a large body of information on chromosome constitution of malignant tumors, both spontaneous and experimentally induced, has been published [3–6]. Certain tumors can be characterized by their chromosome constitution. In terms of chromosome numbers, the tumors can be diploid, pseudodiploid, near triploid, near tetraploid and even near octaploid. Primary tumors, in contrast to serially passed tumors, may show a near-normal chromosome constitution. In fact, in some cases, in virus-induced primary tumors, the majority of cells may show a normal chromosome complement. Thus, the tumor characteristics may not necessarily be reflected by chromosome alterations.

Every cancer appears to be a specific dynamic entity characterized by cellular variability. This is also supported by pathology as well as by other indices of malignancy; invasiveness is a varying cell property or properties dependent

The Molecular Biology of the Mammalian Genetic Apparatus:
edited by P. Ts'o © 1977, Elsevier/North-Holland Biomedical Press

upon a number of different cell sites as well as the degree of progressive evolution. Observations on the variable nature of malignant cells are also compatible with the hypothesis that epigenetic changes may be responsible for neoplastic transformation for which no chromosomal change is found or required.

Because of the diversity of chromosomal changes seen in cancer growth, and in light of the concepts expressed above, one must be extremely cautious in formulating a hypothesis concerning chromosomal changes as a primary event in the malignant disease or as the cause of suppression or extinction of the malignancy. No single theory has been able to encompass all of the experimental data. It is conceivable that while two classes of cancer exist with the same clinical manifestations and pathology, only one involves alterations of chromosome complements. On the other hand, data, particularly when involved with expression and suppression as a result of either special selection techniques or cell hybridization, have not been accompanied with thorough statistical analysis, nor have the protocols included studies such as the Luria Delbrück fluctuation-type of analysis [7]. For example, some neoplastic cell lines can be described as having karyotypic instability in which subpopulations of varying tumorgenicity will exist in the mass culture. As a consequence, subpopulations of low and high tumorgenicity can be selected in the mass culture and the hybridization of these cell lines with low tumorgenicity cells from a low tumorgenicity culture would be expected to yield low tumorgenicity hybrids. This leads to the erroneous conclusion that low tumorgenicity phenotype suppresses the high tumorgenicity phenotype.

Nevertheless, examination of neoplastic or transformed cells of varying ages from different species has resulted in a number of assumptions or hypotheses relating to chromosomal abnormalities identifiable with a light microscope. These include: 1) each agent induces a specific abnormality at the chromosome level for that agent, 2) different agents induce a common specific abnormality, and 3) detectable chromosomal complement changes are the results of secondary events in the neoplastic process and are not essential for malignancy.

Experimental evidence for the assumption that a given agent causes a specific type of chromosome alteration, regardless of the type of cancer induced, has been advanced primarily from Japan and Sweden. The use of 7,12-dimethylbenz(a)anthracene (DMBA) or some of its derivatives in rats resulted in leukemia [8,9], sarcoma [10], or carcinoma [11], which, in some or most cases, possessed a consistent chromosomal abnormality, a trisomy of the largest telocentric chromosome. Dr. Sugiyama has written to me that they now have data indicating that six of 30 leukemias induced by N-nitrosobutylurea in rats had the same non-random cytogenetic change as found with DMBA. No such specific changes were reported for sarcomas induced by Rous virus [12] or by benzo(a)pyrene [13]. Support of the concept that there is a common distinctive abnormality independent of the inducing agent has been suggested by the studies of Sachs and associates [14,15], as well as Benedict [16], who reported complete or partial trisomy or translocation of the same piece of a chromosome designated as a carrier of the gene for expression of malignancy in Syrian hamster cells transformed in vitro by chemicals. Although Sachs and Benedict agreed on the marker responsible for expression, they reported different markers as being responsible for the suppression of malignancy.

In clinical oncology, the neoplasm that unequivocally has a specific and identifiable chromosomal alteration is chronic granulocytic leukemia with a Ph[1] chromosome. When remission occurs as a result of treatment with various agents in CML patients, the bone marrow morphology, as well as a number of other parameters, return to normal, but the percentage of Ph[1] positive ells remains unchanged. Thus, one is confronted with a puzzling situation in which these marrow cells behave like normal cells, despite the continued presence of the visibly recognizable abnormal genome [17]. Recently, banding studies of acute myelocytic leukemia (AML) from two different laboratories have been reported [18,19]. Similar results were obtained. In 50% of the cases all the chromosomes were perfectly diploid by different banding techniques. Nonrandom changes were seen in a small proportion of AML cases that may have appeared after the development of the leukemia.

Another approach to the normalization of the malignancy has been the use of cell hybrids by Wiener, Klein and Harris [20,21]. The somatic cell hybridization techniques have produced some puzzling results, particularly when one considers the work in which hybrids from two highly malignant cells produced tumors, or that of Barski [22], who showed a difference in tumorgenicity between hybrid sister cloned lines that had relatively easy identifiable chromosomes, both of which were karyologically similar but only one of which produced tumors. In fact, the nonmalignant clone even had a slightly higher mean number of chromosomes derived from the normal cells.

For many years, our laboratory has been interested in examining chromosomes in normal and neoplastic cells for visibly recognizable chromosomal changes. We have applied the newer techniques for examination of chromosomes following acute exposure to carcinogens, and for the examination of transformed lines to determine whether alteration in chromosome number, or in banding patterns of chromosomes, would help to answer questions concerning the relationship between chromosome alteration and cancer.

Immediate effects of carcinogens

Our investigations have involved the study of factors that influence the frequency of transformation in vitro. Carcinogenic and non-carcinogenic agents belonging to either physical, chemical, or viral classifications have been utilized alone and in combination to determine their influence on the incidence of transformation [23–25]. Using this approach, consideration has been given to both the immediate and the long-term effect of the various agents on chromosomes of mammalian cells.

Addition particularly of a chemical carcinogen, or exposure of mammalian cells to irradiation results in a certain degree of toxicity. This is best indicated by using the quantitative procedures first described by Puck [26]. The extent of lethality, as indicated by the reproductive capacity of the cells to form visible colonies, is proportional to the kind and concentration of carcinogen used as well as to the type of cell under consideration. For example, benzo(a)pyrene, which requires mixed function oxygenases for its metabolism, when added to Syrian hamster secondary cultures, demonstrates that the population is made up of cells that are sensitive to benzo(a)pyrene-induced lethality, as well as cells resistent to this lethality [27]. The addition of N-methyl-N'-nitro-N-nitrosoguanidine (MNNG), a direct acting carcinogen, to similar Syrian hamster cell cultures may result in a lethality that can approach 100% [28,29]. Nevertheless, by utilizing the proper concentration of a carcinogen, it is possible to study the immediate effect of the carcinogen on the chromosomes of the cells as well as the chromosome profile of populations of cells that have undergone quantitative transformation in vitro.

CARCINOGEN ASSOCIATED CHROMOSOME CHANGES
EARLY EVENTS

TRANSFORMED STATE

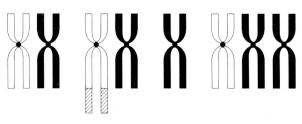

Fig. 1. Summary of possible changes in chromosomes following addition of a carcinogen. Top: the initial effect may cause aberrations that are of the chromatid and chromosome type. Bottom: malignant cells may have chromosomes that are normal, possess a new configuration or are numerically altered.

In general, the initial and the transformed phases may be summarized as indicated in fig. 1. Assuming that the cells to be treated have a normal chromosome complement, the addition of a carcinogen to cells in different phases of the cell cycle may produce chromosome aberrations that can be divided into those associated with both chromatid and chromosome rearrangements. The chromatid rearrangements may involve either single or multiple chromatids. In cells treated prior to DNA synthesis and chromosome replication, the chromosome type of aberration would be expected to occur.

Are these alterations the determinants of transformation and malignancy, or reflections of toxicity? A wide variety of chromosome damage has been observed following addition of chemical carcinogens [30–32]. Benedict [33] concluded that cytotoxicity can be correlated with chromosomal breaks and that the new chromosome rearrangements provide a mechanism for formation of marker chromosomes. Our examination of chromosomes of the cells constituting the transformed state demonstrates that the chromosomes may be normal, following visible examination and comparison to the normal G-band pattern; may have new marker chromosomes; or may be altered in number, resulting in either an excess or loss of chromosomes of a specific type. In any event, the incidence of abnormal chromosomes is very low compared to that obtained immediately after addition of a chemical carcinogen. This leads us to speculate that the result on the chromosome primarily reflects toxicity, since in the transformed state, the incidence and type of marker and non-disjunct chromosomes are significantly lower than that expected on the basis of immediate post-chromosomal damage. In some instances, no variation is observed.

To ascertain whether chemical carcinogen-associated breaks occurred more frequently on certain chromosome(s), Syrian hamster chromosomes were analyzed using the G-band procedure. Secondary Syrian hamster cells grown in Dulbecco's medium were exposed to 0.5 μg of MNNG/ml of medium [34]. Cultures were fixed at 5, 13, and 24 h post-carcinogen treatment; the last four hours prior to fixation of cells also included a 4 h Colcemid treatment. The time intervals were selected to include various phases of cell cycle. On the basis of a previous study [35], it was determined that the first fixation of cells at 5 h would include, for the most part, cells in last S and G2, while the 13 h period would include cells in late G1 and the beginning of S; and the last time period would have permitted the cells to undergo at least one mitosis. The frequency of chromosome aberrations, 20 to 24%, was independent of the time of examination. The type of chromosomal aberration observed following MNNG consisted, for the most part, of chromatid types of aberrations. G-band analysis permitted precise identification of the chromosomes affected by breakage as well as of the origin of the chromosomes involved in chromatid exchanges (fig. 2). The variety of chromosome damage varied from chromatid gaps or breaks located on either the long or short arm of chromosomes, with or without displacement of the chromosome

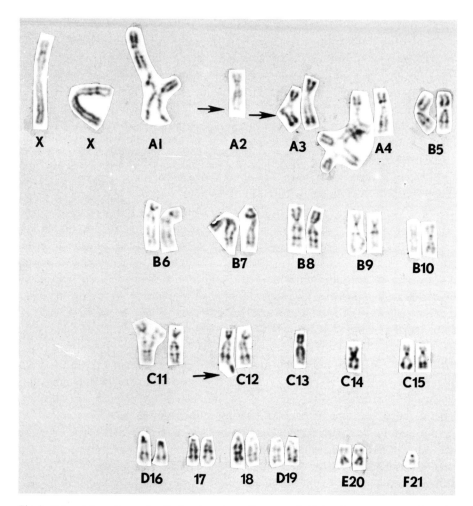

Fig. 2. G-band karyotype of a Syrian hamster cell prepared 24 h after treatment with 0.5 μg MNNG/ml medium. Chromosomes have breaks (arrow): The long arm of A2 lacks the terminal portions; the long arm of A3 has an interstitial break; and chromosome C12 has an acentric piece nearby. Three chromatid exchanges and their derivations are identified: between two A1 chromosomes; between three non-homologous chromosomes, A4 and C13, and C14; and between two non-homologous chromosomes, C11 and F21.

material; in some cases the break involved both chromatids, and in some cases resulted in different types of exchanges. G-band analysis of chromosomes involved in breakage indicated that, with the possible exception of the X chromosome which frequently had breaks in the euchromatic arm, there was no preferential involvement of any chromosomes in the complement. This analysis, furthermore, shows that the number of breaks appears to be related to the size of the chromosome; with the largest chromosomes having most of the breaks.

As suggested in the schematic representation of the immediate effect of a toxic carcinogen, the exchanges observed involved mostly chromatid types. The banding pattern techniques make possible the identification of the chromosomes engaged in the chromatid exchanges and the analysis of the type of exchange between two or more chromosomes. Obviously, depending upon the site of various breaks, a number of rearrangements are possible: for example, between non-homologous chromosomes connected by two chromatids with the two other chromatids remaining open; asymmetrical configuration, in which the chromatids involved are from two chromosome pairs, a configuration which occurs when a break occurs at the telomeric end of one chromosome; and at an interstitial break of a second chromosome. In the event that the breaks occur at the centromeric region, it may be possible to identify displaced chromosome material if the detached short arms are in the vicinity of the chromosomes. In such cases, a rejoining may occur between two homologous and one non-homologous chromosomes. In addition, multiple types of exchanges involving several different chromosomes also occur.

A total of 50 cells were examined for exchanges involving only two chromosomes. Of the 67 interchanges, 33 involved the chromosomes A1 to A4, while the other 34 involved the remaining chromosomes, with the exception of X, Y, and E20. The longer chromosomes were involved in most of the exchanges observed. Furthermore, it is interesting and important to point out that although the X, Y, and E20 chromosomes were affected by chromatid breaks, they were not associated with chromatid exchanges, emphasizing that the heterochromatic areas are not likely to be involved in chromosome exchanges.

The addition of a chemical carcinogen to human leukocyte cultures obtained from peripheral blood or fetal cells freshly derived from therapeutic abortions has not resulted in transformation. Carcinogenic polycyclic hydrocarbons, 4-nitroquinoline-1-oxide, aflatoxin B_1, and N-acetoxy-2-fluorethylacetamide, but not urethane, do cause toxicity as also indicated by chromosomal damage [36]. After a 24 h treatment of human leukocyte cultures or Syrian hamster embryo secondary cultures with concentrations that are known to produce transformation in vitro of hamster cells, the chromosomes exhibit G bands without any further preparation (fig. 3); non-carcinogenic chemicals were ineffective in producing these bands. Examination of cells at 4 or 8 h after treatment failed to produce chromosome bands.

The incidence of metaphases with chromosome bands varied with the carcinogen, and the highest incidence of banded metaphases, 80%, occurred when DMBA was applied to human leukocytes. Although it is possible that subtle variations may have occurred with different agents, it appears that the bands produced by each carcinogen are of the same configuration as that produced by the orthodox G-band procedures on fixed cell preparations. The ability of these known carcinogens to produce bands is a transitory phenomenon. If the

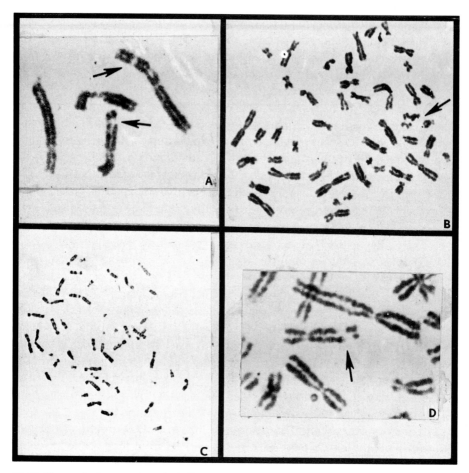

Fig. 3. Human leukocyte culture including a 24 h treatment with benzo(a)pyrene. A and D show chromosomes with gaps and isogaps. B has a radial figure. C demonstrates typical G-bands produced by carcinogen treatment prior to chromosome fixation.

carcinogen was removed and the cells allowed to grow for an additional 24 h, the chromosomes that were prepared were unbanded. The carcinogens capable of producing banding were also capable of causing multiple structural aberrations consisting primarily of chromatid breaks and chromosome exchanges (fig. 3). As expected, the incidence of chromosomal aberrations was dependent upon the carcinogen concentration and length of exposure of the cells. With urethane, not only were no bands produced but neither were chromosome breaks noted nor any indication of inhibition of cell multiplication. These events must be considered as reflections of a non-specific type of chemical toxicity because they may also be caused by other types of chemicals such as hydroxyurea [37] and azure B [38], which are not suspected to be in vivo carcinogens, and because the rearrangements that occurred following carcinogen treatment did not lead to a permanently altered cell. Therefore, these observations are not related to transformation, and the chromosome changes cannot be expected to serve as an index for surveying chemical compounds for possible carcinogenicity.

Radiological insult in conjunction with chemicals has been shown to influence the incidence of cancer in laboratory animals. Important factors in influencing the frequency of tumors are the length of the intervals between the exposures to the two types of agents as well as the nature of the radiation [39,40]. The frequency of the type of cancer produced may be unaffected, decreased, additive or synergistic. In vitro X-irradiation causes enhancement of transformation by benzo(a)pyrene in hamster embryo cells [39]. The addition of carcinogen to cells that had been irradiated and plated to produce discrete colonies resulted in maximum enhancement if the carcinogen was added 48 h after irradiation. The enhancement was greatest after 250R and decreased when higher doses of radiation were used, and when the carcinogen was added 72 h after irradiation. The possibility was pursued that the enhancement caused by the interaction of treatments was due to a specific type of chromosome aberration at the time interval of 48 h. X-ray with 250R produced chromosome damage consisting of chromatid and chromosome aberrations [35] (table 1). Examination of cultures at 24, 48, and 72 h showed that the incidence of cells with aberrations was 29, 26, and 19%, respectively, while the percentages of cells with multiple aberrations were 4, 2, and 0%. If the number of fragments and minute chromosomes are added to the sum of 1-break aberrations, the number of breaks per hundred cells after X-ray increased to 47, 30, and 23% at the different intervals, respectively. The frequency of minute chromosomes was highest at 24 h and decreased to 1 at 48 and 72 h. The only increased incidence in aberrations at 48 h relative to the other time intervals examined was the number of chromatid gaps, which are not considered breaks but probably represent damage to the protein

TABLE 1

Cytogenetic analysis of Syrian hamster cells after 250R*.

Post X-ray (h)	Metaphases			Aberrations						
	% Cells with one aberr.	% Cells with two aberr.	% Cells with multiple aberr.	Gaps	Iso-gaps	Breaks	Iso-breaks	Minutes and fragments	Dicentrics	Trans-locations
24	16	9	4	3	5	4	1	14	13	1
48	19	5	2	15	2	4	1	1	8	4
72	16	3	0	5	1	0	1	1	7	4
Control**	1	0	0	1	0	0	0	0	0	0

* Results based on 100 cells which were randomly selected for examination.
** No irradiation, a secondary culture was examined.

TABLE 2

Ploidy* of Syrian hamster cells after 250R.

Post X-ray (h)	2 N	4 N	8 N	16 N	Endo-	Total**	Percent increase in polyploid
Control***	493	6	1	0	0	500	0
24	475	16	1	0	18	500	5.6
48	452	62	10	1	2	527	12.8
72	537	73	7	1	3	621	11.0

* The values for 4N, 8N, and 16N are approximate only.
** Number of cells counted.
*** No irradiation.

moiety [41]. The examination of metaphases for polyploids indicated a significant increase in polyploids, particularly tetraploids, at 48 and 72 h post-irradiation (table 2). Typical chromosome lesions produced at 48 h post-irradiation included translocations, chromatid gaps and breaks, minute chromosomes and dicentrics. It can be concluded that chromosome aberrations or alterations in ploidy cannot account for the significant increase in enhancement at 48 h post-irradiation.

With human and hamster material, we conclude that the immediate effect of toxic chemicals such as a carcinogen leads to non-specific random changes in terms of breaks, whereas the treatment of rats with materials such as DMBA may lead to a high specific lesion on one chromosome. Either non-radioactive DMBA or [³H]DMBA was added to exponentially growing secondary rat fetal cells for 5, 9, or 24 h, including a 4 h exposure to Colcemid. Chromosome preparations from treated cells had a higher frequency of gaps/breaks associated with the number two chromosome compared to the other chromosomes of the rat complement. With G-band analysis, it was noted that four specific regions of the A2 chromosome were affected, which we refer to as A, B, C, and D. Of these four regions (fig. 4), region B was more susceptible to DMBA, as indicated by the number of chromatid breaks found. In the G-band nomenclature suggested by Levan [42], the segments associated with the breaks or gaps occur in the Q arm; in all four cases, the aberration is located on a negative band. For aberrations A, B, C, and D, the exact locations are Q22, Q24, Q26, and Q34, respectively. However, the vulnerability of region B must be considered non-specific for DMBA treatment since the same aberration occurred with [³H]thymidine at an equivalent concentration and at specific activity as the [³H]DMBA. Chromosome samples collected following either 5 or 9 h exposure to the radioactive carcinogen may show non-random grain accumulation on chromosome A2 in approximately the same area in 1–2% of the metaphases. Following removal of emulsion, treatment with DNase, and reprocessing for

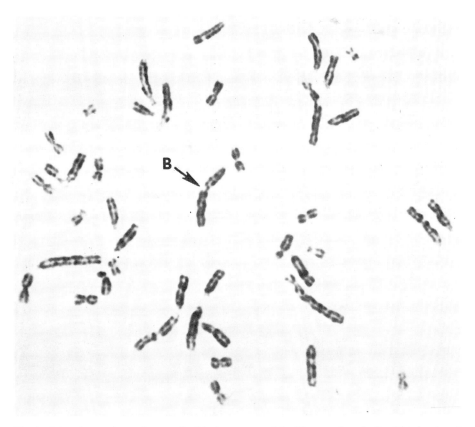

Fig. 4. G-band metaphase of a rat cell with chromosome A2 with a gap found after 5 hrs treatment with 7,12-dimethylbenz(a)anthracene. The gap is at Q2-4.

autoradiography, it was demonstrated that the visible grains represented binding of the radioactive material to the chromosomal DNA. Furthermore, there was no evidence of a chromosome lesion. The non-specificity of cause of this lesion is further supported by evidence obtained by Sugiyama with the food additive, 2-(2-furyl)-3-(5-nitro-2-furyl)acrylamide (AF2) which, in vivo also caused breaks on chromosome A2 [43]. In addition, with the BUDR technique for sister chromatid exchanges [44], chromosome A2 exhibits an increased number of chromatid exchanges including some that are located in what we consider the most vulnerable segment of this chromosome. These studies on rat chromosomes demonstrate that a specific chromosome appears to be involved with several nonspecific reactions: a high frequency of breaks with DMBA, a certain degree of susceptibility to [^{3}H]thymidine, and a susceptibility to [^{3}H]DMBA and to other chemicals such as BUDR and a food additive, AF2.

Cytogenetic alterations in neoplastic cells

Thus far we have reported that mouse, rat, guinea pig, and Syrian hamster cells can be transformed in vitro by chemical carcinogens [45–47]. We have evidence that Chinese hamster cells can also be transformed. In the course of these studies, we have applied the various techniques for examination of chromosomes of early transformed cells, of the serially passed transformed lines, and, in many cases, of the tumor-derived cultures obtained as a result of injecting the transformed cells into the appropriate host.

Syrian hamster chromosomes can be identified by the various crossband techniques and by the C-band technique [48,49]. In all cases, the different pairs of autosomes as well as the sex chromosomes that constitute the normal hamster complement can be separated (fig. 5). The use of the C-band technique for constitutive heterochromatin is considered imperative because hamster chromosomes have a large proportion of heterochromatin. Suggestions have been made

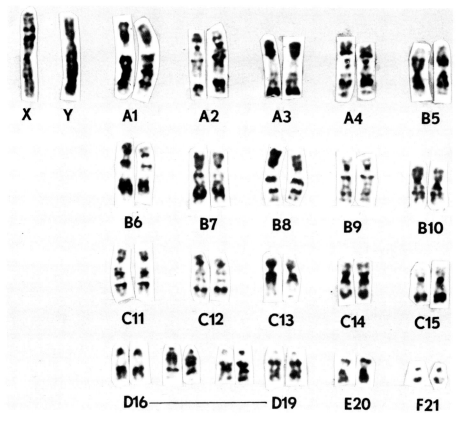

Fig. 5. Trypsin karyotype of a normal male Syrian hamster cell showing typical banding pattern of each chromosome.

that chromosomal aberrations seen in neoplasm might involve heterochromatin, because of its known susceptibility to breakage by chemicals [50]. Thus, when attempting to make conclusions concerning visible changes in chromosomes, both numerical changes and new chromosome markers must be identified in terms of the G-band and the C-band procedures. As pointed out, our data indicate that although a response to carcinogen may occur in areas known to be rich in heterochromatin, in most cases the breaks involve the negative areas of the chromosome, and only in rare cases is the joining of chromosomes observed to occur by reattachment at heterochromatic areas.

A number of transformed lines derived from chemically transformed colonies Simian adeno 7 (SA7) transformed foci or foci obtained by treatment with a chemical carcinogen and SA7 were analyzed for deviations from normal chromosome numbers and structural differences [51,52]. Each colony transformed by chemical carcinogens was derived from a culture that originated from a different hamster embryo culture. A study of the changes in the number of chromosomes associated with each identifiable pair show that the increase and decrease in certain chromosomes is random. A summary of some of the data is shown in table 3. Non-translocational numerical changes found in greater than 25% of the cells analyzed by G-band techniques are included. The increase or decrease in certain chromosomes was random in the chemically transformed lines; however, in three lines trisomy for D17 was present as a result of transformation with benzo(a)pyrene, aflatoxin B1 or 4-nitroquinoline N-oxide. Also as shown, 4-nitroquinoline N-oxide may cause transformation without involvement of the D17 chromosome. Previously, it was shown that chromosomes of cells transformed by polycyclic hydrocarbons need not involve new acrocentrics. A study of some virus-transformed lines again demonstrates the variability in the deviation from normal. In those cases in which the various clones were derived from the same hamster pool, there was the same non-random loss or gain of a chromosome. The foci indicated by the prefix 73 were all derived from the same pool. In the case of A and B, only SA7 was used, whereas with 73C through E the cells were first treated with a chemical carcinogen and then treated with the virus. As with chemical transformation, one must conclude that viral transformation by SA7 was not accompanied by any specific numerical deviation as a result of transformation by either virus alone or from carcinogen pretreatment followed by virus.

From a total of nine different transformed cell lines obtained after treatment with diverse chemical carcinogens, the number of markers ranged from 0 in four lines to as many as three in the remaining lines [51]. Different carcinogens produced transformed lines associated with the same specific marker but not all the transformed lines had the same marker, even with the same chemical carcinogen. Some markers were found in a tumor-derived culture and not in the transformed line, or occurred in later passages of both the transformed lines

TABLE 3

Analysis of numerical changes of individual pairs of chromosomes as indicated by G-banding pattern.

Cell line	X	Y	1	2	3	4	5	6	7	8	9	10	11	12	13	14	15	16	17	18	19	20	21
BP 45																		□	■				
BP 45T										■		□						□	■	■			
AFT 39																			■		■		
AFT 39T							□	■							■	■							
4 NQO 16															□			■					
4 NQO 16T																	□	■					
4 NQO 17							■				■												
4A																				□			
4B																							
4D																							
12G					□						□												□
26A																							
73A	□																				□		
73B	□							□													■		
73V	□							□											□	□			
73D	□																				■		
73E	□																□						

* Numerical increases ■ or decreases □ (25% or greater change in relation to normal) in specific chromosomes indicated by the boxes. Each chemical cell line was isolated from a different animal. Each viral line was isolated from a different Petri dish. In the later cases, the numerical prefix identifies the ancestral embryo cell pool.

and the tumor-derived cultures but not in the early passage of these lines. A summary of the markers found in the various cell lines are indicated by M1 through M6 (fig. 6). These markers demonstrate that different chromosomes of various sizes belonging to different groups in the idiogram may be involved in marker formation. The additional markers MA and MB were obtained as a result of cell progression in vivo, demonstrating that new chromosomes may occur as a response to host pressure. C-band analysis showed that no chemically-transformed line had alterations in chromosome constitutive heterochromatin; but the study of the tumor-derived cultures showed heterochromatic alterations in two different tumor cultures; in one of these the heterochromatic alteration consisted of the addition of a heterochromatic piece which was associated with the reactivation of an ordinarily inert short arm of chromosome number A4 [49].

Chromosomal alterations in mammalian cells

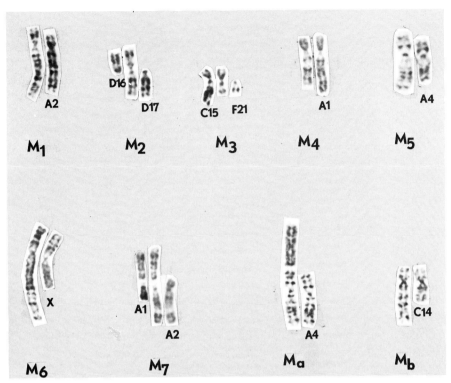

Fig. 6. Representative of all markers, M_1–M_6, found in cell line transformed by chemical carcinogens and/or their tumor derived cultures. M_a and M_b were found in only one tumor. The new chromosomes are either between their normal components or to the left of their normal analog.

The markers found following transformation by virus were different from those found after chemical transformation [52] (fig. 7). SA7 transformation was accompanied by deletions, translocations, and centromeric fusions resulting in new chromosome configurations, all of which differed from those obtained with chemical carcinogens. The addition of chemicals did not result in any marker that had not been observed with virus alone. Therefore, it is again concluded that a common marker did not characterize transformation by either SA7 alone or from carcinogen pretreatment followed by SA7.

SA7 transformation differs from the chemical transformations described in that a specific viral antigen is associated with the transformation. To further determine whether visible alterations had occurred on the chromosomes, even though by banding techniques no specific aberration could be associated with the SA7 transformation, radioactive complementary RNA (cRNA) from SA7 virus was hybridized in situ to fixed transformed cell chromosomes [53]. This represents the first attempt to associate viral integration in intact cells of mammalian transformed cells as opposed to cells which have been merely infected with

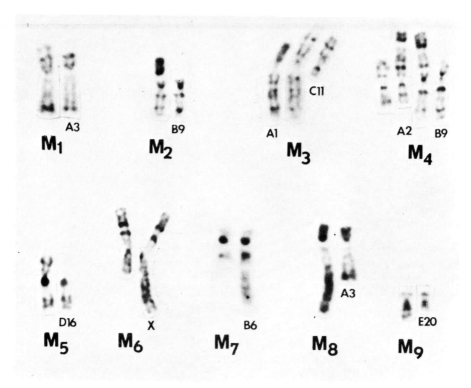

Fig. 7. Marker chromosomes (M) found in 50% or more of the cells of the various virus (SA7) or chemical carcinogen and virus transformed lines. The new chromosomes are either between their normal components or to the left of the normal analog.

the virus. Preliminary evidence indicates that the grain distribution is diffuse over the nucleus as well as the chromosomes.

Guinea pig cells may also be transformed in vitro by chemical carcinogens but the process of transformation differs from that associated with Syrian hamster cell transformation; with hamster cells the progressive events occur during a narrow time range, but with guinea pig cells, successive events are temporally extended so that production of tumors may require four to 18 months of culturing. The normal guinea pig karyotype was established (fig. 8). The examination of chemical carcinogen transformed cells shows that the transformed lines may either have a diploid constitution with no alteration in G or C-bands, as found with benzo(a)pyrene, or near diploid or near tetraploid, as found with 3-methylcholanthrane and N-methyl-N^1-nitro-N-nitrosoguanidine or diethylnitrosamine [54]. The aneuploid cell lines may either have a new metacentric chromosome formed by centromeric fusion of two non-homologous chromosomes, or they may exhibit a new submetacentric or subtelocentric chromosome originating from translocation between two chromosomes, one of which is a number

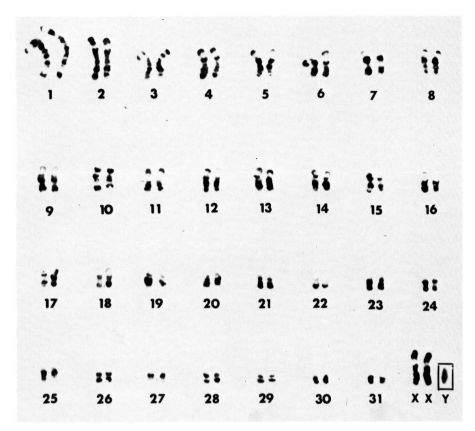

Fig. 8. Normal G-band karyotype of a female guinea pig cell arranged according to chromosome size and banding pattern. The Y chromosome from a male guinea pig cell also is shown at the same magnification.

one chromosome. Although chromosome number one was involved in three of the five lines studied, the involvement is still considered random because G-band analysis shows that the new marker chromosomes involve different regions of the number one chromosome. The vulnerability or the involvement of chromosome number one rather than other chromosomes of the guinea pig may result from association with nucleolar organization [55]. The greater frequency of contact between such chromosomes creates an increased risk of chromatid exchange, possibly explaining their frequent participation in abnormal chromosome formation or non-disjunction. The points at which chromosome number one is involved in the formation of new markers are shown in fig. 9.

Thus far, these studies have been confined to the examination of chromosomes of cells that have been transformed in vitro. The concept that a specific chromosome alteration occurs in rats as a result of treatment with DMBA, regardless of whether a carcinoma, sarcoma or leukemia develops, has been

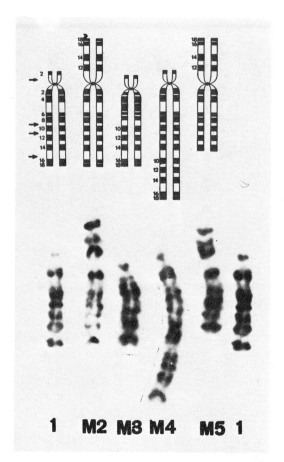

Fig. 9. Idiograms (top) and G banding (bottom) of chromosomes number one, M2, M3, M4, and M5. The positive bands of chromosome number one, but only the translocated segments from number one are numbered. Arrows indicate the point of breakage consistent ith the origin of the various translocated segments.

advanced on the basis of in vivo data [8–11]. Since a high frequency of trisomy of the A2 chromosome was found, the theory has been advanced that the type of chromosomal alteration is dependent upon the etiologic agent. Because of this interesting conclusion, we compared the chromosome banding patterns of rat fibrosarcomas induced by in vitro transformation of rat embryo cells and by in vivo injection of rats with DMBA. A study of the chromosomal constitution of three rat lines transformed in vitro, two non-transformed lines isolated from carcinogen-treated dishes, two untreated control cell lines, and five rat tumors induced in vivo by DMBA revealed numerical and/or structural chromosomal aberrations which preferentially involved chromosomes number one, two or three, whether or not they were transformed (fig. 10). Since no specific alteration in

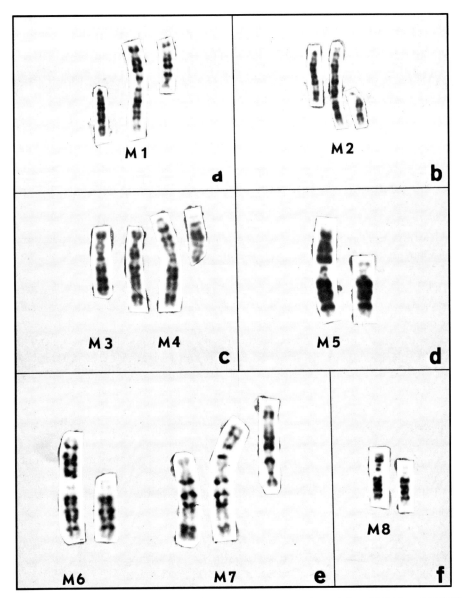

Fig. 10. Marker chromosomes and their derivation in non-transformed lines NT1(a) and NT3(b) isolated from DMBA-treated dishes; DMBA-transformed lines DMBA2(c) and DMBA13(d); and tumors induced in vivo, RT2(e) and RT5(f) (see text).

chromosome distribution nor any specific and constant chromosomal aberrations occurred, it must be concluded that neither in vivo or in vitro malignant transformation by DMBA that leads to formation of fibrosarcomas is associated with specific gross abnormalities of the chromosomes [45]. The vulnerability of the

A2 chromosome may be associated with its function as a nucleolar organizer. It might also be possible that this chromosome contains a specific segment critical for further differentiation since different types of agents may interact at this site. The type of interaction may be at the molecular level since a visible chromosomal lesion need not occur.

Conclusions

Addition of a carcinogen to mammalian cells usually causes an immediate effect on the chromosomes, resulting in a series of events leading to rearrangements of chromatid material.

The immediate effect of a carcinogen on the chromosomes of mammalian cells may be random or non-random and probably reflects species genomic properties.

The correlation between lethality and the frequency of chromosome and chromatid aberrations suggests that the genetic alterations may be responsible for reproductive death.

No universal alteration in specific numbers or visible structure is associated with chemical or viral transformation. In fact, transformation may occur without any visible chromosome alteration.

Chromosome variations are independent of the etiologic agent. Different carcinogens may produce transformation associated with the same specific marker, but not all transformed lines have the same markers even with the same chemical.

Chromosome deviation in cancer is probably species dependent and contributes to the characteristics of malignancy. The presence of identical banding patterns in transformed lines and tumor derived cultures indicates that the transformed cells produce tumors.

References

[1] T. Boveri, Zur Frage der entwicklung maligner tumoren. Jena, Gustav Fischer (1914).
[2] S.A. Makino, A cytological study of the Yoshida sarcoma, an ascites tumor of white rats. Chromosoma, 4, 649 (1952).
[3] T.C. Hsu, Chromosomal evolution in cell populations. Intern. Rev. Cytol. 12, 69 (1961).
[4] P.C. Nowell, Chromosome changes in primary tumors. Progr. Exptl. Tumor Res., 7, 83 (1965).
[5] T.S. Hauschka and A. Levan, Cytologic and functional characterization of single cell clones isolated from the Krebs-2 Ehrlich ascites tumor. J. Natl. Cancer Inst. 21, 77 (1958).
[6] A.A. Sandberg and M. Sakurai, Chromosomes in the causation and progression of cancer and leukemia. Molecular Biology of Cancer (H. Busch, ed.), Academic Press, New York and London, 1976, p. 81.

[7] S.E. Luria and M. Delbrück, Mutations of bacteria from virus sensitivity to virus resistance. Genetics, 28, 491 (1943).
[8] Y. Kurita, Y. Sugiyama and Y. Nishizuka, Cytogenetic studies on rat leukemia induced by pulse doses of 7,12-dimethylbenz(a)anthracene. Cancer Res., 28, 1738 (1968).
[9] E.D. Rees, S.K. Majumdar and A. Shuck, Changes in chromosomes of bone marrow after intravenous injections of 7,12-dimethylbenz(a)anthracene and related compounds. Proc. Natl. Acad. Sci. U.S.A. 66, 1228 (1970).
[10] G. Levan, V. Ahlström and F. Mitelman, The specificity of chromosome A2 involvement in DMBA-induced rat sarcomas. Hereditas 77, 263 (1974).
[11] U. Ahlstrom, Chromosomes of primary carcinomas induced by 7,12-dimethylbenz(a)anthracene in the rat. Hereditas 78, 235 (1974).
[12] F. Mitelman, The chromosomes of fifty primary Rous rat sarcomas. Hereditas 69, 155 (1971).
[13] G. Levan, The detailed chromosome constitution of a benzpyrene-induced rat sarcoma. A tentative model for G-band analysis in solid tumors. Hereditas 78, 273 (1974).
[14] T. Yamamoto, M. Hayashi, Z. Rabinowitz and L. Sachs, Chromosomal control of malignancy in tumors from cells transformed by polyoma virus. Int. J. Cancer 11, 66 (1973).
[15] T. Yamamoto, Z. Rabinowitz and L. Sachs, Identification of the chromosomes that control malignancy. Nature New Biol. 243, 247 (1973).
[16] W. Benedict, N. Rucker, C. Mark and R. Kouri, Correlation between balance of specific chromosomes and expression of malignancy in hamster cells. J. Natl. Cancer Inst. 54, 157 (1975).
[17] A.A. Sandberg and D.K. Hossfeld, Chromosomal changes in human tumors and leukemias, in: Handbuch der allgemeinen Pathologie (W. Altmann et al., eds.), Springer-Verlag, Berlin Heidelberg, 1974, p. 141.
[18] M. Oshimura, I. Hayata, S. Kakati and A.A. Sandberg, Chromosomes and causation of human cancer and leukemia. XVII. Banding studies in acute myeloblastic leukemia (AML). Cancer (in press).
[19] J. Rowley and D. Potter, Chromosomal banding patterns in acute leukemia. Blood 47, 705 (1976).
[20] F. Wiener, G. Klein and H. Harris, The analysis of malignancy by cell fusion. III. Hybrids between diploid fibroblasts and other tumor cells. J. Cell Sci. 8, 681 (1971).
[21] F. Wiener, G. Klein and H. Harris, The analysis of malignancy by cell fusions. IV. Hybrids between tumor cells and a malignancy L cell derivative, J. Cell Sci. 12, 253 (1973).
[22] G. Barski and J. Belehradek, Jr., Expression of malignancy in interspecies cell hybrids, in: Differentiation and Control of Malignancy of Tumor Cells (W. Nakahara, T. Ono, T. Sugimura and H. Sugano, eds.), University Park Press, Tokyo, 1973, pp. 419–441.
[23] J.A. DiPaolo, P.J. Donovan and R.L. Nelson, Quantitative studies of in vitro transformation by chemical carcinogens. J. Natl. Cancer Inst. 42, 867 (1969).
[24] B.C. Casto, W.J. Pieczynski and J.A. DiPaolo, Enhancement of adenovirus transformation by treatment of hamster cells with diverse chemical carcinogens. Cancer Res. 34, 72 (1974).
[25] J.A. DiPaolo, P.J. Donovan and R.L. Nelson, Transformation of hamster cells in vitro by polycyclic hydrocarbons without cytotoxicity. Proc. Nat. Acad. Sci. U.S.A. 68, 2958 (1971).
[26] T.T. Puck, P.I. Marcus and S.J. Cieciura, Clonal growth of mammalian cells in vitro. Growth characteristics of colonies from single HeLa cells with and without a 'feeder' layer. J. Exp. Med. 103, 273 (1956).
[27] J.A. DiPaolo, R.L. Nelson and P.J. Donovan, In vitro transformation of Syrian hamster embryo cells by diverse chemical carcinogens. Nature 235, 278 (1972).
[28] J.A. DiPaolo, P.J. Donovan and R.L. Nelson, In vitro transformation of hamster cells by polycyclic hydrocarbons. Nature New Biol. 230, 240 (1971).
[29] J.S. Bertram and C. Heidelberg, Cell cycle dependency on oncogenic transformation induced by N-methyl-N'-nitro-N-nitrosoguanidine in culture. Cancer Res. 34, 526 (1974).
[30] M.W. Shaw, Human chromosome damage by chemical agents. Ann. Rev. Med. 21, 409 (1970).

[31] M.A. Bender, H.G. Griggs and J.S. Bedford, Mechanisms of chromosomal aberration production. III. Chemicals and ionizing radiation. Mut. Res. 23, 197 (1974).
[32] B.A. Kihlman, Molecular mechanisms of chromosome breakage and rejoining, in: Advances in Cell and Molecular Biology (E.J. Du Praw, ed.), Academic Press, New York 1971, p. 59.
[33] W.F. Benedict, Early changes in chromosomal number and structure after treatment of fetal hamster cultures with transforming doses of polycyclic hydrocarbons. J. Natl Cancer Inst. 49, 585 (1972).
[34] J.A. DiPaolo and N.C. Popescu, Banding pattern analysis of initial structural chromosome alterations induced by MNNG in Syrian hamster cells. Mut. Res. (in press).
[35] J.A. DiPaolo, P.J. Donovan and N.C. Popescu, Kinetics of Syrian hamster cells during X-irradiation enhancement of transformation in vitro by chemical carcinogen. Radiat. Res. 66, 310 (1976).
[36] J.A. DiPaolo and N.C. Popescu, Chromosome bands induced in human and Syrian hamster cells by chemical carcinogens, Brit. J. Cancer 30, 103 (1974).
[37] N.C. Popescu and J.A. DiPaolo, G. and C chromosomal banding produced sequentially in living and fixed human cells. Lancet i, 209 (1974).
[38] T.C. Hsu, S. Pathak and D.A. Shafer, Induction of chromosome cross-banding by treating cells with chemical agents before fixation. Exp. Cell Res. 79, 484 (1973).
[39] J.A. DiPaolo, In vitro transformation: Interactions of chemical carcinogens and radiation, in: Biology of Radiation Carcinogenesis (J.M. Yuhas, R.W. Tennant and J.B. Regan, eds.) Raven Press, New York, 1976, p. 335.
[40] A.C. Upton, Somatic and genetic effects of low-level radiation, in: Recent Advances in Nuclear Medicine, Vol. IV, (J.H. Lawrence, ed.) Grune & Stratton, Inc., New York, 1974, pp. 1–40.
[41] A. Brogger, Is the chromatid gap a folding defect due to protein change? Evidence from mercaptoethanol treatment of human lymphocyte chromosomes. Hereditas 80, 131 (1975).
[42] G. Levan, Nomenclature for G-bands in rat chromosomes. Hereditas 77, 37 (1974).
[43] T. Sygiyama, K. Goto and H. Uenara, Acute cytogenetic effect of 2-(2-furyl)-3-(5-nitro-2-furyl)-acrylamide (AF-2) a food preservative on rat bone marrow cells in vivo. Mutat. Res. 31, 241 (1975).
[44] P. Perry and S. Wolff, New Giemsa method for the differential staining of sister chromatids. Natura 251, 156 (1974).
[45] J.A. DiPaolo, Quantitative aspects of in vitro chemical carcinogenesis, in: Chemical Carcinogenesis (P.O.P. Ts'o and J.A. DiPaolo, eds.) Marcel Dekker, Inc., New York, 1974, pp. 443–455.
[46] C.D. Olinici and J.A. DiPaolo, Chromosome banding patterns of rat fibrosarcomas induced by in vitro injection of rats by 7,12-dimethylbenz(a)anthracene. J. Natl. Cancer Inst. 52, 1627 (1974).
[47] C.H. Evans and J.A. DiPaolo, Neoplastic transformation of guinea pig fetal cells in culture induced by chemical carcinogens. Cancer Res. 35, 1035 (1975).
[48] N.C. Popescu and J.A. DiPaolo, Identification of Syrian hamster chromosomes by acetic-saline-Giemsa (ASG) and trypsin techniques. Cytogenetics 11, 500 (1972).
[49] J.A. DiPaolo and N.C. Popescu, Distribution of chromosome constitutive heterochromatin of Syrian hamster cells transformed by chemical carcinogens. Cancer Res. 33, 3259 (1973).
[50] T. Sugiyama, Chromatid rearrangements and carcinogenesis, in: Recent Topics In Chemical Carcinogenesis (S. Odashima, S. Takayama and H. Sato, eds.) University Park Press, Tokyo, 1975, p. 393.
[51] J.A. DiPaolo, N.C. Popescu and R.L. Nelson, Chromosomal banding patterns and in vitro transformation of Syrian hamster cells. Cancer Res. 33, 3250 (1973).
[52] N.C. Popescu, C.D. Olinici, B.C. Casto and J.A. DiPaolo, Random chromosome changes following SA7 transformation of Syrian hamster cells. Int. J. Cancer 14, 461 (1974).

[53] M.L. Pardue and J.G. Gall, Chromosomal localization of mouse satellite DNA. Science 170, 1356 (1970).
[54] N.C. Popescu, C.H. Evans and J.A. DiPaolo, Chromosome patterns (G and C-band) of in vitro chemical carcinogen transformed guinea pig cells. Cancer Res. 36, 1404 (1976).
[55] S. Ohno, C. Weiler and C. Stenius, A dormant nucleolus organizer in the guinea pig *cavia cobaya*. Exp. Cell Res. 25, 498 (1961).

Chapter 19

THE IMPORTANCE OF CHROMOSOMAL CHANGES IN THE EXPRESSION OF MALIGNANCY

WILLIAM F. BENEDICT

Division of Hematology-Oncology, Department of Medicine, Childrens Hospital of Los Angeles, USC School of Medicine, Los Angeles, California 90027, U.S.A.

Summary

The role of the specific chromosomal changes associated with malignant transformation and/or tumorigenicity in mammalian model systems, including those of hamster, mouse, and human origin, is discussed. Criteria to enable the adequate evaluation of the relationship between specific chromosomal imbalances and expression of malignancy is also outlined. Finally, a general hypothesis is presented to illustrate how viral, chemical, and physical carcinogens can each alter the chromosomal genetic balance which in turn may be directly related to oncogenesis.

Introduction

For the last several years we have felt that chromosomal aberrations may play an important role in the expression of human malignancy, particularly solid tumors. This belief initially evolved from an awareness that abnormal chromosomal patterns occur not only in invasive solid tumors but also, more significantly, before the tumor has shown evidence of invasion [1]. Thus, aneuploidy is found as early as malignancy can be recognized histologically. Abnormal chromosomal patterns, in fact, have been used as a diagnostic criterion for malignant cells [2] and as an indication of the presence of malignant components of solid tumors long before they could be diagnosed using histological criteria [3]. In contrast, approximately 50% of the initial bone marrow samples from patients with acute lymphoblastic leukemia show a diploid chromosomal pattern [4,5]. Although it is highly speculative, this may imply a basic difference in the molecular events responsible for the development of human solid tumors compared to the evolution of certain human leukemias.

Early evidence that chromosomal aberrations may play a role in malignant expression came from the study of mammalian in vitro transformation systems. It was shown that chromatid breaks per se were correlated with cytotoxicity in hamster embryo diploid cells following treatment with carcinogenic polycyclic aromatic hydrocarbons [6] as well as with the cancer chemotherapeutic agent ara-C [7], whereas the production of aneuploidy was closely associated with transformation [6]. Polycyclic hydrocarbons produced aneuploidy in cells within 24 to 72 h after exposure, and it was suggested that some of these rapidly appearing chromosomal changes may be important in cell transformation [8].

In this paper some of the recent studies on chromosomal changes associated with the expression of malignancy will be reviewed, including new information on the transformation of human diploid cells. I shall then present a general hypothesis in an attempt to show how viral, chemical and physical carcinogens can change the chromosomal genetic balance which, in turn, is directly associated with the expression of malignancy.

Chromosomal changes associated with malignant expression in mammalian cells

Diploid hamster embryo cells

With the advent of various chromosomal banding techniques which allowed individual chromosomes to be identified unequivocally [9,10] (fig. 1), it became possible to study the relationship between the presence of specific chromosomes and the expression of malignancy. The first paper suggesting that specific chromosomal imbalances could be involved in the expression of malignancy following

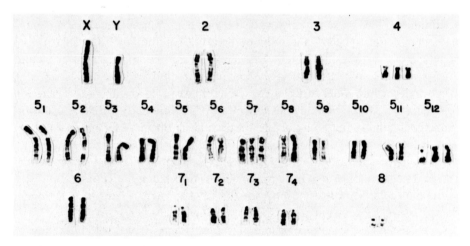

Fig. 1. Karyotype of the original ara-C transformed hamster cell line A [13] using the trypsin-Giemsa technique [9] for chromosomal banding.

chemical transformation of hamster cells came from the laboratory of Dr. Leo Sachs [11]. Hamster cells were transformed by treatment with dimethylnitrosamine. Chromosomes 5_7 and 5_{10} were implicated as having genes for the expression of transformation and/or malignancy and chromosomes 7_2 and 7_3 were thought to have genes for the suppression of the same properties [11]. Similarly, in cells transformed by polyoma virus it was reported that genes for the expression of malignancy were located on the 5_6 chromosome and genes for suppression of malignancy were located on the 5_3 and 7_2 chromosomes [12].

During the same period we were also studying the importance of specific chromosomal imbalances to the expression of malignancy in hamster embryo cells [13]. The hamster cells were transformed by exposure to ara-C. Variants were then isolated from two tumors produced by subcutaneous injection of the original transformed cell line (fig. 1) which had either high or low malignant potential. Chromosomal analysis was performed rapidly at a very early passage both on the cells injected and on certain tumors that were produced after the tumor cells were placed back into culture [13]. In all cases, tumorigenicity was associated with an increased number of 5_7 chromosomes compared to the number of 7_3 chromosomes, i.e. the ratio of 5_7 to 7_3 chromosomes was greater than one. In this study a marker chromosome, M_2, was often counted as one of the 5_7 chromosomes (fig. 2). This marker chromosome was formed by a translocation of a 7_4 chromosome onto the long arm of a 5_7 chromosome. Thus, since only the long arm of the 5_7 chromosome was involved in these calculations, we suggested that information for the expression of malignancy could be assigned, in part, to the long arm of the 5_7 chromosome in these cells. However, the M_2 chromosome did not necessarily have to be present in the transformed cells to correlate an excess of 5_7 to 7_3 chromosomes with the expression of malignancy [13].

Additional independent studies recently published on a hamster embryo cell line, transformed following treatment with dimethylnitrosamine, again suggest that genes for the expression of malignancy and transformation are located on the 5_7 chromosome and, to a lesser extent, on the 5_{12} chromosome [14]. The 7_2 chromosome was again reported to contain genes for the suppression of

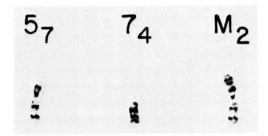

Fig. 2. Marker chromosome M_2 is shown to be formed from a translocation of a 7_4 chromosome onto the long arm of a 5_7 chromosome.

malignancy and transformation in these cells. However, the 7_3 chromosome was not shown to have any genetic information for the suppression of transformation, in contrast to earlier studies [11,13]. These variations in the specificity of chromosomes implicated in expression or suppression of malignancy suggest that there may be several chromosomes which contain genes regulating malignancy. One or another of these chromosomes could, however, be more important in a given transformed clone for the phenotypic expression of transformation and/or malignancy than in a different, independently derived malignant clone. Nevertheless, the similarity between the studies from separate laboratories would also imply that the chromosomes which contain genetic information involved in oncogenesis are far from random.

We have previously commented [13] on the fact that in a recent study by DiPaolo et al. [15] no specific chromosomal imbalance associated with oncogenicity could be found. It may be that chromosomal changes related to the expression or suppression of malignancy are more subtle in certain cell lines [15] than in others [11–14]. However, it must be emphasized that in order to investigate the relationship between specific chromosomal imbalances and the expression of malignancy without the use of cell hybridization techniques, one must be able to identify every chromosome or chromosomal portion unequivocally in each metaphase examined. Unless the exact chromosomal complement of each analyzed metaphase is reported [13,14] it is difficult to compare the results from one study to another.

Mouse cells

There has been evidence also in mouse cells that specific chromosomes can be important in the suppression of malignancy. Codish and Paul reported that in tumor cells initially induced in vivo by methylcholanthrene malignancy appeared to be related to whether or not one or two of the same marker chromosome, named M_8, was present in the cell [16]. In the mass culture, cells with either one or two M_8 chromosomes were found. However, when the mixed population was injected, only the cells containing one M_8 chromosome formed tumors [16]. This M_8 marker chromosome was formed by a translocation of a number 7 chromosome onto a number 19 chromosome. Therefore, it was concluded that either the 7 or 19 chromosome contained genetic information for the suppression of malignancy in these cells since an addition of an extra M_8 chromosome appeared to prevent the expression of tumorigenicity.

Diploid human cells

Under given experimental conditions, specific human chromosomes have also been shown to be involved in the expression of malignancy. Croce and

Koprowski along with their colleagues first reported that in Simian Virus 40 (SV40) transformed human cells, the tumor antigens and SV40 genome could be assigned to the human chromosome number 7 [17]. They later showed in somatic hybrids between mouse peritoneal macrophages and the SV40 transformed human cells that the presence of the human chromosome 7 was necessary to retain the transformed phenotype in vitro [18]. More recently, it was found that the presence of this same human chromosome containing the SV40 genome was responsible for the tumorigenicity of the hybrid cells when injected into 'nude' mice [19].

These observations are of fundamental importance and may explain why in certain cases no chromosomal abnormalities are found in specific human malignancies, such as in many instances of acute lymphocytic leukemia. Utilizing chromosomal banding methods, no cytogenetic difference between normal and SV40 transformed human diploid cells can be seen. However, by using cell hybridization techniques it could be shown that, in the SV40 treated cells, there was additional genetic material incorporated into the number 7 chromosome which was associated with not only the expression of transformation but also the development of malignant tumors. Thus, this or similar techniques might be needed to implicate certain chromosomes as containing genetic material associated with the expression or suppression of malignancy.

There is also great promise that human diploid cells will soon become available to study the relationship of specific chromosomal changes and the expression of malignancy following chemical transformation, although numerous attempts to transform human diploid cells with chemicals have been unsuccessful. Igel et al. [20] recently reported on human cells which were apparently transformed by urethane. These two transformed cell lines were later further characterized [21]. The control cell line was completely diploid and had none of the characteristics associated with transformed cells such as growth in soft agar, plasminogen activator production, or tumor formation in immunosuppressed animals. The two cell lines studied, in contrast, were markedly aneuploid with unique chromosomal patterns. The two lines also each had distinct marker chromosomes [21].

Additional characteristics of transformed cells such as growth in soft agar, increased plasminogen activator and production of fibrosarcomas in immunosuppressed animals were shown for one of these lines. Increased plasminogen activator was also found in the second transformed cell line but soft agar and tumorigenicity studies were not done. Since, however, the diploid non-transformed cells could not be developed into an established cell line, only a limited number of chemical transformation experiments could be carried out with these cells.

A recent study by Kakunaga appears to be particularly significant, since it is reported that an established human diploid cell line can be transformed with the chemical carcinogens, N-methyl-N'-nitro-n-nitrosoguanidine and 4-nitroquinoline-1-oxide [22]. Five transformed lines were isolated and each had abnormal chromosomal patterns, grew in soft agar, and produced malignant tumors when injected into 'nude' mice. Consequently, it seems that a reproducible diploid human transformation system has finally been developed. This cell line or a similar human diploid system for transformation should now enable us to look at the relationship between the balance of specific human chromosomes and the expression of malignancy using techniques which previously have been described for the hamster transformation studies [11–14]. Theoretically, variants with decreased malignant potential could be isolated from the initial transformed cell lines and their complement of specific chromosomes studied with respect to the changes in tumorigenicity.

In addition, Freeman et al. [23] have been able to reproducibly obtain aneuploid conversion of human diploid cells when skin epithelial cells were grown on pigskin and then treated with carcinogenic aromatic polycyclic hydrocarbons. From these studies it was concluded that aneuploid conversion may be one of the early changes leading to malignant transformation of human diploid cells following treatment with chemical carcinogens.

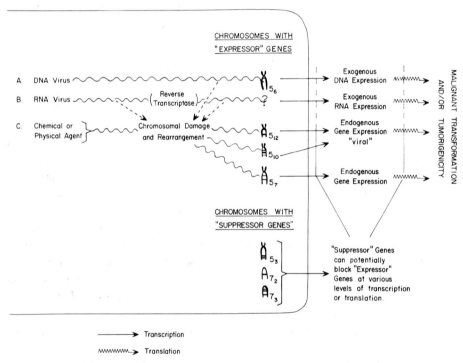

Fig. 3. Outline of general hypothesis on possible specific chromosomal changes produced by all carcinogens and how these changes could lead to the expression of malignant transformation and/or tumorigenicity (see text). The Syrian hamster system has been used as an example [11–14] for the illustration of specific chromosomal involvement.

A general hypothesis on specific chromosomal changes and malignant expression produced by all carcinogens

I should finally like to present an hypothesis to explain how diverse carcinogens such as DNA and RNA viruses or chemical and physical agents could produce chromosomal imbalances directly related to the expression of malignancy. These chromosomal imbalances can be of two types: 1) Gross changes involving abnormalities in either chromosomal number or structure; 2) small insertions which may cause specific imbalances at genetic loci rather than at a chromosomal or subchromosomal level. The pathways are outlined in fig. 3.

DNA oncogenic viruses

The DNA of viruses such as SV40 or polyoma which are known to produce malignant transformation may be integrated into specific chromosomes. This integrated DNA might contain oncogenic information which in turn could be transcribed and translated. In this case, one would *not necessarily* be able to find any specific chromosome involved in the expression of malignancy unless cell hybridization studies were done [18–19]. One main reason for this difficulty is that a diploid transformed cell would contain one or more chromosomes with exogenous genetic information for the expression of malignancy [18–19]. If this transformed cell were to 'revert' to a non-transformed state, the most common mechanism might be to delete the chromosome with the oncogenic DNA, although it is also possible to delete only the specific gene(s) associated with the transformed phenotype. This chromosomal loss most likely would be lethal since the hypodiploid cell formed would have lost too much genetic material to be viable. The DNA virus, however, might also cause chromosomal rearrangements which could induce the expression of endogenous viral or non-viral oncogenic information that is normally 'switched off' in non-transformed cells. Evidence for specific chromosomal imbalance and the expression of malignancy could then be found more readily using chromosomal banding techniques [12].

RNA oncogenic viruses

In a similar manner, genetic information for the expression of malignancy could be integrated into specific chromosomes following malignant transformation by certain RNA viruses. The viral RNA and its associated reverse transcriptase enzyme could give rise to DNA sequences which become integrated into chromosomal DNA. Again, *no specific* chromosome necessarily would be found associated with the expression of malignancy unless cell hybridization techniques were used, since chromosomal banding patterns cannot distinguish between a normal diploid cell and a malignant cell which has incorporated a small oncogenic segment into specific chromosomes. Such a situation might occur, for example, in some cases of human leukemia, as alluded to initially in this paper. An RNA virus, theoretically, could also produce chromosomal rearrangements which would yield specific chromosomal changes leading to a 'switch on' of endogenous oncogenic information.

Chemical or physical agents

Chemical carcinogens have been shown to produce chromosomal damage in mammalian systems [24–26]. Since in most instances massive chromosomal damage is lethal to the cell, marked chromosomal breakage in general leads to cytotoxicity rather than oncogenic transformation [6,7]. However, in a small percentage of cells, a significant specific chromosomal rearrangement could occur to allow for the expression of endogenous oncogenic genes, whether viral or nonviral in origin. If careful studies are performed using chromosomal banding techniques, specific chromosomal rearrangements associated with the expression of malignancy should be found in most chemically transformed cells [11,13,14]. The only exception to this generality would occur if the chromosomal rearrangements were below the resolution of those chromosomal aberrations which can be detected by banding techniques. This could happen, for instance, if a chemical carcinogen bound to regions of the chromosomal DNA, producing specific small frame shift mutations or base pair substitutions which allow the expression of oncogenic information while maintaining a diploid chromosomal pattern. I do not, however, believe that this is a common mechanism for chemically induced transformation in vitro or tumorigenesis in vivo, since aneuploidy is found almost without exception in chemically induced transformed cells or tumors.

Certain physical agents which are carcinogenic, such as X-rays, also produce considerable chromosomal damage [24], and thus the mechanism by which they initiate malignancy may be similar to chemical carcinogens, as shown in fig. 3. No studies have been undertaken thus far, however, to identify specific chromosomal changes related to the expression of malignancy following exposure to physical agents.

Role of chromosomes with suppressor genes

Since much of the hypothesis to explain how specific chromosomal changes could result in malignancy involves 'switching on' endogenous oncogenic genes, it must be assumed that these genes are normally 'switched off'. Thus, if one concludes that there are specific chromosomes which contain genes for the suppression as well as expression of malignancy, then the various carcinogens could also cause a cell to be oncogenic by deleting specific chromosomal material with genetic information for the suppression of malignancy [11–14,16].

Such specific chromosomes with genes for the suppression of malignancy could theoretically block the expression of malignancy at numerous steps between the initial transcription and final translation of the oncogenic information as shown in fig. 3. It might, however, be less likely (broken lines) for endogenous suppressor genes to block exogenous oncogenic information since these exogenous genes may have no endogenous counterpart. Therefore, at times, only specific chromosomes with genes for the expression of malignancy following viral transformation with such agents as SV40 [18,19] can be found. On other occasions evidence for chromosomes with suppressor genes can be found following viral transformation [12], perhaps by allowing for the expression of endogenous oncogenic information (also illustrated in fig. 3).

Our hypothesis states that carcinogens can produce malignant transformation by 'switching on' both endogenous non-viral as well as viral oncogenic information. Non-viral factors might include the regulation and induction of plasminogen activator [27–29] which appears to be an early event in malignant transformation [30].

Conclusion

We now have the technical ability to test the role of specific chromosomes in the expression of malignancy, particularly in human systems. Since human solid tumors as well as transformed human cells can be grown in immunosuppressed animals, especially 'nude' mice, we have the capacity to investigate the tumorigenicity of human cells. It would appear that a human diploid transformation system will soon be available [22], and thus we shall be able to examine the relationship between specific chromosomal changes in these cells and the expression of malignancy. Similar studies could also be done using established human tumor lines in which the initial chromosomal pattern contains only one or two changes from the normal diploid complement. The latter chromosomal abnormalities would obviously need to be easily recognized using banding techniques.

If one could then isolate malignant cell clone from the original highly human 'revertant' clones with a markedly decreased oncogenic potential, it would be possible to see whether or not there were specific chromosomal changes associated with the differences in tumorigenicity. It is also necessary to analyze cytogenetically the tumor cells produced by the various cell lines with high or low malignant potential since significant dissimilarities may occur between the chromosomal complement of the cells injected and those isolated from tumors produced by these cells [13]. Diploid cells obtained from the tumors may also reflect the presence of a considerable number of normal fibroblasts derived from the tumor. It is my belief that these investigations will be important to our understanding of the mechanism(s) of malignant transformation in various mammalian systems, and in particular, to our knowledge of the molecular changes which are fundamental to the development of human cancer.

Acknowledgments

W.F.B. is the recipient of Career Development Award CA-70996 from the National Cancer Institute. I wish to thank Dr. Margery Nicolson for her critical review of this manuscript and especially Dr. Richard Kouri for his advice and encouragement through many years of collaborative efforts.

References

[1] I.H. Porter, W.F. Benedict, C.D. Brown and B. Paul, Recent advances in molecular pathology. Exp. Mol. Pathol. 11, 340 (1969).
[2] W.F. Benedict, I.H. Porter and C.D. Brown, The cytogenetic diagnosis of malignancy in effusions. Acta Cytol. 16, 304 (1972).
[3] W.F. Benedict, I.H. Porter, C.D. Brown and R.A. Florentin, Cytogenetic diagnosis of malignancy in recurrent meningioma. Lancet 971 (1970).
[4] A.A. Sandberg, N. Takagi, R. Sofuni and L.H. Chrosswhite, Chromosomes and causation of human cancer and leukemia. V. Karyotypic aspects of acute leukemia. Cancer 22, 1268 (1968).
[5] J. Whang-Peng, T. Knutsen, J. Ziegler and B. Leventhal, Cytogenetic studies in acute lymphocytic leukemia, special emphasis in long-term survival. Med. Ped. Onc. 2, 333 (1976).
[6] W.F. Benedict, J.E. Gielen and D.W. Nebert, Polycyclic hydrocarbon-produced toxicity, transformation, and chromosomal aberrations as a function of aryl hydrocarbon hydroxylase activity in cell cultures. Int. J. Cancer 9, 435 (1972).
[7] M. Karon, W.F. Benedict and N. Rucker, Mechanism of 1-ε-D-arabinofuranosylcytosine induced cell lethality. Cancer Res. 32, 2612 (1972).
[8] W.F. Benedict, Rapid production of numerical and structural abnormalities in hamster cells by transforming doses of polycyclic hydrocarbons. J. Natl. Cancer Inst. 49, 585 (1972).
[9] T. Casperson, S. Farber, G.E. Foley, J. Kudynowski, E.J. Modest, E. Simonsson, U. Wagh and L. Zech, Chemical differentiation along metaphase chromosome. Exp. Cell Res. 49, 219 (1968).
[10] M. Seabright, A rapid banding technique for human chromosomes. Lancet 2, 971 (1971).
[11] T. Yamamoto, Z. Rabinowitz and L. Sachs, Identification of the chromosomes that control malignancy. Nature New Biol. 243, 247 (1973).
[12] T. Yamamoto, M. Hayashi, Z. Rabinowitz and L. Sachs, Chromosomal control of malignancy in tumors from cells transformed by polyoma virus. Int. J. Cancer 11, 555 (1973).
[13] W.F. Benedict, N. Rucker, C. Mark and R. Kouri, Correlation between the balance of specific chromosomes and the expression of malignancy in hamster cells. J. Natl. Cancer Inst. 54, 157 (1975).
[14] N. Bloch-Shtacher and L. Sachs, Chromosome balance and the control of malignancy. J. Cell Physiol. 87, 89 (1976).
[15] J.A. DiPaolo, N.C. Popescu and R.L. Nelson, Chromosomal banding patterns and in vitro transformation of Syrian hamster cells. Cancer Res. 33, 3250 (1973).

[16] S.D. Codish and B. Paul, Reversible appearance of a specific chromosome which suppresses malignancy. Nature 252, 610 (1974).
[17] C.M. Croce, J. Girardi and H. Koprowski, Assignment of the T antigen gene of simian virus 40 to human chromosome C-7. Proc. Natl. Acad. Sci. U.S.A. 70, 3617 (1973).
[18] C.M. Croce and H. Koprowski, Somatic cell hybrids between mouse peritoneal macrophages and SV40-transformed human cells. I. Positive control of the transformed phenotype by the human chromosome 7 carrying the SV40 genome. J. Exp Med. 140, 1221 (1974).
[19] C.M. Croce, D. Aden and H. Koprowski, Somatic cell hybrids between mouse peritoneal macrophages and simian-virus-40-transformed human cells: II. Presence of human chromosome 7 carrying simian virus 40 genome in cells of tumors induced by hybrid cells. Proc. Natl. Acad. Sci. U.S.A. 72, 1397 (1975).
[20] H.J. Igel, A.E. Freeman, J.E. Spiewak and K.L. Kleinfeld, Carcinogenesis in vitro. II. Chemical transformation of diploid human cell cultures: A rare event. In vitro 11, 117 (1975).
[21] W.F. Benedict, P.A. Jones, W.E. Laug, H.J. Igel and A.E. Freeman, Characterization of human cells transformed in vitro by urethane. Nature 256, 322 (1975).
[22] T. Kakunaga, Transformation of human diploid cells by chemical carcinogens. Proc. Nat. Acad. Sci. (submitted).
[23] A.E. Freeman, H.J. Igel, R.S. Lake, L. Gernand, M.R. Pezzutti, C. Mark and W.F. Benedict, Euploid to heteroploid conversion of human skin cells by methylcholanthrene. Proc. Nat. Acad. Sci. (submitted).
[24] B.A. Kihlman, Actions of Chemicals on Dividing Cells, Vol. 1 (Prentice Hall, Inc., Englewood Cliffs) 1966.
[25] S.M. Sieber and R.H. Adamson, The Clastogenic, mutagenic, teratogenic and carcinogenic effects of various antineoplastic agents, in: Pharmacological Basis of Cancer Chemotherapy (Williams and Wilkins, Baltimore) 1975, p. 401.
[26] W.F. Benedict, A. Banerjee, A. Gardner and P.A. Jones, Induction of morphological transformation in mouse C3H/10T$\frac{1}{2}$Cl-8 cells and chromosomal damage in hamster A(T_1)C1-3 cells by cancer chemotherapeutic agents. Cancer Res (submitted).
[27] P.A. Jones, W.E. Laug and W.F. Benedict, Clonal variation of fibrinolytic activity in a human fibrosarcoma cell line and evidence for the induction of plasminogen activator secretion during tumor formation. Cell 6, 245 (1975).
[28] W.E. Laug, P.A. Jones and W.F. Benedict, Relationship between fibrinolysis of cultured cells and malignancy. J. Natl. Cancer Inst. 54, 173 (1975).
[29] W.E. Laug, P.A. Jones, C. Nye and W.F. Benedict, The effect of cyclic AMP and prostaglandins on the fibrinolytic activity of mouse neuroblastoma cells. Biochem. Biophys. Res. Commun. 68, 114 (1976).
[30] J.C. Barrett, B.D. Crawford, D.L. Grady, L.D. Hester, P.A. Jones, W.F. Benedict and P.O.P. Ts'o, The temporal acquisition of enhanced fibrinolytic activity by syrian hamster embryo cells following treatment with benzo(a)pyrene. Cancer Res. (submitted).

Chapter 20

THE RELATIONSHIP BETWEEN NEOPLASTIC TRANSFORMATION AND THE CELLULAR GENETIC APPARATUS

PAUL O.P. TS'O, J. CARL BARRETT and ROBERT MOYZIS

Division of Biophysics, School of Hygiene and Public Health, The Johns Hopkins University, Baltimore Md., U.S.A.

Introduction

In addition to its serious impact on health and socio-economic activities, cancer is a unique category of diseases which provides both great challenge and opportunity in biological research. In this paper, we shall concentrate on the cellular aspects of cancer initiation which can be studied in vitro and temporarily ignore those complex and medically important aspects of cancer which are related to the physiology of the host, such as hormonal control, immune response, metastasis, and vascularization of tumors. Attention is focused on the problem of carcinogenesis as a cellular event.

One of the fundamental and most frequently posed questions in research on carcinogenesis is whether neoplastic transformation is genetic in nature, epigenetic in nature, or both. Superficially, this question appears to be well-formulated and clear-cut, but in actuality, it is predetermined by one's concept of the term 'genetic' and, therefore, 'epigenetic'. Many of the experimental approaches adopted in genetics research today center on unicellular microorganisms, such as phage, bacteria, and yeast. With the onset of molecular biology, the term 'genetic' appears to be used to describe the function and the structure of DNA. In this framework, and with the notion that genetic information can be explained by a one-dimensional arrangement, most hereditary changes of single-celled organisms can be traced to changes in DNA sequence. This concept in microbial and Mendelian genetics is useful in explaining the single-gene event. However, this concept is inadequate when applied to the problem of differentiation.

From a fertilized zygote, a multicellular organism develops with every cell having the same genetic endowment (such as the same DNA content and arrangement) but with different physical and physiological properties, which are heritable in vivo or in vitro. In analogy to carcinogenesis, the question has been raised whether differentiation is either a genetic or an epigenetic phenomenon,

The Molecular Biology of the Mammalian Genetic Apparatus:
edited by P. Ts'o © 1977, Elsevier/North-Holland Biomedical Press

or both. Therefore, the study of neoplastic transformation of cells (unregulated growth) provides very valuable information concerning the regulated development and maintenance of differentiated cells. The terms 'genetic', 'epigenetic' and 'heritable properties' defined by concepts adopted from microbial genetics are not necessarily very useful in describing the properties of differentiated cells from a metazoan. In short, while the molecular mechanism of the control of gene expression in mammalian cells may be similar to that in bacteria, the inheritance of such control may be more complex in the mammalian cells than in the bacteria.

Another major issue concerning the research on in vitro neoplastic transformation is the basic recognition and characterization of this phenomenon. At present, even though much effort has been expended with a certain degree of success (vide infra), the phenomenon of cancer is only recognized and characterized in a universally accepted manner through the intercellular, social properties of cells. On the other hand, many of the successful investigations in molecular biology have centered on *intracellular* events such as nucleic acid synthesis, protein synthesis, and enzyme function. Intercellular reactions and responses have so far not been fully described in molecular terms, although the science of cell surfaces has received much deserved attention during the last five years. Therefore, one major challenge in the research on neoplastic transformation is to attempt to relate the *intercellular* events (usually described as growth properties) to the *intracellular* events, which in this paper particularly are concerned with the functions and properties of the genetic apparatus.

Concepts

In this section we shall describe our general concepts in dealing with the two most fundamental questions raised above, i.e. genetic nature versus epigenetic nature and intercellular events versus intracellular events.

In considering the unique properties of differentiated cells in metazoans, we have formulated a concept which states that the genetic and epigenetic processes are mutually interdependent and inter-reactive. The statement of this concept in biology would be: the genetic control of the epigenetic process; the statement of this concept in molecular biology would be: the DNA control of the control of DNA; the statement of this concept in biochemistry would be: the effect of DNA structures on the DNA-regulator interaction. The regulators can be protein, RNA, lipo-membranes, or other structures; the effects can be direct or remote. These statements can be expressed in the following diagram (chart 1). In this diagram, DNA is still considered a central component of the genetic apparatus which is most easily identifiable. However, there is another equally important component in the genetic apparatus which is termed 'regulator'. The regulator can be composed of various substances in performing various roles,

Relationship between neoplastic transformation and genetic aparatus

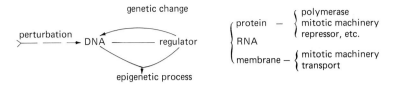

as exemplified in the chart. This chart also indicates that the process of DNA-regulator interaction can be considered an epignetic process from the viewpoint of microbial genetics since this process usually does not cause any change in DNA sequence. However, the DNA-regulator interaction controls the expression of DNA. Therefore, this chart depicts the DNA control of the control of DNA. A perturbation administered directly to the DNA (a genetic event by definition) can lead to a substantial effect on the DNA-regulator interaction which is an epigenetic process; on the other hand, a direct perturbation to the regulator (by definition in microbial genetics an epigenetic event) can also lead to a change in the DNA under certain specific circumstances, an outcome which can then be correctly termed a genetic change. A perturbation to the genetic apparatus, either to DNA or to regulator, can lead to death, mutation (usually recognized as intracellular events) or neoplasia (currently recognized as an intercellular event).

Experimental design

The experimental design can be divided into three sections: system, perturbation and responses of the genetic apparatus.

System

Our choice of experimental animal is the Syrian hamster and the first phase of the experiment is done with the embryonic fibroblast. Important pioneering work with this system has been done by Sachs and collaborators [1,2], by DiPaolo and collaborators [3–6] and by others [7–10]. The Syrian hamster system has the following important characteristics: in terms of economy, a rodent with a short life span must be used. Syrian hamsters have 44 chromosomes, very similar to the number of chromosomes in human cells and, in fact, this animal has about 10% more DNA per cell than the human cells. The cells from these animals can go through a senescent stage and have a very low spontaneous transformation frequency upon serial passage. In this respect, the Syrian hamster fibroblasts are very similar to the human fibroblasts. The karyotypic patterns of the cells grown in culture are more or less constant and can be analyzed carefully (see W. Benedict's chapter). The animals are relatively tame in captivity (in contrast to the Chinese hamster). Also, they can be induced to form tumors upon injection of transformed cells into the newborn animals. In addition, they have been used extensively as model animal systems for various types of cancer study, particularly lung cancer. In our laboratory, we have been successful in the study of transformation of the Syrian hamster embryonic fibroblasts in the absence of feeder layer [11]. In this system, all of the cells in the Petri dishes are potentially target cells. Such a system can be used for mechanistic studies of transformation and somatic mutation, as well as for the metabolism of carcinogens. Currently, we have begun to extend our research to epithelial cells from liver and to lymphocytes.

Perturbation to the genetic apparatus

In this paper, we shall temporarily ignore the perturbation delivered by a biological agent such as a virus and concentrate on chemical or physical perturbations which can be formulated in the following three categories.

1) Decrease in the genetic information. In general, this has been considered as the role of chemical or physical carcinogens/mutagens. In our experiments, we have used an important environmental carcinogen, benzo(a)pyrene (B(a)P) and a well-known mutagen, N-methyl-N^1-nitro-nitrosoguanidine (MNNG). The former compound requires metabolic activation, while the latter compound has no such requirement and these two compounds attack the cells with very different mechanisms. However, these two carcinogens/mutagens do suffer from one serious drawback, i.e. they attack many types of target macromolecules inside the cells: nucleic acids, proteins, and so forth. It is therefore vital to develop a system in which the chemical attack is delivered mainly to one type of macromolecule or one specific area of the cell. In this situation, two types of specific perturbations are being developed, one of which is based on the incorporation of bromodeoxyuridine (BrdU) into the cell, subsequently followed by near UV irradiation (approximately 300–400 nm). It is known that BrdU is incorporated only into DNA, and the incorporated BrdU can be photoactivated by irradiation in the near UV region and cause DNA breakage. Such a perturbation is therefore specific to DNA alone. When such a study is carried out on synchronized cell cultures with BrdU supplied in pulse at different periods of the S phase, the destruction of DNA can be further specific to the DNA synthesized during the pulse period. These procedures, however, will not provide perturbations to other types of target macromolecules in the cells.

The second type of perturbation can be described as internal irradiation by incorporated isotopes; for example, ^{3}H-thymidine, ^{3}H-uridine, or ^{3}H-amino acids. The decay of the radioactive precursors within a cell results in a specific perturbation to the macromolecules which contain these precursors under defined conditions. For instance, incorporation of ^{3}H-thymidine will most likely destroy DNA, while the incorporation of ^{3}H-uridine will probably destroy RNA. Such investigations have been carried out effectively by the researchers on phages and bacteria [12–19]. However, in eukaryotic cells the position effect of the beta irradiation can also be utilized. ³H-Thymidine will probably have an effect on the entire nucleus, and under certain circumstances, ³H-RNA will preferentially effect the cytoplasmic structures. When used judiciously, this approach can provide perturbations having a certain degree of specificity to cellular macromolecules and internal structures.

2) Increase in genetic information. In this category genetic information is introduced into the target cell in a series of gradations in complexity, i.e. from fusion with viable cells to uptake of purified DNA. These experiments are related to the hypothesis of the imbalance of genetic material as the cause of carcinogenesis (see chapter by W. Benedict).

3) Change in the ploidy state of the genetic apparatus. In these types of experiments, tetraploid cells (or even octoploid cells) are artificially created. Their response to various types of perturbations, including the mutagens and carcinogens can be compared with the response of their original diploid cells.

The perturbations described in [2] and [3] above are very informative in the study of neoplastic transformation. They are basically biological in nature and will not be discussed further.

Responses

Experimental measurements of the responses can be divided into two categories: the biological responses and the cytological and molecular responses.

Biological responses

1) Somatic mutation of single genes. The recent progress in this area has removed much doubt concerning the genetic nature of the heritable changes observed in the cells grown in culture after treatment with various mutagens [20]. In our laboratory, a system utilizing Syrian hamster embryonic fibroblasts has been developed which can be used to investigate somatic mutation and neoplastic transformation simultaneously. With such a system, a comparative study of these two biological responses can now be made.

In this paper only two genetic markers are described, the HGPRT locus which depends on the selections of the 6-thioguanine/8-azaguanine resistant cells, and the K^+/Na^+ membrane ATPase locus which depends on the selection of the ouabain resistant cells. Other markers are currently being developed in our laboratory for the Syrian hamster cells, such as resistance to α-amanitin (Leavitt and Morry, unpublished results).

In these experiments, single mutations due to various perturbations are measured. The measurements in these experiments include: i) the forward and reverse frequencies of mutation, and ii) biochemical characterization of somatic mutation. In addition, these mutants may serve as cell markers for somatic hybridization and experiments concerning the transfer of genetic material.

2) Neoplastic transformation. Only tumor formation is recognized universally as a reliable phenotypic marker of the neoplastic state in cultured cells. Other 'abnormal' properties of cells in vitro have been associated with neoplastic transformation. Some of these phenotypic abnormalities are listed in table 1. In order to be useful in a mechanistic study, these phenotypic abnormalities have to be measured quantitatively. In our laboratory, four phenotypic characteristics and their relationship with each other have been investigated extensively. These are: alteration in clonal morphology, appearance of fibrinolytic activity, growth in soft agar, and, finally, tumorigenicity. The results of this investigation (vide infra) show that the neoplastic transformation process appears to be progressive in nature, not a single-step phenomenon, as is the case with a single-gene somatic mutation. The progression of neoplastic transformation was recognized in the in vivo tumor formation by the pathologists a long time ago [21]. However, only recently has this phenomenon been recognized in in vitro

TABLE 1

Phenotypes of transformed cells.

1. Tumorigenicity
2. Loss of anchorage, dependency for growth
    a) soft agar
    b) methylcellulose
    c) teflon
3. Morphological changes
4. Enhanced fibrinolytic activity
5. Chromosomal changes
6. Indefinite life span
7. Lack of density dependent inhibition of replication (saturation density)
8. Growth at low serum concentrations
9. Changes in membrane properties
    a) altered glycoproteins and glycolipids
    b) increased rate of transport
    c) loss of LETS protein
    d) increased agglutinability by plant lectins
    e) increased mobility of membrane protein
    f) altered surface structure (SEM)
10. Changes in microtubules and actin cables
11. Decreased cAMP levels
12. Growth in low calcium medium

carcinogenesis. The reason for such a delay in recognition of this important feature of neoplastic transformation is due to two aspects: 1) Such a feature may be less obvious when aneuploid, but contact-inhibited non-tumorigenic cell lines are used in neoplastic transformation studies (such as BHK cells), particularly with oncogenic viruses. 2) Such a feature does not appear to be a predominant characteristic of single-gene somatic mutation of cells in culture. Nevertheless, this is a very important aspect of neoplastic transformation in vitro both from the theoretical consideration of the mechanism of this process and also from practical considerations as to why such experiments are difficult to repeat and quantitate. Therefore, it should now be noted that the measurement of neoplastic transformation can be a time-dependent phenomenon.

Cytological and molecular responses

1) Loss of DNA sequences. Abnormalities in karyotypes during transformation have long been recognized, as discussed in two chapters in this volume (chapter by DiPaolo and chapter by Benedict). However, the loss or rearrangement of DNA sequences of the neoplastic cell has never been systematically and quantitatively measured by the techniques of nucleic acid hybridization. The following section describes the current findings in the search for missing DNA sequences in a highly malignant transformed cell line which possesses a subdiploid set of chromosomes. It is of fundamental importance to know whether any such loss occurs in other types of transformed cells which have more chromosomes than the diploid cells. Under that condition, it is still possible that these aneuploid cells may have a loss of specific DNA sequences, although the total mass of DNA in the cell is even greater than in normal cells. In the event that the malignant aneuploid cells are found to contain all of the DNA sequences in the normal diploid cells, this finding can indicate that the imbalance of genetic material may be the important factor in neoplastic transformation. However, the significance of these experiments depends on the level of detection of the missing DNA sequences. Also, in these experiments an alteration of nucleotide sequences leading to mutation of one or several genes presumably can never be detected by such techniques. Therefore, such a conclusion must be reinforced by positive results obtained from experiments in which foreign DNA or foreign genetic material prepared from non-neoplastic cells is introduced into normal host cells. The imbalance of genetic material can come about due to a loss of control in mitotic processes leading to an uneven distribution of genetic material between the two daughter cells.

2) Change in transcription. Conceptually since neoplastic cells are morphologically and functionally different from the normal cells, it would be logical to predict a change in the transcription program. Such a change can be both quantitative and qualitative in nature. The gain or the loss of transcription of an

entire class of RNA (particularly mRNA) would be much more easily detectable than merely the change in number of RNA strands being transcribed in all categories. Therefore, experiments have been performed to first analyze whether there is any appearance of new types of mRNA in the highly tumorigenic cells as compared to the normal daughter cells. Similar experiments can be done to examine the disappearance of a class of mRNA in the malignant cells as compared to the normal cells.

Karyotypic analysis has been made in conjunction with the molecular analysis of nucleic acid sequences. The missing DNA sequences and possibly the tumorigenicity-related mRNA sequences can be located in the metaphase chromosome through in situ hybridization. Such a study could then correlate cytological analysis with the molecular analysis.

Current results

Somatic mutation

As mentioned above, the somatic mutation assays developed are for two loci, HGPRT (8-azaguanine/6-thioguanine resistance) and Na^+/K^+ ATPase (ouabain resistance). After exposure to B(a)P (1 μg or 10 μg/ml, 24 h) or MNNG (1×10^{-6} M or 5×10^{-6} M, 2 h) and allowing a 4-day growth period for recovery, early passage Syrian hamster embryonic fibroblasts (SHE) were grown in medium containing the following selective agents: 40 μg/ml azaguanine, 1 μg/ml 6-thioguanine, or 1 mM ouabain. These cells were grown in the selective media for 1–2 weeks before scoring or clonal isolation. 8-azaguanine ($8AG^r$), 6-thioguanine ($6TG^r$) and ouabain (oua^r) resistant cells have been isolated, and the characteristics of the isolated clones are shown in table 2. These properties are consistent with the 8AG/6TG resistant cells having a mutation in the structural gene or a regulatory gene for HGPRT and the ouabain resistant cells having an altered Na^+/K^+ ATPase.

TABLE 2

Characterization of mutants from Syrian hamster embryonic fibroblasts.

1. Isolated clones have stable resistance to selective drugs
2. $8AG^r$ clones are cross-resistant to 6TG
3. The reversion frequency of $6TG^r$ and $8AG^r$ clones as measured by HAT resistance is less than 10^{-6}
4. HGPRT activity of $8AG^r/6TG^r$ clones is <1% of parental cells
5. Ouabain resistant clones show decreased sensitivity to ouabain inhibition of rubidium uptake
6. Frequency of resistant clones is increased by known mutagens in dose-dependent manner

Quantitative measurements of the mutation frequency of each loci following mutagenesis with different doses of MNNG and B(a)P have been determined. The spontaneous mutation frequency is $< 10^{-6}$ per cell for Oua^r, 6TG^r and 8AG^r. Mutagenesis with B(a)P (10 μg/ml, 24 h) or MNNG (5×10^{-6} M, 2 h) increases the frequency to about 10^{-5}–10^{-4}.

A comparative study has been made of the mutation frequency of normal diploid SHE cells and transformed subdiploid hamster fibroblasts (BP6T) mutagenized by MNNG (1×10^{-6} M and 5×10^{-6} M). The spontaneous and induced mutation frequencies of these two cell types are comparable to each other (within 2-fold) for the ouabain resistant and 6-thioguanine resistant assays; however, transformed BP6T cells appear to afford higher frequencies (both spontaneous and induced) for the 8-azaguanine resistance assays.

The Syrian hamster embryo (SHE) fibroblast system has been used for studies on neoplastic transformation by a number of investigators [1–10]. The system has been quantitated by measurement of the morphological transformation frequency of the cells following treatment with carcinogens. Our laboratory has further developed this system by adoption of the clonal morphological trans-

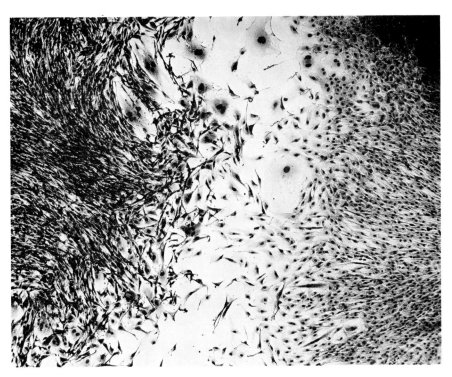

Fig. 1. Edge of a morphologically transformed colony (left) contrasted with the edge of a normal colony (right) of Syrian hamster embryo cells.

formation assay in the absence of a feeder cell layer [11] which has enabled us to use the system for somatic mutation studies.

Fig. 1 illustrates the morphological alterations observed in clones of SHE cells following treatment with B(a)P. The normal clone has a characteristic organized arrangement of cells while the transformed clone displays a disorganized, criss-cross arrangement of spindle shaped cells. A survey of about 6000 clones of SHE cells exposed to B(a)P, B(e)P (10 μg/ml) and pyrene (20 μg/ml), as well as another 6000 clones exposed to DMSO as control, has been made [11]. This experiment indicated a morphological transformation frequency of about 1 to 3% (or about 60 transformed clones in ~5000 clones) when exposed to 1–10 μg of B(a)P for 24 h, while zero or near zero was found for cultures exposed to B(e)P (0/520), pyrene (0/584) or DMSO (4/~6000). A number of such transformed clones have been isolated and shown to produce fibrosarcomas when injected into newborn hamsters [11].

In addition to morphological transformation, we have studied neoplastic transformation by examining the properties of enhanced fibronolytic activity, growth in soft agar and tumorigenicity of normal, transformed and B(a)P treated SHE cells. Emphasis has been placed on these phenotypes because they can be measured quantitatively. Fibrinolytic activity can be measured quantitatively either by the release of labeled fibrin which measures the activity of a population of cells [22–27] or by the fibrin overlay method of Jones et al. [28] which enables one to detect the enhanced activity of individual colonies of transformed cells in the presence of normal cells. With the latter method, colonies of cells are overlaid with agarose containing fibrin, giving the plate an opaque appearance. Clones which possess fibrinolytic activity lyse the fibrin, resulting in a clear zone of lysis [28]. Quantitatively, the data can be expressed as a percentage of

TABLE 3

Cloning efficiency of various cell types in soft agar.

Cell line	Cell number	0.3% Agar	0.3% Agar + bactopeptone
SHE	10^5	0%	0 %
BP6	10^4	20%	21 %
	10^3	0%	41 %
	10^2	0%	51 %
	10^1	0%	75 %
	2	N.D.	80 %
BP6T	10^4	TNTC	TNTC
	10^2	0%	96 %
	10^1	0%	90 %

cells in culture exhibiting lysis-zones at a given time, or the time of incubation required for 100% of the cells to exhibit lysis-zones. As shown in table 3, the soft agar assay is also a quantitative assay if one supplements the agar suspension with bactopeptone [9]. In the absence of bactopeptone, three different transformed cell lines failed to grow in soft agar unless very high cell numbers were used. However, by adding 0.1% bactopeptone to the agar medium enabled very small numbers, as few as 2 cells, to form colonies in soft agar with a high degree of efficiency. In another experiment addition of normal SHE cells, which do not grow in agar (table 3), to the transformed cell lines did not inhibit but, in fact, enhanced colony formation. Thus, by this method, 1–2 transformed cells can be detected in the presence of 10^5 normal cells. Tumorigenicity of the chemically transformed cells can also be tested quantitatively in newborn Syrian hamsters with a high degree of sensitivity. Upon injection of 10 to 10^7 cells into the newborn hamster, tumors (fibrosarcomas) can be examined after a period of 2–4 months. With normal cells, injection of 10^7 cells/animal produced no tumors in 12 months, while with the highly malignant BP6T, injection of 10 cells/animal produced tumors in 100% of the animals in 2–3 months.

TABLE 4

A comparison of the tumorigenicities and other phenotypes in normal SHE cells and three benzo(a)-pyrene transformed lines.

Cell type	Normal SHE[a]	BP12[b]	BP6[b]	BP6T[b]
Tumorigenicity[c]	0% at 10^7	50% at 10^4	75% at 10^2	100% at 10^1
Fibrinolytic activity[d]	neg. at 24 h	100% pos. at 24 h	100% pos. at 3 h	100% pos. at 6 h
Cloning efficiency in[e] soft agar	none (10^5 cells)	3%	51%	96%
Cloning efficiency at[f] low cell density in liquid medium	<10%	73%	84%	85%

[a] Syrian hamster embryo cells, passage 3–7.
[b] Transformed cell lines derived from SHE cells following exposure to benzo(a)pyrene.
[c] The percent of tumor bearing animals 6–12 months after subcutaneous injection of 3 day old newborn Syrian hamsters with the number of cells indicated.
[d] Fibrinolytic activity as measured by the percent of colonies producing positive lysis zones in the fibrin agarose overlay after incubation for the specified time.
[e] Percent of cells forming macroscopic colonies in 0.3% agar supplied with complete growth medium and 0.1% bactopeptone following incubation at 37°C for 14–21 days.
[f] Percent of plated cells forming colonies on plastic petri dishes after 7 days incubation at 37°C.

A comparison of the growth in soft agar, fibrinolytic activity and tumorigenicity of three cell lines has been performed (table 4). The cell lines differed in their tumorigenicity, with the least tumorigenic line requiring 10^4 cells to produce tumors. The tumorigenicity data paralleled the growth of these cells in agar quantitatively, while they had similar cloning efficiencies in liquid media. All of the lines possess enhanced fibrinolytic activity when compared to the early passage SHE cells. However, the fibrinolytic activity of these cells did not correlate quantitatively with their tumorigenicity. A cell line which produced tumors with 10^2 cells had a higher fibrinolytic activity than a cell line which required only 10 cells. This may indicate that fibrinolysis, while perhaps necessary for tumorigenicity, is not limiting in these cell lines.

To further determine the significance of fibrinolysis and growth in soft agar to the transformation process, we have examined the temporal acquisition of these parameters following treatment of Syrian hamster embryo cells with B(a)P. Cells were seeded overnight, treated with B(a)P and then allowed to clone for 8 days. After this time, the plates were either fixed and stained or overlaid by the agarose fibrin technique for fibrinolytic activity detection. Treatment with B(a)P resulted in morphological transformations but no fibrinolytic activity was detected. These cells were also trypsinized and suspended into soft agar. No clones in agar resulted. These data suggest that morphological transformation, fibrinolytic activity and soft agar growth are not related quantitatively at this point in time in the transformation process.

To determine when fibrinolytic activity appears, a mass culture experiment was performed. Tertiary passage SHE cells were treated in 75 cm² flasks, washed and allowed to grow to confluency. They were trypsinized and a subculture of $1-5 \times 10^5$ cells was initiated. Cells were also seeded at low density for cloning to examine morphological transformation and fibrinolytic activity. Also, 10^6 cells were tested at each passage for soft agar growth. The passages are numbered from the 1st passage after treatment called post-treatment passage one or PTP-1. The process is repeated at each PTP.

At the first or second passage after treatment with benzo(a)pyrene, cells are present which give rise to fibrinolytically active colonies. The number of such colonies is quite high, up to 10%, and appears to occur with a dose response. The controls have no such activity. The presence of these cells persists with passaging in 3 of the 4 treated cultures. In the fourth culture, fibrinolytically active clones started to disappear after the fourth passage. This 4th culture senesced at a later passage. In three treated cultures, cells capable of colony formation in soft agar appeared but not until PTP6. The 4th culture which lost its fibrinolytic activity did not produce any colonies capable of growing in soft agar and senesced at a later PTP, as mentioned. Control cultures did not give rise to soft agar positive clones even at PTP30. After PTP6, the number of clones (per 10^6 cells seeded) grown in soft agar slowly increased with each passage while the percentage of fibrinolytically-active clones, and the percentage of morphologically transformed clones remained about constant until PTP9 to PTP11. At that time, the percentage of fibrinolytically active clones and the morphologically transformed clones increased rapidly. Within 2 passages, these cultures were completely dominated by morphologically transformed, fibrinolytically active cells with the cloning efficiency in soft agar approaching 1–2% at PTP13.

The question which then arises is why such long periods of growth (greater than forty population doublings) are required before cells have the ability to grow in agar. Two explanations can be given for this:

First, fully transformed cells are present at an earlier point but at a level below the sensitivity of our assay, which is ~ 1 in 10^5, and further growth selects for these cells, or second, the transformation process in vitro is not a one-step process but a multistep, progressive change through many qualitatively different stages. In providing a determination of these two alternative explanations, nine fibrinolytically-positive clones were isolated at PTP3. At this timepoint, there were 17.5% of the fibrinolytically-active colonies in the culture while no clone capable of growth in soft agar were found when 10^6 cells were tested. Subsequently, five fibrinolytically-active clones developed into cell lines which can grow in soft agar (the other four clones senesced). This result implies that the culture at the stage of PTP3 might contain up to 9% of the total number of cells potentially capable of growth in soft agar, or 0.9% of the population should grow in soft agar, giving an expected cloning efficiency of 10%. Yet none of these cells could grow in soft agar at PTP3; this was achieved only at PTP6 or later. This observation and reasoning strongly endorses the second alternative as the correct explanation. In these cultures exposed to B(a)P, a substantial proportion of these cells was capable of growth in soft agar at a late passage but not at an early passage, and this phenomenon is due to *progression* and *not to selection*. At a later stage (say PTP11 onward) these cultures were dominated by the transformed cells, presumably through a selective mechanism, but the underlining mechanism of progression can be clearly observed at an earlier period.

Relationship between neoplastic transformation and somatic mutation

Development of quantitative assays for the processes of neoplastic transformation and somatic mutation which can be measured concomitantly with the same cell system enables us to study in depth the relationship of these two processes. As has been shown, the frequency of morphological transformation of the SHE cells is $\sim 10^{-2}$ per surviving colonies after treatment with B(a)P or MNNG. The apparent somatic mutation frequency as measured by 8AG^r, 6TG^r or Oua^r clones is less than 10^{-4} per surviving cell. Thus, the morphological transformation frequency is about two orders of magnitude greater than the mutation frequency. This could represent a much greater target size for the morphological transformation genes or it may represent an enhanced sensitivity of this portion of the genome to chemical carcinogens.

However, there is another more important distinction between neoplastic transformation and single-gene somatic mutation which can be drawn from our studies. The single gene mutations were found to occur rapidly, within one week, and appear to be manifested in a one-step mechanism with two stages, i.e., wild type → mutant. Transformation, on the other hand, appears to require longer periods of time and occurs in a multi-step progressive manner. Neoplastic transformation may therefore be initiated by a mutational event but cannot be completely explained by such a mutation.

In summary, there exists a variety of altered, heritable phenotypic changes which are considered to be closely associated with neoplastic transformation. Three such phenotypic alterations, morphological transformtion, enhancement of fibrinolytic activity and efficient growth in soft agar, were investigated here in conjunction with tumorigencity in the host animal. The data clearly indicate that these three heritable alterations, while all occuring on exposure to carcinogen, appeared in different time sequences after treatment and have no direct relationship in a 1 : 1 correspondence with each other. Temporally, morphological transformation appears first (7–10 days), enhancement of fibrinolytic activity is next (10–14 days, or 4–10 total population doublings), efficient growth in soft agar then occurs (30–60 days or 30–40 total population doublings), followed by tumorigenicity in the host animal. These observations suggest that the appearances of these heritable phenotypic alterations may reflect various facets of the progressive and cascading process of neoplastic transformation due to an initial perturbation and subsequent, continued loss of proper function of the genetic apparatus.

Changes in DNA sequence organization and gene expression in neoplastic transformation

The above biological studies point to the need for a molecular, intracellular characterization of the transformation process, particularly at the nucleic acid level. Ultimately, we hope to answer two fundamental questions in the investigations described in this section. First, is DNA sequence loss or DNA sequence organizational change, either specific or non-specific, causally related to neoplastic transformation? Second, what is the extent of new transcription in chemically transformed cells; are these neoplastic phenotypes related to the 'turning on' of 10 genes or 10,000? In this study, we wish to develop the strategy and the methodology of answering these two fundamental questions by comparing a newly transformed subdiploid cell line (18 Cl 10), transformed in vitro by B(a)P, and its highly malignant tumor cell derivative (BP6T) to normal diploid SHE

TABLE 5

Analyses of karyotypes from eight transformed hamster fibroblasts (BP6T)[a].

Cell.	X	A1	A2	A3	A4	B5	B6	B7	B8	B9	B10	C11	C12	C13	C14	C15	D16	D17	D18	D19	E20	F21	*	Total no. chromosomes
1	1	2	M$_1$ 1(2)	M 1(1)	3	2	2	0	2	M$_2$ (2)	2	M 1(1)	2	2	2	M (1)	M$_3$ 2(1)	0	2	M$_3$ 2(1)	2	2		41
2	1	1	M$_1$ 1(1)	2	3	2	3	0	M 2(1)	M$_2$ (1)	1	2	2	3	3	1	2	0	2	2	2	1		39
3	1	2	M$_1$ 1(1)	2	2	2	1	0	2	M$_2$ 1(2)	1	2	3	1	2	3	M 1(1)	0	2	2	2	2	(M)	40
4	2	2	M$_1$ 1(1)	2	2	2	3	0	2	M$_2$ (2)	2	2	2	2	1	M? (0)		8 total	D group		2	2		39
5	1	M 1(1)	M$_1$ 1(1)	2	2	2	2	1 0	2	M$_2$ 0(1)	2	2	1	2	1	M 1(1)		8 total	D group		2	1		35
6	1	2	M$_1$ 1(1)	M 2(1)	2	2	3	0	2	M$_2$ (2)	1	M 2(1)	2	2	2	2	M 2(1)	(1)?	2	1	2	2		43
7	1	2	1	M 1(2)	M 1(1)	2	3	0	2	1	1	2	2	2	4	1		9 total	D group		2	2		40
8	1	2	M$_1$ 1(1)	2	2	2	2	0	2	M$_2$ (2)	2	2	2	2	1	2	M$_3$ 2(1)	0	2	M$_3$ 2(1)	2	2	(4M)	43
									Averages of Columns															
	1.125	1.875	2.0	2.25	2.25	2.0	2.375	0	2.125	1.75	1.5	2.125	2.0	2.0	2.0	1.5	2.4	0-0.2	2.0	2.2	2.0	1.75		40

[a] The procedures for karyotyping and classification of the chromosomes are based on the work of Popescu and DiPaolo.
* Unidentifiable-atypical G-banding.
M$_1$, M$_2$, M$_3$ – refer to specific chromosome markers present in most cells.
M – refers to nonspecific marker chromosomes infrequent in the population.

cells. As described above, as few as ten BP6T cells produce tumors in 100% of the injected newborn hamsters. Karyotypic studies indicate that both 18 Cl 10 and BP6T have a model chromosome distribution of 40 in comparison to the normal 44 chromosomes of cultured diploid SHE cells [11]. This loss in chromosome number appears to be the result of the loss of actual chromosome pieces and is not predominantly due to extensive Robertsonian fusions and translocations (table 5). Preliminary flow-microfluorometric analyses of normal and tumor cells indicate a 5–10% decrease in DNA content in BP6T. There is additional information which can be learned from the results shown in table 5. As shown by this preliminary analysis of the karyotypes of 8 BP6T cells, there exists a substantial variation in the metaphase chromosome patterns within the tumor cell population, yet the range of distribution of the DNA content in the BP6T cell population shown by the flow-microfluorometric analyses is similar to that of the normal SHE cell population which has an uniform 44 metaphase chromosomal pattern. This preliminary observation suggests that for a given amount of DNA content in a tumor cell type, there may exist a substantial variation in metaphase chromosomal pattern. This phenomenon may reflect a basic malfunction of the genetic apparatus of the tumor cells. This observation also warns against the selection of a *single* metaphase chromosomal pattern as the sole representative of the entire tumor cell population. Obviously, this phenomenon deserves further investigation and close attention.

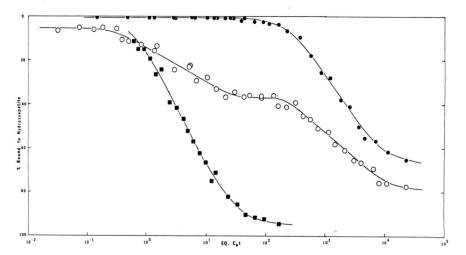

Fig. 2. Reassociation kinetics of 250 nucleotide ^{3}H-SHE cell DNA driven by a 1,000 fold excess of hamster organ DNA, ○——○. Isolated single-copy ^{3}H-SHE DNA driven by a 1,000 fold excess of hamster organ DNA, ●——●. ^{32}P *E. coli* DNA standard, ■——■. Reactions were carried out in 0.12 M PB 60°C or 0.48 M PB 67°C at various DNA concentrations and times. The extent of reassociation was assayed by hydroxyapatite chromatography [92].

It should be noted that this variation of karyotypes was present in the population which was derived from a single cell shortly after its isolation. Thus, this variation appears to be an intrinsic property of the transformed line.

These karyotypic studies are being extended to the molecular level by analysis of DNA-DNA reassociation kinetics (fig. 2). If one isolates DNA from normal SHE cells, denatures it, and allows the sheared 250 nucleotide long DNA to reassociate at 60°C in 0.12 M phosphate buffer, at least three kinetic components are apparent. Approximately 5% of the genome consists of highly repetitive and/or fold-back sequences, with a $C_0t_{1/2} < 10^{-3}$; 35% behaves as moderately repetitive sequences with a $C_0t_{1/2}$ value of 2.5. These sequences consist of between 250 and 2,000 families repeated 500 to 5,000 times per genome; 60% of the genome behaves like non-repetitive sequences present only once per genome. Reassociation kinetics of tumor cell DNA, and various mixing experiments using ^{3}H-normal cell DNA and ^{14}C-tumor cell DNA driven by either a 40,000 excess of normal or tumor cell DNA indicate three things. First, some single-copy and repetitive sequences appear to be present in greater or lesser copies on the average in tumor cells than in normal cells. This is consistent with the chromosomal imbalance observed cytologically and lends molecular support to the balance hypothesis of carcinogenesis [29–31]. Second, certain single-copy DNA sequences appear to be missing from tumor cell DNA. Again, this is consistent with the cytological finding that certain chromosome segments are missing in this particular transformed cell line. Thirdly, no net loss in the kinds of mid-repetitive DNA sequences has been found. This is consistent with chromosomal interspersion (in addition to DNA interspersion) of mid-repetitive DNA sequences. In addition to the above, other probes of specific DNA sequences such as the RNA cistrons [32–34], poly (A) tracts [35–36] and the sequences adjacent to fold-back DNA [37–39] are currently being analyzed to explore the general relevance of these intriguing DNA changes to neoplastic transformation.

With respect to the problem of possible disturbances in transcription patterns, we have initially investigated the extent of new transcription by examining polysome-associated poly (A) mRNA, a cellular component whose role in protein synthesis is well defined [40–45]. The sizes of poly (A) mRNAs in SHE cells, 18 Cl 10, and BP6T appear similar, having a heterogeneous distribution with a maximum at approximately 2,000 nucleotides. The gross turnover rate, measured by the approach to steady-state labeling with ^{3}H-uridine in logarithmically growing cells [46–50] indicates a similar turnover rate in all three cell types, with a half-life approximately equal to the doubling time of the cells.

There is also no apparent gross qualitative change in the types of mRNAs produced in these three cell types. Trace amounts of steady-state ^{3}H-labeled polysomal poly (A) mRNA were hybridized to a vast excess of hamster organ DNA, and hybridization was assayed with RNase [51–56]. Under these conditions, the probability of forming a hybrid is related only to the concentration of the complementary sequence in the DNA. Repetitive sequence transcripts will hybridize at lower $C_o t$ values than will single-copy sequence transcripts. Fourteen to fifteen percent of the mRNA in all three cell types is complimentary to mid-repetitive DNA sequences, while the remainder of the hybridizable mRNA appears to be complementary to single-copy DNA sequences. No significant differences are apparent between normal and transformed cells [57].

Early investigation of gene transcription changes during transformation [58–62] were limited to changes in repetitive sequence transcripts [63–66]. Recent advances in hybridization technology, however, have made reactions with single-copy transcripts possible [67–79], also see McCarthy et al. and Davidson et al., this volume). Under conditions of mRNA excess, the sequences complimentary to the mRNA (mDNA) can be isolated from the remaining sequences (nmDNA) by hydroxyapatite chromatography. The hybridization conditions used are sensitive enough to detect down to one mRNA per cell. As shown in table 6, normal, transformed and tumor cell mRNAs saturate 0.64% of the ^{3}H-normal cell single-copy DNA probe, indicating a complexity of 2.5×10^7 nucleotides or approximately 12,500 genes of average size. Total poly(A) mRNA from 13 day hamster embryos saturates at least twice this amount of DNA. If one mixes normal and transformed cell poly(A) mRNAs, the percent saturation is a measure of the difference in sequence complexity of the two cell types. If all the mRNAs were different, the saturation value would be additive. If all the mRNAs are the same, no increase would be observed. When SHE and BP6T cell mRNAs

TABLE 6

The hybridization level at saturation of single-copy ^{3}H-SHE DNA (400,000 cpm, spec. act. 4.82×10^5 cpm/µg) in excess of poly (A)-mRNA[b].

Source of poly (A)-mRNA	DNA-RNA hybridization level (%)	
	At mRNA $C_o t$ value near 200[a]	At mRNA $C_o t$ value near 400[a]
SHE Cell	0.68 (194)	0.595 (403)
18 Cl 10	0.60 (180)	0.68 (370)
BP6T	0.63 (216)	0.65 (459)
BP6T + SHE	0.66 (220)	0.68 (450)

[a] The exact $C_o t$ values are in parentheses.
[b] The hybridization was carried out in 0.48 M phosphate buffer, 1 mM EDTA, 0.4% SLS, pH 6.8, 67%, up to 4 days at various mRNA levels. The mixture was analyzed by a hydroxyapatite column at 60°C. Single-stranded nucleic acid was eluted at 0.12 M phosphate and double-stranded nucleic acid was eluted in 0.4 M phosphate. DNA-DNA hybrid was distinguished from the DNA-RNA hybrid by the method of Galau et al. [89]. Poly A-mRNA was prepared from isolated polysomes [40,90] by two recyclings through oligo dT-columns according to a published method [91].

are mixed, no increase is observed (table 6). Taking into consideration the accuracy of the technique, no more than 4% of 12,500 mRNAs are different between normal and transformed cells [57].

Thus, the maintenance of malignant transformation can be accompanied by relatively small changes in the number of polysome-associated poly(A)mRNA sequences and massive gene activation need not be invoked to account for the transformed state. Recycling experiments, using isolated mDNA and nmDNA, are in progress to detect, isolate and identify any small amounts of new mRNA in transformed cells which have escaped notice by the simple additive experiments.

A direct demonstration of the involvement of DNA damage in neoplastic transformation

The research in the former section leads to the realization of the urgent need for a new experimental approach which can directly demonstrate the DNA involvement in neoplastic transformation. The analytical data obtained as described above cannot be used readily to show a direct, causative involvement of DNA damage in neoplastic transformation, since the activated carcinogens can react with many types of molecules inside the cells. Thus, the change in DNA sequences could well be a secondary effect and not an initial, crucial event.

In order to induce direct damage to DNA, we made use of the fact that 5-bromodeoxyuridine (BrdU) can only be incorporated into DNA and subsequently can be specifically activated by near UV irradiation, leading to DNA strand scission or cross-linkage with nearby protein [80].

2.5×10^{-6} M BrdU was introduced into an unsynchronized SHE cell culture for 24 h together with 2×10^{-4} M deoxycytidine to reduce toxicity. Under this condition, the BrdU was about 50% cytotoxic. The culture was irradiated with BLB fluorescent bulbs (~ 330 erg/mm²/sec); irradiation at this wavelength for 20 min is non-toxic to the cells. However, when cells have been incubated with BrdU, subsequent irradiation with the near UV light for 5 min resulted in a 90% reduction in cloning efficiency. No morphologically transformed clones could be found in untreated cultures, cultures receiving irradiation alone, or cultures treated with BrdU alone. Morphological clones appeared at a frequency of 1.2–5% per surviving colonies in cultures treated with BrdU and subsequently irradiated for 1–5 min. Experiments with cells grown in mass culture indicated that cells treated with BrdU and irradiated subsequently, developed the ability to grow in soft agar while the cells in control cultures did not. The tumorigenicity of the treated and irradiated cells is currently being examined.

In order to probe this sytem further pulse experiments were conducted in collaboration with Dr. T. Tsutsui using synchronized SHE cells. The synchronized cultures of SHE cells were obtained by the following procedure. After growing in 1% serum for a period of 36 h, cells were cultured in 3.2×10^{-4} M hydroxyurea with 10% serum for 12 h. After this period, the inhibitor was removed and the cells were released to grow in 10% serum. The cultures were greater than 80% synchronous as shown by the labeling index and thymidine incorporation. These cultures show no morphological transformation or gross chromosomal abnormalities. The synchronous cultures were then treated with BrdU (10^{-5} M) for 1 h pulse at hourly intervals at the start of the S-phase after being released from hydroxyurea. Five 1-h pulses were administered during the entire S phase and a 1-h pulse was administered before the S-phase. Irradiation alone or treatment of BrdU alone produced no morphological transformation in these synchronized cultures. However, when the 1-h pulse with BrdU was followed by a 5-min near UV irradiation period, morphological transformation at 1.5% frequency of the survival clones was observed for the pulses administered in the early S phase (1–2 h after hydroxyurea release) but were not observed for pulses in the late S phase or pulse in the G_1/S interphase. Under these conditions, the cytotoxic effects of the treatment were minimal, around 20%.

Somatic mutation induced by BrdU incorporation followed by near UV irradiation has been previously reported [81]. We have also observed somatic mutations at the HGPRT locus and the Na+/K+ ATPase locus of the SHE cells in asynchronous culture as induced by BrdU incorporation followed by irradiation. Currently, the inducement of somatic mutation of SHE cells in synchronized culture by this procedure is under active investigation.

Potentially, this same approach can be used to incorporate highly radioactive ^{3}H-thymidine into cells. Dr. M. Takii of our laboratory has undertaken this investigation, and has been able to induce somatic mutation of the HGPRT locus in the tumor cell line, BP6T, by the incorporation of ^{3}H-thymidine. The effects of the incorporation of ^{3}H-thymidine on in vitro morphological and neoplastic transformation is being investigated.

These results strongly suggest that direct damage to the DNA in the cell can initiate the process of neoplastic transformtion.

Concluding remarks

Four years ago, our laboratory presented an in-depth discussion of the mechanism of chemical carcinogenesis [82]. Based on the available data at that time [83–86], we compared the frequency of somatic mutation in Chinese hamster cells with the frequency of morphological transformation in Syrian hamster embryo cells, induced in each case by 2 μg/ml of N-acetoxy-N-2 fluoresylacetamide. This dosage of carcinogen resulted in about 30–40% cell death and about 0.08% mutation in 8-azaguanine resistance assay in Chinese hamster cells, and about 50% cell death and 2% morphological transformation in Syrian hamster cells [86]. Based on this comparison and other data in the literature, we stated: 'This observation suggests that the genetic target size for transformation appears to be considerably larger (25- to 50-fold) than that for mutation' [82].

This statement has been strongly confirmed by the work reported above, where a comparison of single-gene somatic mutation and morphological transformation can be made simultaneously in the same in vitro system (the Syrian hamster embryonic fibroblasts) perturbed by the same mutagen (MNNG) or carcinogen (B[a]P). The most recent work of Huberman, Mager and Sachs [87] shows that a ratio of about 20 can be observed in a comparison between morphological transformation (clonal assay) and somatic mutation (ouabain resistance assay) in Syrian hamster cells when induced by B(a)P 7,8-transdihydrodiols. This ratio is slightly lower than our values of somewhere between 50–100 based on our observations.

In the previous discussion, we have also commented on the possibility that the 'reversion' rate of transformed cells (at least for the virally transformed cells [30]), may be quite high in comparison to the 'reversion' rate of single-gene mutated cells. At present, while reversion of chemically transformed cells has been observed (see chapter by W. Benedict), there is a real need for information about the frequency of 'reversion' of chemically transformed cells. An extensive comparison of 'reversion frequency' between single-gene mutation and neoplastic transformation is highly warranted.

In explaining the high forward rate (or large gene target size) of morphological (possibly neoplastic) transformation as opposed to that of single-gene mutations, we postulate that a battery of genes and their proper functioning are involved in the cellular machinery of 'stimuli-response-regulation', which is a multi-channel network [82]. Damage to any one of these genes or disturbance of any part of their operation in the network may cause the malfunction of the 'stimuli-response-regulation' machinery, and these malfunctions can have a cascading effect felt throughout the network. This postulate can also be used to explain a high 'reversal rate' to normalcy (or pseudo-normalcy), if this is indeed what occurs. We stated, 'the reversion of the transformed cell, due to the repair of the damaged process in this multi-channel network, can be achieved by (a) restoration of the original path, (b) bypassing the damaged step in the original path, or (c) building a totally new path. The reverted cell, while functioning normally, may not be the same in all respects as the original cell'. Similarly, owing to the large number of genes vulnerable as targets, the transformed cells themselves are different in molecular details although they have similar overall physiological processes.

In support of this postulate, we cited evidence [82] that the transformed cells tend to lose their sensitivity to environmental controls, such as hormones, serum factors, and cell-contact inhibition. The membrane properties of the transformed cells are different from those of non-transformed cells. In addition, we pointed out the importance of the finding about the individuality of the resulting transformed cells [82].

Upon losing the sensors to the environmental stimuli or due to the malfunctioning of the response mechanism to the stimuli, the transformed cells which originated from a metazoan gradually acquire the characteristics and lifestyle of single-celled organisms. This point has also been emphasized in the chapter of Dr. T.T. Puck in this volume. The selective advantage of a single-celled organism in survival and growth depends on *multiplication* and adaptability, areas in which the transformed cells excel. Malfunction of the regulation process in the 'stimuli-response-regulation' machinery evidently can lead to genetic instability, readily observable in the abnormalities and instability of the karyotypes. A high level of spontaneous single-gene mutations could be predicted from this situation. Additionally, this situation may lead to a relatively high death rate among the individual cells, at the same time that immortality has been conferred upon the population as a whole, a situation similar to that in a population of bacteria. These predictions should be investigated and documented. The questions of genetic instability and clonal evolution of tumor cell populations were extensively discussed by Nowell [88]. He proposed that 'most neoplasms arise from a single-cell of origin, and tumor progression results from acquired genetic variability within the original clone allowing sequential selection of more aggressive sublines. Tumor cell populations are apparently more genetically unstable than normal cell populations. The acquired genetic instability and associated selection process, most readily recognized cytogenetically, result in advanced human malignancies being highly individual karotypically and biologically.' We are in agreement with the assessment of this aspect of neoplasia.

In returning to the concept stated in the first section of this chapter, there remains the enigma of how a particular pattern of the 'stimuli-response-regulation' network becomes heritable and survives many cell divisions. Thus, according to Chart I, the DNA-regulator complex, or the genetic apparatus, is heritable through cell divisions. The biochemists are searching for a type of DNA-regulatory molecule interaction which is tissue specific and once established can become semi-permanent through cell divisions. From the biophysical viewpoint, our postulate would be that perhaps the hereditary information of the genetic apparatus in mammalian (or other eukaryotic) cells is three-dimensional and not one-dimensional, as in phage and possibly in bacterium. In other words, the relative spatial relationship and arrangement among the genes and their regulators are *informational* and *heritable*. We know that at least at the mitotic phase of the cell cycle, the arrangement of genes is two-dimensional in nature, i.e. genes are grouped in various chromosomes for their assembly and precise partition. The question is whether or not the interphase chromatin in the cell nucleus has a definite three-dimensional structure which is related to its function. This hypothesis proposes that, indeed, the genetic apparatus during interphase does have a definite three-dimensional structure specific for its function, and is semi-permanent through many cell divisions. A change in the spatial arrangement of the genetic apparatus in the nucleus, with or without a necessary change in the DNA primary sequence, can lead to different types of function and this change can be heritable. This change can arise as a result of a normal process, as in differentiation, or it can be a perturbation as in carcinogenesis. This is a bold and challenging hypothesis for the investigation of the molecular biology of the mammalian genetic apparatus.

References

[1] Y. Berwald and L. Sachs, In vitro transformation of normal cells to tumor cells by carcinogenic hydrocarbons. J. Natl. Cancer Inst. 35, 641 (1965).
[2] E. Huberman and L. Sachs, Cell susceptibility to transformation and cytotoxicity by the carcinogenic hydrocarbon, benzo(a)pyrene. Proc. Natl. Acad. Sci. U.S.A. 56, 1123 (1966).
[3] J.A. DiPaolo and P.J. Donovan, Properties of Syrian hamster cells transformed in the presence of carcinogenic hydrocarbons. Exp. Cell Res. 48, 361 (1967).
[4] J.A. DiPaolo, P. Donovan and R. Nelson, Quantitative studies of in vitro transformation by chemical carcinogens. J. Nat. Cancer Inst 42, 867 (1969).
[5] J.A. DiPaolo, P.J. Donovan and R.L. Nelson, In vitro transformation of hamster cells by polycyclic hydrocarbons: factors influencing the number of cells transformed. Nature New Biology 230, 240 (1971).
[6] J.A. DiPaolo, N.C. Popescu and R.L. Nelson, Chromosomal banding patterns and in vitro transformation of Syrian hamster cells. Cancer Res. 33, 3250 (1973).
[7] W.F. Benedict, Rapid production of numerical and structural abnormalities in hamster cells by transforming doses of polycyclic hydrocarbons. J. Natl. Cancer Inst. 49, 585 (1972).
[8] W.F. Benedict, N. Rucker, C. Mark and R. Kouri, Correlation between the balance of specific chromosomes and the expression of malignancy in hamster cells. J. Natl. Cancer Inst. 54, 157 (1975).
[9] T. Kakunaga and J. Kamahora, Properties of hamster embryonic cells transformed by 4-nitroquinoline-1-oxide in vitro and their correlations with the malignant properties of the cells. Biken J. 11, 313 (1968).
[10] T. Kakunaga and J. Kamahora, Analytical studies in the process of malignant transformation of hamster embryonic cells in culture with 4-nitroquinoline-1-oxide. Symposia Cell Chem. 20, 135 (1969).
[11] L.M. Schechtman, J.C. Barrett, R.K. Moyzis and P.O.P. Ts'o, In vitro transformation of Syrian hamster embryonic fibroblasts in the absence of a feeder layer. Submitted to Cancer Research (1977).
[12] F. Krasin, S. Person, R.D. Ley and F. Hutchinson, DNA crosslinks, single-strand breaks and effects on bacteriophage T4 survival from tritium decay of [2-^{3}H]adenine, [8-^{3}H]adenine and [8-^{3}H]guanine. J. Mol. Biol. 101, 197 (1976).
[13] F. Krasin, S. Person, W. Snipes and B. Benson, Local effects for [5-^{3}H] decay: production of a chemical product with possible mutagenic consequences. J. Mol. Biol. 105, 445 (1976).
[14] F. Krasin, S. Person and W. Snipes, DNA strand breaks from tritium decay: a local effect for cytosine-6-^{3}H. Int. J. Radiat. Biol. 23, 417 (1973).
[15] A.L. Koch, A distributional basis for the variation in killing efficiencies by different tritiated compounds incorporated in E. coli. Radiation Res. 24, 398 (1965).

[16] P.N. Rosenthal and M.S. Fox, Effects of disintegration of incorporated ^{3}H and ^{32}P on the physical and biological properties of DNA. J. Mol. Biol. 54, 441 (1970).
[17] Donald E. Wimber, Effects of intracellular irradiation with tritium. Adv. Rad. Res. 1, 85 (1964).
[18] F. Funk, S. Person and R.C. Bockrath, Jr., The mechanism of inactivation of T_4 bacteriophage by tritium decay. Biophysical J. 8, 1037 (1968).
[19] S. Person, Comparative killing efficiencies for decay of tritiated compounds incorporated into *E. coli*. Biophysical J. 3, 183 (1963).
[20] L. Siminovitch, On the nature of heritable variation in cultured somatic cells. Cell 7, 1 (1976).
[21] L. Foulds, Neoplastic development. Academic Press, New York (1969).
[22] J. Unkeless, A. Tobia, L. Ossowski, J.P. Quigley, D.B. Rifbin and E. Reich, An enzymatic function associated with transformation of fibroblasts by oncogenic viruses I. Chick embryo fibroblast cultures transformed by avian RNA tumor viruses. J. Exp. Med. 137, 85 (1973).
[23] L. Ossowski, J. Unkeless, A. Tobia, J.P. Quigley, D.B. Rifbin and E. Reich, An enzymatic function associated with transformation of fibroblasts by oncogenic viruses II. Mammalian fibroblast cultures transformed by DNA and RNA tumor viruses. J. Exp. Med. 137, 112 (1973).
[24] J. Unkeless, K. Dano, G. Kellerman and E. Rich, Fibrinolysis associated with oncogenic transformation: partial purification and characterization of the cell factor, a plasminogen activator. J. Biol. Chem. 249, 4295 (1974).
[25] J. Quigley, L. Ossowski and E. Reich, Plasminogen, the serum proenzyme activated by factors from cells transformed by oncogenic viruses. J. Biol. Chem. 249, 4306 (1974).
[26] L. Ossowski, J. Quigley and E. Reich, Fibrinolysis associated with oncogenic transformation: morphological correlates. J. Biol. Chem. 249, 4312 (1974).
[27] J.C. Barrett, B.D. Crawford and P.O.P. Ts'o, Quantitation of fibrinolytic activity of Syrian hamster fibroblasts using ^{3}H-fibrinogen prepared by reductive alkylation. Submitted to Cancer Res. (1977).
[28] P. Jones, W. Benedict, S. Strickland and E. Reich, Fibrin overlay methods for the detection of single transformed cells and colonies of transformed cells. Cell 5, 323 (1975).
[29] T. Yamamoto, Z. Rabinowitz and L. Sachs, Identification of the chromosomes that control malignancy. Nature New Biol. 243, 247 (1973).
[30] T. Yamamoto, M. Hayashi, Z. Rabinowitz and L. Sachs, Chromosomal control of malignancy in tumors from cells transformed by polyoma virus. Int. J. Cancer 11, 555 (1973).
[31] W.F. Benedict, N. Rucker, C. Mark and R. Kouri, Correlation between the balance of specific chromosomes and the expression of malignancy in hamster cells. J. Natl. Cancer Inst. 54, 157 (1975).
[32] M.L. Birnstiel, M. Chipchase and J. Speirs, The ribosomal RNA cistrons, Progr. Nucl. Acid Res. Mol. Biol. 11, 351 (1971).
[33] D.E. Wimber, D.M. Steffensen, Localization of gene function. Ann. Rev. Genetics 7, 205 (1973).
[34] T.C. Hsu, S.E. Spirito and M.L. Pardue, Distribution of 18S and 28S ribosomal genes in mammalian genomes. Chromosoma 53, 25 (1975).
[35] A. Shenkin and R.N. Burdon, Deoxyadenylate-rich and deoxyguanylate-rich regions in mammalian DNA. JMB 85, 19 (1974).
[36] J.O. Bishop, M. Rosbash and D. Evans, Polynucleotide sequences in eukaryotic DNA and RNA that form ribonuclease-resistant complexes with polyuridylic acid. J.M.B. 85, 75 (1974).
[37] C.W. Schmid, J.E. Manning and N. Davidson, Inverted repeat sequences in the *Drosophila* genome. Cell 5, 159 (1975).
[38] S. Perlman, C. Phillips and J.O. Bishop, A study of foldback DNA. Cell 8, 33 (1976).
[39] W.G. Flamm, Highly repetitive sequences of DNA in chromosomes. Int. Rev. Cytol. 32, 1 (1972).
[40] G. Brawerman, Eukaryotic messenger RNA. Ann. Rev. Biochem. 43, 621 (1974).

[41] B. Lewin, Units of transcription and translation: the relationship between heterogeneous nuclear RNA and messenger RNA. Cell 4, 11 (1975).
[42] B. Lewin, Units of transcription and translation: sequence components of heterogeneous nuclear RNA and messenger RNA. Cell 4, 77 (1975).
[43] J.R. Greenberg, Messenger RNA metabolism of animal cells. J. Cell Biol. 64, 269 (1975).
[44] M. Rosbash, M.S. Campo and K.S. Gummerson, Conservation of cytoplasmic poly(A)-containing RNA in mouse and rat. Nature 258, 682 (1975).
[45] M. Noll and M.M. Burger, Membrane-bound and free polysomes in transformed and untransformed fibroblast cells. JMB 90, 215 (1974).
[46] J.R. Greenberg, High stability of messenger RNA in growing cultured cells. Nature 240, 102 (1972).
[47] R.H. Singer and S. Penman, Messenger RNA in HeLa cells: kinetics of formation and decay. JMB 78, 321 (1973).
[48] R.P. Perry and D.E. Kelley, Messenger RNA turnover in mouse L cells. JMB 79, 681 (1973).
[49] B.P. Brandhorst and E.H. McConkey, Relationship between nuclear and cytoplasmic poly-(adenylic acid). PNAS 72, 3580 (1975).
[50] T.E. Sensky, M.E. Haines and K.R. Rees, The half-life of polyadenylated polysomal RNA from normal and transformed cells in monolayer culture. BBA 407, 430 (1975).
[51] J.O. Bishop and G.P. Smith, The determination of RNA homogeneity by molecular hybridization. Cell 3, 341 (1974).
[52] J.R. Greenberg and R.P. Perry, Hybridization properties of DNA sequences directing the synthesis of mRNA and hnRNA. J. Cell Biol. 50, 774 (1971).
[53] R.B. Goldberg, G.A. Galau, R.J. Britten and E.H. Davidson, Nonrepetitive DNA sequence representation in sea urchin embryo messenger RNA. PNAS 70, 3516 (1973).
[54] W.H. Klein, W. Murphy, G. Attardi, R.J. Britten and E.H. Davidson, Distribution of repetitive and non-repetitive sequence transcripts in HeLa mRNA. PNAS 71, 1785 (1974).
[55] M.S. Campo and J.O. Bishop, Two classes of messenger RNA in cultured rat cells: repetitive sequence transcripts and unique sequence transcripts. JMB 90, 649 (1974).
[56] E.H. Davidson, B.R. Hough, W.H. Klein and R.J. Britten, Structural genes adjacent to interspersed repetitive DNA sequences. Cell 4, 217 (1975).
[57] R.K. Moyzis, J. Bonnet, M. Melville, P.O.P. Ts'o and M. Zajac, Changes in DNA sequence organization and gene expression in neoplastic transformation. J. Cell Biol. 70, 324a (1976).
[58] V.P. Chiarugi, Changes in nuclear RNA in hepatomas as revealed by RNA/DNA hybridization. BBA 179, 129 (1969).
[59] R.B. Church, S.W. Luther, B.J. McCarthy, RNA synthesis in taper hepatoma and mouse liver cells. BBA 190, 30 (1969).
[60] P.E. Neuman and P.H. Henry, Ribonucleic acid, deoxyribonucleic acid hybridization and hybridization-competition studies of the rapidly labeled ribonucleic acid from normal and chronic lymphocyte leukemia lymphocytes. Biochemistry 8, 275 (1969).
[61] R.W. Turkington and D.J. Self, New species of hybridizable nuclear RNA in breast cancer cells. Cancer Res. 30, 1833 (1970).
[62] A.S. Levine, M.N. Oxman, H.M. Eliot and P.H. Henry, New species of rapidly hybridizing RNA in contact-inhibited as well as transformed hamster cell lines. Cancer Res. 32, 506 (1972).
[63] B. McCarthy and R.B. Church, The specificity of molecular hybridization reactions. Ann. Rev. Biochem. 39, 131 (1970).
[64] D.E. Kennel, Principles and practices of nucleic acid hybridization. Prog. Nuc. Acid Res. 11, 259 (1971).
[65] P. Szabo, R. Elder and O. Uhlenbeck, The kinetics of in situ hybridization. Nucleic Acids Res. 2, 647 (1975).

[66] M.L. Birnstiel, B.H. Sells and I.F. Purdom, Kinetic complexity of RNA molecules. JMB 63, 21 (1972).
[67] J.O. Bishop, The gene numbers game. Cell 2, 81 (1974).
[68] J.O. Bishop, J.G. Morton, M. Rosbah and M. Richardson, Three abundance classes in HeLa cell messenger RNA. Nature 250, 199 (1974).
[69] G.U. Ryffel and B.J. McCarthy, Complexity of cytoplasmic RNA in different mouse tissues measured by hybridization of polyadenylated RNA to complementary DNA. Biochemistry 14, 1379 (1975).
[70] G.U. Ryffel and B.J. McCarthy, Polyadenylated RNA complementary to repetitive DNA in mouse L-cells. Biochemistry 14, 1385 (1975).
[71] J.C. Williams and S. Penman, The messenger RNA sequences in growing and resting mouse fibroblasts. Cell 6, 197 (1975).
[72] R. Axel, P. Feigelson and G. Schutz, Analysis of the complexity and diversity of mRNA from chicken liver and oviduct. Cell 7, 247 (1976).
[73] M.J. Getz, P.K. Elder, E.W. Benz, Jr., R.E. Stephens and H.L. Moses, Effect of cell proliferation on levels and diversity of poly(A),containing mRNA. Cell 7, 255 (1976).
[74] G.A. Galau, W.H. Klein, M.M. Davis, B.J. Wold, R.J. Britten and E.H. Davidson, Structural gene sets active in embryos and adult tissues of sea urchin. Cell 7, 487 (1976).
[75] J.A. Bantle, W.E. Hahn, Complexity and characterization of polyadenylated RNA in the mouse brain. Cell 8, 139 (1976).
[76] B.D. Young, G.D. Birnie and J. Paul, Complexity and specificity of polysomal poly (A+) RNA in mouse tissues. Biochemistry 15, 2823 (1976).
[77] L.J. Grady and W.P. Campbell, Non-repetitive DNA transcription in mouse cells grown in tissue culture. Nature New Biol. 243, 195 (1973).
[78] L.J. Grady and W.P. Campbell, Non-repetitive DNA transcripts in nuclei and polysomes of polyoma-transformed and non-transformed mouse cells. Nature 254, 356 (1975).
[79] L.J. Grady and W.P. Campbell, Transcription of the repetitive DNA sequences in polyoma-transformed and nontransformed mouse cells in culture. Cancer Res. 35, 1559 (1975).
[80] F. Hutchison, The lesions produced by ultraviolet light in DNA containing 5-bromouracil. Quart. Rev. Biophys. 6, 201 (1973).
[81] E.H.Y. Chu, N.C. Sun and C.C. Chang, Induction of auxotrophic mutation by treatment of Chinese hamster cells with 5-bromodeoxyuridine and black light. Proc. Nat. Acad. Sci. U.S.A. 69, 3459 (1972).
[82] P.O.P. Ts'o, W.J. Caspary, B.I. Cohen, J.C. Leavitt, S.A. Lesko, Jr., R.J. Lorentzen and L.M. Schechtman, Basic mechanisms in polycyclic hydrocarbon carcinogenesis, in: Chemical Carcinogenesis (ed. P.O.P. Ts'o and J.A. DiPaolo) Marcel Dekker, Inc. New York, 1974, p. 113.
[83] E. Huberman, L. Aspiras, C. Heidelberger, P.L. Grover and P. Sims, Mutagenicity to mammalian cells of epoxides and other derivatives of polycyclic hydrocarbons. Proc. Nat. Acad. Sci. 68, 3195 (1971).
[84] P.L. Grover, P. Sims and E. Huberman, In vitro transformation of rodent cells by K-region derivatives of polycyclic hydrocarbons. Proc. Nat. Acad. Sci. 68, 1098 (1971).
[85] J.A. DiPaolo, P.J. Donovan and R.L. Nelson, Quantitative studies of in vitro transformation by chemical carcinogens. J. Nat. Cancer Inst. 42, 867 (1969).
[86] E. Huberman, P.J. Donovan and J.A. DiPaolo, Mutation and transformation of cultured mammalian cells by N-acetoxy-N-2-fluorenylacetamide. J. Nat. Cancer Inst. 48, 837 (1972).
[87] E. Huberman, R. Mager and L. Sachs, Mutagenesis and transformation of normal cells by chemical carcinogens. Nature 264, 360 (1976).
[88] P.C. Nowell, The clonal evolution of tumor cell populations. Science 194, 23 (1976).
[89] G.A. Galau, R.J. Britten and E.H. Davidson, A measurement of the sequence complexity of polysomal messenger RNA in sea urchin embryos. Cell 2, 9 (1974).

[90] S. Penman, RNA metabolism in the HeLa cell nucleus. J. Mol. Biol. 17, 117 (1966).
[91] H. Aviv and P. Leder, Purification of biologically active globin messenger RNA by chromatography on oligothymidylic acid-cellulose. Proc. Natl. Acad. Sci. 69, 1408 (1972).
[92] R.J. Britten, D.E. Graham and B.R. Neufeld, Analysis of repeating DNA sequences by reassociation, Methods in Enzymology XXIX (ed. Lawrence Grossman and Kivie Moldave, Nucleic Acids and Protein Synthesis, Part E, p. 363 (1974).

Chapter 21

CELLULAR MODIFICATIONS OF TRANSFECTING SV40 DNA

DANIEL NATHANS and CHING-JUH LAI

Department of Microbiology, Johns Hopkins University School of Medicine, Baltimore, Md. 21205, U.S.A.

Virus infection leads to the introduction of a new nucleic acid molecule into a cell, often followed by expression of the viral genome and consequent development of new virus particles. In addition to expressing its encoded developmental program, the viral genome is also subject to cellular enzymes which modify nucleic acids; for example, enzymes concerned with recombination and repair of DNA. In the case of bacteria, many of these enzyme systems were first detected and characterized by the use of relatively simple bacteriophage DNA's, suitably marked genetically and/or biochemically. With the availability of mutants and fairly detailed physical maps of the genomes of certain animal viruses, it has been possible to extend this approach to animal cells. In this report we describe experiments with Simian Virus 40 (SV40) which have revealed DNA modifying reactions in infected monkey cells analogous to those described in bacteria. Much of the data to be presented has already been reported, though often in a different context.

Cyclization of linear SV40 DNA

It has been known for some time that nicked SV40 DNA circles are fully infectious, thus implying that phosphodiester bond breaks in one strand are efficiently sealed in animal cells. In addition, full length linear SV40 DNA molecules with cohesive single-stranded termini are infectious [1]. When the termini result from cleavage by endo $R \cdot EcoR_1$, which generates 5' tails with the sequence AATT [2], the efficiency of infection is about 10% that of form I SV40 DNA [1]. Since the progeny DNA contain the original R_1 site, one can infer that the linear molecules cyclized via their cohesive ends, followed by ligase sealing of the strand discontinuities. Similar linear molecules of SV40 DNA of less than unit length generate deleted genomes within the cell [3,4], and the resulting deletion mutants can be cloned by complementation with helper virus [5,6].

A more general and mechanistically more obscure DNA cyclization reaction has also been found in SV40-infected cells. Linear molecules of SV40 DNA containing even ends or single-stranded tails can cyclize by an intramolecular recombination event near their termini, thus producing terminal deletions of DNA [3,7]. Based on physical mapping of such 'extended deletions' it appears that the recombination event does not require extensive base pairing, i.e., it is 'illegitimate', since the limits of deletions are highly variable.

Although the mechanism of recombinational cyclization is not known, a plausible hypothesis is shown in fig. 1: exonuclease creates single-stranded termini; limited base pairing between single-stranded tails leads to cyclization (possibly stabilized by specific proteins); and finally, enzymatic trimming of single-stranded tails, gap filling and ligation result in a covalently closed short genome. Whether any of the enzymes involved is virus-coded is not yet known. This reaction has allowed the construction and cloning of a large number of deletion mutants in all segments of the genome of SV40 [3,7].

Repair of partial heteroduplex molecules

As in the case of bacteria [8], animal cells can be transfected by partial heteroduplexes of viral DNA consisting of a single-stranded circle and a complementary single-stranded fragment [9]. Such SV40 heteroduplexes are infectious for monkey cells, thus allowing application of marker rescue techniques to mapping ts mutants of SV40 [9,10,11]. Based on these facts, one can infer that the partial heteroduplex is converted to a fully duplex structure by cellular enzymes (fig. 2). Since the efficiency of infection by partial heteroduplexes containing a large DNA fragment is up to 10% that of circular SV40 DNA, it appears that this conversion reaction is rather efficient. Whether in the case of wt/ts heteroduplexes, correction of mismatched bases occurs (prior to DNA replication),

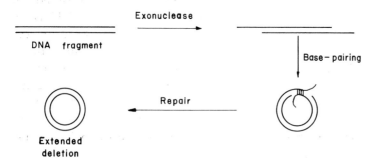

Fig. 1. Cyclization of linear SV40 DNA within infected cells by 'illegitimate' type recombination [3].

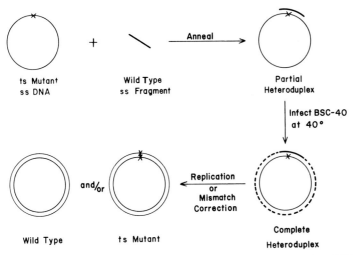

Fig. 2. Inferred modifications of transfecting heteroduplex molecules of SV40 DNA. X indicates a mutational site.

as illustrated in fig. 2, is not clear from the above observations alone. However, as indicated below, other experiments suggest that this is the case.

Evidence for correction of mismatched bases in SV40 DNA

The marker rescue experiments diagrammed in fig. 2 raised the possibility that in SV40 infected cells mismatched bases may be corrected with high efficiency. To test this hypothesis monkey cells were infected at non-permissive temperature with complete SV40 heteroduplex DNA's, each strand of which was derived from the DNA of a different ts mutant in the same complementation group [10,12] (table 1). In this instance, replication cannot 'correct' the mutational defect. As seen in the table, many such heteroduplex molecules are infectious, including those in which the strands were derived from two different non-complementing tsA mutants, each of which is defective in initiation of viral DNA replication at 40° (B, C, and D mutants are not defective in DNA replication). Based on the specific infectivity of these intracistronic heteroduplexes, it appears that the correction process is again rather efficient.

Several of the ts/ts heteroduplex preparations indicated in table 1 had both mutations within a small, localized segment of the SV40 genome (see legend to table 1). Some of these were not infectious at high temperature, suggesting that the two mutational sites were corrected by the same excision-repair event, thus generating only a mutant genotype (fig. 3). In other instances wild type SV40 was generated, and from the approximate distance between mutations in this category we infer that two mismatched base pairs separated by more than

TABLE 1

Infectivity of ts/ts heteroduplexes (pfu per dish). Heteroduplex molecules were formed by melting and reannealing singly nicked SV40 DNA Form II from each of the ts mutant pairs noted in the table. Results are expressed as number of plaques (pfu) in each of two dishes of BSC-40 monkey cells infected with 20 ng of DNA at 40°. (Data from [11].) All the A mutants used map within the Hin-I segment of the SV40 genome (about 230 base pairs); all D mutants map within Hin-E (about 430 base pairs); all C mutants map within Hin-J (about 220 base pairs).

tsA				tsA		
	207	209	239	241	255	276
207	1,0	71,58	0,0	0,0	52,59	44,66
209		0,0	74,58	69,64	34,35	37,60
239			0,0	1,0	72,79	45,39
241				0,0	81,83	58,63
255					1,0	33,16
276						0,0

tsD				tsD			
	101	202	222	238	263	270	275
101	0,0	28,29	20,16	24,28	0,0	11,9	30,20
202		0,0	16,17	12,18	15,6	9,16	21,33
222			0,0	0,0	22,13	7,10	0,0
238				0,0	17,29	40,37	0,0
263					0,0	10,15	16,34
270						0,0	31,21
275							0,0

ts	B228	B201	B221	C240	C260
B228	0,0	80,82	65,60	91,100	60,49
B201		0,0	22,27	30,56	47,75
B221			0,0	38,35	29,50
C240				0	44,31
C260					0

about 200 nucleotides can be corrected independently. The minimal distance still allowing independent correction is not known.

Although the experiments cited in table 1 were done with mixtures of hetero- and homoduplexes, the basic observations have been repeated with purified ts/ts heteroduplexes constructed as diagrammed in fig. 4 and isolated as open circular Form II DNA by agarose gel electrophoresis. Results with one tsA/tsA heteroduplex and one tsB/tsB heteroduplex prepared in this way are given in table 2. As seen in the table, the A209/A255 heteroduplex and the B218/B228

a. Two separated mis-matched base pairs

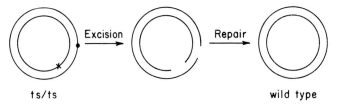

b. Two close mis-matched base pairs

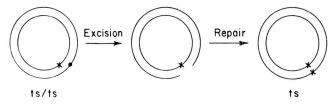

Fig. 3. Scheme for correction of mismatched bases in ts/ts heteroduplex molecules by excision-repair. X and ● indicate mutational sites.

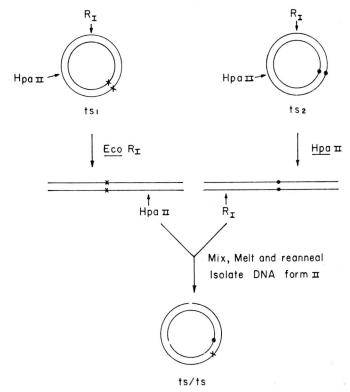

Fig. 4. Preparation of pure ts/ts heteroduplexes. X and ● indicate mutational sites. Sites cut by endo R·$EcoR_I$ and Hpa II are shown at the top.

TABLE 2

Specific infectivity of purified ts/ts heteroduplexes at 32° and 40°. Heteroduplex molecules were constructed and purified as diagrammed in fig. 4. Results are expressed as number of plaques (pfu) in each of three dishes of BSC-40 monkey cells infected with 5, 0.5, or 0.05 ng of DNA at 32° (permissive for ts mutants) or at 40° (non-permissive for ts mutants). A209 and A255 map in the *Hin*-I segment of the SV40 genome (about 230 base pairs); B218 and B228 map in the *Hin*-K segment (about 200 base pairs).

Heteroduplex		pfu per dish			pfu/ng
		5 ng	0.5 ng	0.05 ng	
A209/A255	40°	10,20,15	9,2,2	0,0,1	3
	32°		19,24,18	4,3,1	40
B218/B228	40°	16,16,23	0,2,2	1,1,1	3.6
	32°		41,20,24	3,0,4	56

heteroduplex each had about the same specific infectivities at 40° and 32°. Since A mutants are defective in viral DNA replication at 40°, whereas B mutants are not, these results again suggest that correction is not dependent on SV40 DNA replication. Further, since A mutants do not express early or late viral functions at non-permissive temperature, it is likely that correction is dependent on cellular enzymes rather than virus-coded enzymes.

In an attempt to exclude the possibility that tsA/tsA heteroduplexes are leaky enough to lead to some viral DNA replication followed by high frequency recombination between daughter molecules to generate wild type virus, we analyzed the progeny from the 32° A209/A255 heteroduplex infection for ts and wild type virus. At 32° there should be little or no selection for wild type progeny. Three of 25 plaques selected randomly at 32° contained only wild type virus on subsequent testing. The presence of plaques of this class is best explained by correction of both pairs of mismatched bases in the infecting heteroduplex (as illustrated in fig. 3a). rather than recombination.

Conclusion

The genome of SV40 is now sufficiently understood to allow its use as a rather precise probe for reactions in mammalian cells that modify DNA in various ways. A few examples, studied in our own laboratory, have been discussed in this report: cyclization of linear DNA, repair of partial heteroduplex molecules, and correction of mismatched bases. The cyclization reaction has been exploited to construct deletion mutants, and mismatch correction provides a means for constructing point mutants throughout the genome. Extension of these studies should help characterize the cellular reactions involved in these modifications of DNA and may also aid in defining enzymatic blocks in mutant human cells defective in these functions.

Acknowledgements

The authors' research summarized in this report has been supported by the American Cancer Society (VC-132A) and the National Cancer Institute (CA16519).

References

[1] J.E. Mertz and R.W. Davis, Cleavage of DNA by R_1 restriction endonuclease generates cohesive ends. Proc. Nat. Acad. Sci. 69, 3370 (1972).

[2] J. Hedgpeth, H.M. Goodman and H.W. Boyer, DNA nucleotide sequence restricted by the RI endonuclease. Proc. Nat. Acad. Sci. 69, 3448 (1972).

[3] C.-J. Lai and D. Nathans, Deletion mutants of Simian virus 40 generated by enzymatic excision of DNA segments from the viral genome. J. Mol. Biol. 89, 179 (1974).

[4] J.E. Mertz, J. Carbon, M. Herzberg, R.W. Davis and P. Berg, Isolation and characterization of individual clones of Simian virus 40 mutants containing deletions, duplications and insertions in their DNA. Cold Spring Harbor Symp. Quant. Biol. 39, 69 (1974).

[5] W.W. Brockman and D. Nathans, The isolation of Simian virus 40 variants with specifically altered genomes. Proc. Nat. Acad. Sci. 71, 942 (1974).

[6] J.E. Mertz and P. Berg, Defective Simian virus 40 genomes: Isolation and growth of individual clones. Virology 62, 112 (1974).

[7] J. Carbon, T.E. Shenk and P. Berg, Biochemical procedure for production of small deletions in simian virus 40 DNA. Proc. Nat. Acad. Sci. 72, 1392 (1975).

[8] C.A. Hutchison III and M.H. Edgell, Genetic assay for small fragments of bacteriophage øX174. J. Virol. 8, 181 (1971).

[9] C.-J. Lai and D. Nathans, Mapping temperature-sensitive mutants of Simian virus 40: Rescue of mutants by fragments of viral DNA. Virology 60, 466 (1974).

[10] C.-J. Lai and D. Nathans, A map of temperature-sensitive mutants of Simian virus 40. Virology 66, 70 (1975).

[11] N. Mantei, H.W. Boyer and H.M. Goodman, Mapping Simian virus 40 mutants by construction of partial heterozygotes. J. Virol. 16, 754 (1975).

[12] J.Y. Chou and R.G. Martin, Complementation analysis of Simian virus 40 mutants. J. Virol. 13, 1101 (1974).

Chapter 22

THE TRANSFORMING GENE OF AVIAN SARCOMA VIRUS

J.M. BISHOP, D. STEHELIN, J. TAL, D. FUJITA, D. SPECTOR,
D. ROULLAND-DUSSOIX, T. PADGETT and H.E. VARMUS

University of California. Department of Microbiology. San Francisco, California 94143, U.S.A.

Introduction

Avian sarcoma viruses (ASV) are C-type RNA tumor viruses capable of inducing sarcomas in animals and transforming cultured fibroblasts to a neoplastic state [1]. Several features make these viruses particularly valuable for the study of viral oncogenesis. First, the viruses arose in a natural setting and can be found as oncogenic agents in the field. Second, strains of ASV are available which require no helper virus to replicate. Finally, both conditional mutants and deletions of the transforming function in ASV have been isolated. At present, no other species of RNA sarcoma virus offers this useful set of properties.

The transforming function of ASV is encoded in a single gene (denoted *onc*; see [2] whose product is required for cellular transformation [3]. Three other viral genes have also been identified [2]: *env*, the glycoprotein(s) of the viral envelope; *pol*, the viral reverse transcriptase; and *gag*, which encodes a polyprotein containing the structural proteins found in the interior of the virus. These genes, together with *onc*, may account for the entire coding potential of the ASV genome [2].

ASV genes are not inevitably expressed in infected cells. When a culture of mammalian cells is infected with ASV, only a small proportion (ca. 10^{-6}) of the cells are transformed, yet most or all of the untransformed cells contain the entire ASV genome [4] integrated into the chromosomal DNA of the host [5]. The cells which are transformed do not usually produce virus and may or may not contain viral antigens, yet these cells also contain the ASV genome in an integrated state [5]. Finally, ASV-transformed mammalian cells can segregate phenotypically normal cells [6] which retain the entire ASV genome [7] as an integrated DNA provirus [8].

In order to analyze the mechanisms by which the expression of ASV genes are modulated, we have developed a procedure to prepare radioactive DNA

The Molecular Biology of the Mammalian Genetic Apparatus:
edited by P. Ts'o © 1977, Elsevier/North-Holland Biomedical Press

(cDNA) complementary to specific regions of the ASV genome. We began by preparing DNA (cDNA$_{sarc}$) complementary to part or all of the nucleotide sequence which encodes *onc* [9] and have now also prepared cDNA (cDNA$_{gp}$) for *env*. In this communication, we review our experience in the preparation of specific cDNAs and illustrate how the use of these reagents can provide information on the origin and nature of viral genes.

Preparation of specific cDNAs

Our procedure to prepare specific cDNAs exploits the existence of deletion mutants which affect single ASV genes. For example, propagation of ASV gives rise to transformation-defective (td) variants in which approximately 15% of the viral genome has been deleted (fig. 1). The td strains of ASV replicate normally but cannot induce sarcomas in animals or transform fibroblasts in culture [1]. Moreover, td strains cannot complement conditional mutants of transformation in ASV [10], and analysis by recombination has mapped the known conditional mutants for transformation in the deleted region (personal communication from G.S. Martin). Hence, we assume that the nucleotide sequences contained in the deletion (or *sarc*) encode part or all of *onc*. We have prepared cDNA$_{sarc}$ as outlined in figs. 1 and 2 and described in detail elsewhere [9]. Briefly, cDNA is transcribed from the ASV genome with reverse transcriptase; the DNA specific to *sarc* is then selected by virtue of its failure to hybridize to RNA from the td

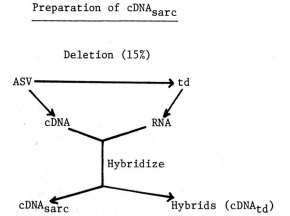

Fig. 1. Scheme for the preparation of DNA complementary to the transforming gene of ASV. The figure outlines the strategy for preparation of cDNA$_{sarc}$; details are published elsewhere [9]. Fractionation of the hybridized cDNAs is illustrated in fig. 2. Abbreviations: ASV, avian sarcoma virus; td, transformation-defective variant; cDNA$_{sarc}$, cDNA specific for the nucleotide sequences deleted in the genesis of td strains; cDNA$_{td}$, DNA representing part or all of the td genome but devoid of *sarc* nucleotide sequences.

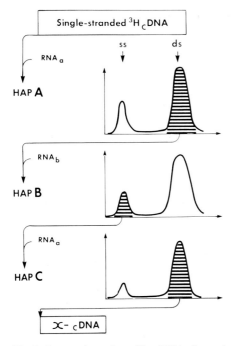

Fig. 2. Preparation of specific cDNAs by molecular hybridization and fractionation on hydroxyapatite. DNA transcribed from the genome of ASV was first subjected to a positive selection by hybridization of ASV 70S RNA (RNA a). DNA which hybridized to the viral RNA was selected by fractionation on hydroxyapatite (HAP A), then hybridized with a large excess of 70S RNA from a td variant (RNA b). The DNA which failed to hybridize with td RNA was selected on hydroxyapatite (HAP B) and subjected to a final positive selection by hybridization with ASV 70S RNA. DNA which hybridized to ASV RNA was selected on hydroxyapatite (HAP C) and then used as cDNA$_{sarc}$.

deletion mutant. The purified cDNA$_{sarc}$ is composed of DNA chains 100–200 nucleotides in length representing ca. 16% of the ASV genome (i.e., a genetic complexity of ca. 1600 nucleotides) when analyzed by molecular hybridization[9]; we conclude that cDNA$_{sarc}$ represents most or all of the deletion which defines *sarc* and could represent the entire transforming gene (*onc*) of ASV.

Since deletion mutants for *env* in ASV also exist[1], we have been able to use the above procedure to prepare cDNA$_{gp}$ specific for nucleotide sequences encoding glycoprotein in the ASV genome (unpublished work of the authors). We used a strain of ASV which is deficient only in the *env* gene product[11] and which lacks ca. 20% of the parental ASV genome[12]; our preparation of cDNA$_{gp}$ represented approximately 20% of the ASV genome when analyzed by molecular hybridization (unpublished observation of J.T.).

Specificity of the nucleotide sequences in the transforming gene of ASV

We tested the ability of various viral genomes to hybridize with cDNA$_{sarc}$ and found reactions only with RNA from ASV (table 1). There was no hybridization with RNA from two mammalian sarcoma viruses (murine and feline), both of which induce sarcomas and transform fibroblasts, nor with RNA from mouse mammary tumor virus, which induces carcinoma of the breast. The DNA of viruses (SV-40, adenovirus-5, and Herpes simplex-1) capable of transforming fibroblasts and/or epithelial cells contained no homology with cDNA$_{sarc}$.

Genomes from a series of avian leukosis viruses (RAV-2, RAV-6, RAV-50 and AMV) were also tested with negative results. These viruses induce a variety of tumors, including lymphoid leukosis, myeloblastosis and renal carcinoma;

TABLE 1

Nucleotide sequences of cDNA$_{sarc}$: distribution among viral genomes. Nucleic acids prepared from the viruses listed were incubated with cDNA$_{sarc}$ under conditions for hybridization described elsewhere [9]. The extent of hybridization was measured by hydrolysis with S1 nuclease [9].

Viral RNA or DNA	Hybridization to: [^{3}H]cDNA$_{sarc}$ (Pr-C ASV) %
Pr-C ASV	100
B77-C ASV	95
Carr-Zilber ASV	100
Schmidt-Ruppin ASV	93
Bryan ASV	94
Fujinami ASV	92
td Pr-C ASV	1
td B77-C ASV	1
Rous-Associated Virus-0	1
Rous-Associated Virus-2	0
Rous-Associated Virus-6	0
Rous-Associated Virus-50	1
Avian Myeloblastosis Virus	2
Myelocytomatosis Virus (MC-29)	1
Murine Sarcoma-Leukosis Virus (MOL)	0
Feline Sarcoma-Leukosis Virus (RIC)	0
Murine Mammary Tumor Virus (RIII)	0
SV40	0
Adenovirus Type 5	0
Herpes Simplex Type 1	0

in certain instances, sarcomas may occur [12a], but none of these viruses can transform fibroblasts in culture. Previous analyses indicated that the genomes of these viruses were no larger than the genome of td variants of ASV [13] and that the missing portion corresponds to *sarc* in ASV[14]; our data confirm to these conclusions and further implicate the nucleotide sequences of *sarc* in the transformation of fibroblasts by ASV.

We have tested one strain of avian leukosis virus (MC-29) which can transform fibroblasts in culture [15]. The genome of MC-29 is no larger than the RNAs of other leukosis viruses [16] and has no homology with *sarc* (table 1). We conclude that genetic determinants other than *onc* can induce transformation of avian fibroblasts; to date, transformation of mammalian cells by MC-29 has not been reported, a fact which distinguishes this virus from the conventional ASV which bears *sarc*.

RAV-0 is an endogenous virus produced spontaneously by certain inbred lines of chickens [17]; production of an identical or similar virus by chicken cells can also be induced with a variety of chemical and physical agents [18]. To date, no pathogenicity has been described for RAV-0 [19], and the viral genome is devoid of both *sarc* (table 1) and nucleotide sequences which comprise ca. 20% of the genome of pathogenic avian leukosis viruses [20]; it is not known whether the latter nucleotide sequences are responsible for the pathogenicity of the leukosis viruses, but the RAV-0 genome is otherwise closely related to the leukosis virus genome [20].

We conclude that the nucleotide sequences of *sarc* include part or all of *onc* of ASV and that they are unique to the genome of ASV. Although genetic determinants other than *sarc(onc)* can also facilitate transformation of fibroblasts by avian C-type RNA tumor viruses (e.g., leukosis virus MC-29), all of the ASV isolated from sarcomas in a natural setting contain *sarc*. The restriction of *sarc* to avian viruses is explicable by the origin of the constituent nucleotide sequences, an issue which we now address.

The origin of nucleotide sequences in the transforming and glycoprotein genes of ASV

Prior to our studies with $cDNA_{sarc}$, the only viral genes known to be endogenous to normal chicken cells were the genes contained in the genome of RAV-0, and these genes do not include *onc*. However, by using $cDNA_{sarc}$ in molecular hybridization, we discovered that nucleotide sequences homologous to part or all of *sarc* are present in the normal DNA of all the avian species we have tested (table 2) [21]. Analysis of the kinetics of reassociation between $cDNA_{sarc}$ and avian DNAs indicated that the *sarc* nucleotide sequences are present as a single copy or as a few copies in avian DNAs [21].

TABLE 2

The phylogeny of ASV_{sarc}. Denatured DNAs were annealed with [^{3}H]cDNA$_{sarc}$ and the resulting duplexes adsorbed to columns of hydroxyapatite in 0.12 M sodium phosphate [9]. The columns were then exposed to a thermal gradient and the denaturation of the duplex was monitored by continuously washing the columns with 0.12 M sodium phosphate (0); the temperatures at the midpoints of the denaturations are expressed as T_m. The estimates of phylogenetic distance, deduced from fossil records and antigenic relationships among proteins [3], were provided by Professor Allan Wilson.

DNA	% cDNA$_{sarc}$ in duplex	T_m (°C)	ΔT_m (°C)	Phylogenetic distance from chicken (years × 10^{-6})
Provirus (XC)	56	81	0	?
Chicken	52	77	−4	0
Quail	37	74	−7	20–40
Turkey	30	72	−9	40
Duck	16	71	−10	80
Emu	15	70	−11	100
Mouse	<2	–	–	–
Calf	<2	–	–	–

The extent of duplex formation between cDNA$_{sarc}$ and cellular DNAs from different species varied roughly as a function of the phylogenetic distance among the species (table 2) [21]; a similar proportional variation was observed with the melting temperatures of the duplexes (table 2) [21]. We also investigated the duplex between cDNA$_{sarc}$ and DNA from rat cells (XC) transformed by ASV; since mammalian DNAs normally contain no homology with cDNA$_{sarc}$ (table 2), the DNA homologous to *sarc* in XC cells must represent viral DNA integrated at the time of infection by ASV and can therefore be used to generate 'perfectly matched' duplexes with cDNA$_{sarc}$. These duplexes have a T_m 4° higher than the T_m of duplexes between cDNA$_{sarc}$ and chicken DNA (table 2), indicating that the nucleotide sequences in *sarc* of the viral genome are appreciably diverged from the homologous sequences in the chicken genome. We conclude that the nucleotide sequences from which *sarc* of ASV was derived arose and diverged during the course of avian speciation and appeared subsequently in the genome of ASV (fig. 3).

The acquisition of *sarc* by ASV could be due to a process akin to transduction, facilitated by either a DNA tumor virus or recombination between an avian leukosis virus and the host genome (fig. 3). Alternatively, as proposed by Temin [22], the entire genome of ASV may have been generated by modification and mobilization of cellular genes into an antecedent 'protovirus' (fig. 3). The nucleotide sequences of *sarc* in ASV are now perceptibly diverged from their alleged progenitor sequences in the normal avian cell (table 2); this could be

POSSIBLE ORIGINS OF SARC IN ASV

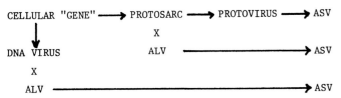

Fig. 3. Several mechanisms for the mobilization of a cellular gene into the genome of ASV. Abbreviations and terms used: X, a recombination; ALV, avian leukosis virus; ASV, avian sarcoma virus; protosarc, the cellular gene prior to mobilization into a viral genome by mechanisms proposed by Temin [23].

the consequence of either the process which generated viral genes from cellular genes or mutations during the course of repeated viral propagation.

Our finding of nucleotide sequences homologous to *sarc* in chicken cells demonstrates that normal cells may harbor genes at least closely related, if not identical to viral genes responsible for transformation of fibroblasts. The failure of *sarc* to appear in the genome of the endogenous chicken virus RAV-0 conforms to the general experience that endogenous viruses, when pathogenic at all, induce leukemia rather than sarcomas and cannot transform fibroblasts [23].

Using $cDNA_{gp}$, we have found in preliminary experiments that chicken DNA contains nucleotide sequences homologous to the *env* gene, whereas DNAs from quail, ring-necked pheasant and duck do not (table 3). Since all of these DNAs contain homology with *sarc*, it appears that the nucleotide sequences for the two viral genes may have been generated in substantially different ways. Moreover, we cannot yet conclude that the gene for *env* evolved during avian speciation; its presence in one species could as easily be accounted for by horizontal virus infection of the germ line, as has been described recently for primate and murine

TABLE 3

Detection of nucleotide sequences homologous to genes encoding *onc* and *env* in normal avian DNAs. DNA was extracted from 9 or 10-day old embryos and hybridized with either $cDNA_{sarc}$ or $cDNA_{gp}$; hybridization was detected by hydrolysis with S1 nuclease. Phylogenetic distances were derived as for table 2.

DNA	sarc	gp	Phylogenetic distance (years $\times 10^{-6}$)
Chicken	+	+	0
Ring necked pheasant	+	−	20
Japanese quail	+	−	20–40
Duck	+	−	80

viruses [24,25]. The presence of nucleotide sequences for *env* in chicken conforms to previous demonstrations that this species harbors genes for viral glycoprotein [1].

Expression of *sarc* nucleotide sequences in normal cells

We have found RNA homologous to *sarc* in a variety of embryonic avian cells, as summarized in table 4. We have tested these cells by bioassay and/or molecular hybridization and have found no evidence for contamination by ASV; hence, the presence of *sarc* in RNA in normal avian cells is due to transcription from the *sarc* nucleotide sequences in normal avian DNA. The concentrations of *sarc* RNA are low (ca. 10^{-6} of total cellular RNA) and comparable to the concentration of virus-specific RNA in mammalian cells transformed by ASV (unpublished data of the authors). Since the amount of viral RNA in the mammalian cells is sufficient to facilitate viral gene expression (synthesis of viral antigens; transformation of the cells), we conclude that the cellular gene transcribed into *sarc* RNA in normal cells may also be functionally active.

The *sarc* nucleotide sequences are part of the unique fraction of cellular DNA and could represent either a structural or a regulatory gene. Neither the gene product of *onc* in ASV nor the product of *sarc* in avian cells has been identified; moreover, we do not know if *sarc* RNA is translated in avian cells. Consequently, our deduction that transcription from *sarc* is equivalent to gene expression is only speculative.

TABLE 4

Avian embryos and embryonic fibroblasts containing *sarc* RNA. RNA was extracted and tested for the presence of *sarc* nucleotide sequences by molecular hybridization with cDNA$_{sarc}$, using techniques described elsewhere [9]. Nucleotide sequences homologous to *sarc* generally comprised ca. 1×10^{-6} of the total cellular RNA (unpublished data of the authors).

Chick embryos (3–16 days)
Chicken embryo fibroblasts (10th passage)

Duck embryo fibroblasts (10th passage)

Quail embryos (9–10 days)
Quail embryo fibroblasts (1st, 2nd and 5th passages)

Conclusion

'... death dwelled in the cell though the cell be looked in on at its most quick.' (T. Pynchon, in *The Crying of Lot Forty Nine*)

The nucleotide sequences of *sarc* apparently encode a normal cellular gene which has been highly conserved during evolution and is transcribed in growing cells. Since the same (or a very similar) gene in the genome of ASV can transform fibroblasts to a neoplastic state, we suggest that *sarc* in avian cells is involved in cell growth. *Onc* of ASV is apparently a trigger for cellular DNA synthesis [26]; *sarc* in the avian genome might serve the same purpose. If this account is correct, ASV has fortuitously selected a gene vital to cellular growth and division and has made this gene available for close analysis. The mobilization of *sarc* from the cell as part of a viral genome appears to be a rare event: field isolates of ASV are extremely uncommon and most avian sarcomas do not contain detectable transforming virus. Provisional claims to have mobilized *sarc* experimentally with an avian leukosis virus require further evaluation [27].

Expression of *sarc* would be regulated in normal cells in order to effect ordered cell growth. Mutation of the controlling element, releasing *sarc* from its normal regulation, could account for the oncogenic effects of either somatic mutation or environmental carcinogens. Integration of *sarc* into the host genome as part of the proviral DNA of ASV could insert the gene at a site not susceptible to regulation, thereby leading to a constant stimulus for cellular DNA synthesis and consequent neoplastic transformation. These suggestions all treat *sarc* as a normal cellular gene rather than as an original constituent of a viral genome and require no mandatory connection between oncogenesis and genes of endogenous viruses, in contrast to the oncogene hypothesis of Huebner and Todaro [28]. We see no reason to view *sarc* as a viral gene; it is not required for viral replication [3], in contrast to other known transforming genes of viruses [29], and it is not linked to any endogenous viral genome which can be mobilized from the cell [9]. We are now attempting to test the role of *sarc* in cell growth and oncogenesis by searching for *sarc* RNA in cells arrested in G_o, in fully differentiated adult tissues, and in avian tumors induced by chemical carcinogens.

Acknowledgements

This work was supported by USPHS grant #CA 12705-05, contract NO1 CP 33293 within the Virus Cancer Program of the National Cancer Institute, NIH, PHS, and grant #VC70 from the American Cancer Society.

References

[1] H. Hanafusa, Avian RNA tumor viruses, in: Cancer: A Comprehensive Treatise, vol. 2 (F.F. Becker, ed) Plenum Press, N.Y., p. 49.

[2] D. Baltimore, Tumor Viruses. Cold Spring Harb. Symp. Quant. Biol. 39, 1187 (1974).

[3] J.A. Wyke, Temperature sensitive mutants of avian sarcoma viruses. Biochim. Biophys. Acta 417, 91 (1975).

[4] D. Boettiger, Virogenic nontransformed cells isolated following infection of normal rat kidney cells with B77 strain Rous Sarcoma Virus. Cell 3, 71 (1974).

[5] H.E. Varmus, J.M. Bishop and P.K. Vogt, Synthesis and integration of Rous sarcoma virus-specific DNA in permissive and non-permissive hosts, in: Virus Research, Second ICN-UCLA Symposium on Molecular Biology (C.F. Fox and W.S. Robinson, eds.) Academic Press, N.Y. and London, pp. 373–383 (1973).

[6] I.A. Macpherson, Reversion in cells transformed by tumour viruses. Proc. Roy. Soc. (Biol.) 177, 41 (1971).

[7] D. Boettiger, Reversion and induction of Rous sarcoma virus expression in virus-transformed baby hamster kidney cells. Virology 62, 512 (1974).

[8] C.T. Deng, D. Boettiger, I. Macpherson and H.E. Varmus, The persistence and expression of virus-specific DNA in revertants of Rous sarcoma virus-transformed BHK-21 cells. Virology 62, 522 (1974).

[9] D. Stehelin, R.V. Guntaka, H.E. Varmus and J.M. Bishop, Purification of DNA complementary to nucleotide sequences required for neoplastic transformation of fibroblasts by avian sarcoma viruses. J. Mol. Biol., in press (1976).

[10] J.A. Wyke, J.G. Bell and J.A. Beamand, Genetic recombination among temperature-sensitive mutants of Rous sarcoma virus. Cold Spring Harb. Symp. Quant. Biol. 39, 897 (1974).

[11] S. Kawai and H. Hanafusa, Isolation of a defective mutant of avian sarcoma virus. Proc. Nat. Acad. Sci. U.S.A. 70, 3493 (1973).

[12] P. Duesberg, S. Kawai, L.-H. Wang, P.K. Vogt, H.M. Murphy and H. Hanafusa, RNA of replication-defective strains of Rous sarcoma virus. Proc. Nat. Acad. Sci. U.S.A. 72, 1569 (1975).

[12a] H.G. Purchase and B.R. Burmester, in: Diseases of Poultry (M.S. Hofstad et al., eds.) pp. 502–567, 6th edition. Iowa State University Press, Ames.

[13] P.H. Duesberg, P.K. Vogt, J. Maisel, M.M.-C. Lai and E. Canaani, Tracking defective tumor virus RNA, in: Virus Research, the Proceedings of the 2nd ICN-UCLA Symposium on Molecular Biology (C.F. Fox and W.S. Robinson, eds.). Academic Press, N.Y. and London, pp. 327–338 (1973).

[14] M.M.-C. Lai, P.H. Duesberg, J. Horst and P.K. Vogt, Avian tumor virus RNA: A comparison of three sarcoma viruses and their transformation-defective derivatives by oligonucleotide fingerprinting and DNA-RNA hybridization. Proc. Nat. Acad. Sci. U.S.A. 70, 2266 (1973).

[15] A.J. Langlois, S. Sankaran, P.H.-L. Hsuing and J.W. Beard, Massive direct conversion of chick embryo cells by strain MC29 avian leukosis virus. J. Virol. 1, 1082 (1967).

[16] P.H. Duesberg and P.K. Vogt. RNA species obtained from clonal lines of avian sarcoma and avian leukosis virus. Virology 54, 207 (1973).

[17] P.K. Vogt and R.R. Friis, An avian leukosis virus related to RSV(0): properties and evidence for helper activity. Virology 43, 223 (1971).

[18] R.A. Weiss, R.R. Friis, E. Katz and P.K. Vogt, Induction of avian tumor viruses in normal cells by physical and chemical carcinogens. Virology 46, 920 (1971).

[19] J.V. Motta, L.B. Crittenden, H.G. Purchase, H.A. Stone, W. Okazaki and L. Witter, Low oncogenic potential of avian endogenous RNA tumor virus infection or expression. J. Nat. Cancer Inst. 55, 685 (1975).

[20] P.E. Neiman, S.E. Wright, C. McMillin and D. MacDonnell, Nucleotide sequence relationships

of avian RNA tumor viruses: Measurement of the deletion in a transformation-defective mutant of Rous sarcoma virus. J. Virol. 13, 837 (1974).
[21] D. Stehelin, H.E. Varmus, J.M. Bishop and P.K. Vogt, DNA related to the transforming gene(s) of avian sarcoma viruses is present in normal avian DNA. Nature, in press (1976).
[22] H.M. Temin, On the origin of RNA tumor viruses. Ann. Rev. Genetics 8, 155 (1974).
[23] J.R. Stephenson, J.S. Greenberger and S.A. Aaronson, Oncogenicity of an endogenous C-type virus chemically activated from mouse cells in culture. J. Virol. 13, 237 (1974).
[24] R.E. Benveniste and G.J. Todaro, Evolution of C-type viral genes: inheritance of exogenously acquired viral genes. Nature 252, 456 (1974).
[25] R.E. Benveniste and G.J. Todaro, Evolution of type C viral genes: preservation of ancestral murine type C viral sequences in pig cellular DNA. Proc. Nat. Acad. Sci. U.S.A. 72, 4090 (1975).
[26] J.G. Bell, J.A. Wyke and I.A. Macpherson, Transformation by a temperature-sensitive mutant of Rous sarcoma virus in the absence of serum. J. Gen. Virol. 27, 127 (1975).
[27] R.A. Weiss, L.B. Crittenden, H.G. Purchase and P.K. Vogt, Genetic control and oncogenicity of endogenous and exogenous avian RNA tumor viruses, in: Chemical and Viral Oncogenesis, Proceedings of the XI International Cancer Congress, pp. 248–253, Excerpta Medica, Amsterdam (1974).
[28] G.J. Todaro and R.J. Huebner, The viral oncogene hypothesis: new evidence. Proc. Nat. Acad. Sci. U.S.A. 69, 1009 (1972).
[29] R. Dulbecco, Cell transformation by viruses and the role of viruses in cancer. J. Gen. Microbiol. 79, 7 (1973).
[30] E.M. Praeger, H.H. Brush, R.A. Noland, M. Nakaniski and A.C. Wilson, Slow evolution of transferrin and albumin in birds according to microcomplement fixation analysis. J. Mol. Evol. 3, 243 (1974).

Chapter 23

SOMATIC CELL GENETIC INVESTIGATIONS OF CLONAL SENESCENCE

GEORGE M. MARTIN, THOMAS H. NORWOOD and HOLGER HOEHN

Department of Pathology, University of Washington, Seattle, Washington, U.S.A.

In 1961, Hayflick and Moorhead [1] first clearly delineated 2 major classes of cultivated cell lines from mammalian tissues (table 1) [2]. For technical reasons, biochemists have largely utilized that class, typified by HeLa and L cells, which, at least in mass cultures, appears to have an infinite replicative life-span. We refer to these as *neoplastoid* cell lines, since they may be regarded as in vitro models of certain aspects of the neoplastic process. In contrast, those cell lines which have finite replicative life-spans and which maintain the donor karyotype in vitro may be regarded as in vitro models of certain aspects of hyperplastic processes such as wound healing. Prototypes of such *hyperplastoid* cell lines are the WI-38 lines from human fetal lung and the fibroblast-like cells derived from human skin biopsies. There is now a large body of literature which shows that mass cultures and clones from such cultures invariably cease replicating

TABLE 1

A comparison of some properties of hyperplastoid cell lines with neoplastoid cell lines [2].

Property	Hyperplastoid cell lines	Neoplastoid cell lines
Prototype cultures	Human skin 'fibroblasts' and WI-38	HeLa and L cells
In vivo analogs	Healing wounds	Ca of cervix; subcut. sarcoma
Replicative life-span	Finite	Apparently infinite
Cloning efficiency	Decreases with age of culture	May approach 100% in some lines
Karyotype	Same as donor's normal somatic cells	Usually different from donor's normal somatic cells
Tumor assay	Negative	Sometimes positive
Growth in suspension culture	No	Frequently
Metabolic cooperation	Easily demonstrable	Less readily demonstrable

The Molecular Biology of the Mammalian Genetic Apparatus:
edited by P. Ts'o © 1977, Elsevier/North-Holland Biomedical Press

after various numbers of cell doublings. That this *clonal senescence* has something to do with in vivo aging is supported by 4 lines of evidence. First of all, the replicative life-spans of cultures are inversely related to the ages of the donors. In unpublished work, we have recently confirmed this point for human skin fibroblast cultures in the critical age range from 20–90 years. There is a highly significant ($P = <0.001$) regression of cumulative cell doublings on age of -0.18 ± 0.05 S.E. per year. This supports previous studies by ourselves and others [3–6] and unpublished studies by E. Snyder. Perhaps of more significance to the phenotype of aging is the fact that the growth potentials of cells in and around the vascular wall (fig. 1) clearly diminish as a function of donor age [7]. A second line of evidence is that the growth potential of somatic cells is a function of the precise tissue of origin [2,5,7]. For example, evidence that the growth potential of cells from the thoracic aorta exceeds that from the abdominal aorta [2,7] fits with a pathogenetic theory which we have proposed for atherosclerosis [8,2,7], a disease which preferentially involves the abdominal aorta. A third line of evidence is that, in certain genetically determined syndromes of premature aging, the growth potential of cultivated somatic cells is substantially diminished in comparison with those of age-matched controls. This has been very well substantiated for the case of Werner's Syndrome [5] but has also been suggested in the case of at least certain patients with the Hutchinson–Gilford Syndrome [9], in which there is an onset of progeria during childhood. Although we do not believe that either of these presumptive point mutations results in what might be called a 'global progeria' (for example, there is no evidence of accelerated aging of the central nervous system), they result in striking caricatures of the phenotype of aging, especially as it is manifested in mesenchymal tissues. Finally, there is some evidence of a correlation between the replicative life-span of cultured somatic cells and the life-span of the donor species [10]

Proposals as to mechanisms of clonal senescence generally fall into one of three broad categories, although the distinction is not as clearcut as is generally believed: 1) Some investigators believe that there is a primacy of deteriorative stochastic events, with or without modulation by the genome. An example would be the original error catastrophe theory of Leslie Orgel [11], in which a cascade of abnormal cell proteins accumulates because of mistakes in transcription and/or translation; the consequences of such errors could be modulated by genetically controlled scavenger enzymes [12,13]. Another example is the somatic mutation theory; rates at which these develop or are rectified by DNA repair enzymes might also be under genetic control, as has recently been discussed by Burnet [14,15]. 2) Other investigators emphasize some type of genetic program controlling the transition from cycling to noncycling states. For example, normal diploid mammalian cells in vitro may be doing what they often do in vivo – undergoing a series of differentiation steps towards a post-replicative, terminally differentiated state [16–18]. Although this process must ultimately be under genetic control, the switch from a proliferative to a non-proliferative state might itself be a stochastic event subject to certain probability laws [18,19]. It is of course possible that deteriorative stochastic events might then ensue in the terminally differentiated cells as secondary epiphenomena [18].

Finally, one could think about clonal senescence in terms of a 'running out of genetic program'. This view incorporates elements of the first two categories, but is sufficiently distinctive conceptually to warrant delineation.

In this communication, I wish to illustrate how somatic cell genetic experiments can be used to evaluate these theories on the mechanisms of clonal senescence. While the results to date provide no proof of any given hypothesis, I believe that, in aggregate, they are more consistent with genetic programming ideas than they are with error theories. I suspect, however, that there will be a sub-set of error-prone biologists who will take issue with this conclusion. Therefore, to make everyone happy, at the end of this communication, I shall briefly describe the results of two more direct approaches to the problem – one of which argues against error theory and one of which argues against our own pet genetic programming theory – namely that normal cells in vitro stop proliferating because they 'differentiate themselves to death' [18].

There are three types of cell fusion experiments which could help us with a genetic analysis of the phenotype of the post-replicative, 'senescent' fibroblast:
1) *Synkaryosis* – in which two different nuclei are fused within a common pool

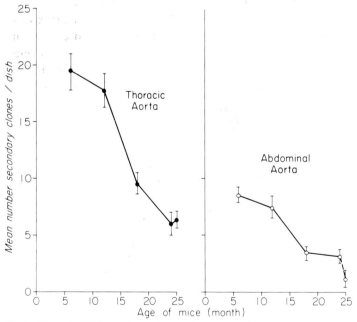

Fig. 1. Mean ($\pm$ S.E.) numbers of 'secondary' clones from strain ICR male mouse aortas plotted as a function of age of donor. Methods are given in [7]; reproduced by permission of the publishers.

of cytoplasm; 2) *heterokaryosis* – in which two different cells are fused short of nuclear fusion and 3) *heteroplasmosis* [21], *cybridization* [22] and variants thereof, in which two different cells are fused, one of which is enucleate or one of which is a karyoplast retaining only remnants of cytoplasm. Unfortunately, we and others are still at a comparatively early stage in the exploitation of these methods for the study of cell aging, but I shall share with you the results to date, mainly those from our own laboratory.

We shall begin with synkaryon experiments. The somatic mutation hypothesis of clonal senescence and the more recent version of the Orgel hypothesis (in which somatic mutations are thought to result from erroneous DNA polymerases [23–25]), both predict complementation in synkaryons (assuming most mutations are random and recessive), giving greater growth in hybrids than in parentals. However, if the putative genetic lesions are very numerous, only rare synkaryons might be expected to complement sufficiently to sustain cellular replication. John Littlefield [26] could not detect such complementation in crosses between pairs of different senescent fibroblast cultures and between senescent fibroblasts and young fibroblasts. We have had a similar experience. More recently [27], we have sought evidence for complementation in crosses involving actively replicating fibroblasts, in which we compared the longevities, in terms of cumulative population cell doublings, of parental diploids and hybrid tetraploids. For this purpose, we had to develop a new approach, since no one had ever before isolated *proliferating* hybrids between pairs of *normal* diploid somatic cells (fig. 2). A technique which proved quite useful was the visual comparison of the lengths of the equatorial plates of mitotic metaphases of emerging tetraploid clones with control diploids growing in the same culture dishes [27,28]. Using rapid electrophoretic screening for glucose-6-phosphate dehydrogenase (G6PD) polymorphic markers, we could rather efficiently screen a series of isolated presumptive tetraploids and to differentiate hybrids from tetraploids which resulted from selfing. These experiments, incidentally, show that proliferating synkaryons

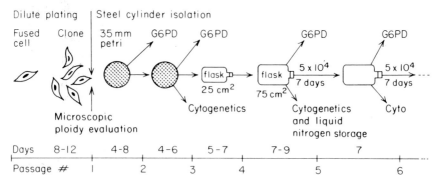

Fig. 2. Diagram of the protocol for the synkaryon experiments between lines of human diploid skin fibroblasts. Methods are given in [27].

between euploid cells can be obtained without the use of biochemical selective techniques. They open the door to the analysis of complementation between any pair of euploid somatic cells for which there are no available selective markers but for which there are differentiating markers. Table 2 summarizes the types of

TABLE 2

Summary of synkaryon experiments between lines of human diploid skin fibroblasts.

Type of fusion	# of separate experiments	Fusion protocol*	Fusion index**	Cloning efficiency***
G6PD A male × G6PD B male	8	7 I 1 M	5–18	8–16
G6PD A/B × self	7	3 I 4 M	9–12	3–16
G6PD A male × Werner G6PD B	1	I	8	14

* I = interphase cells fused; M = mitotic shake-off cells fused.
** Percent bi- and multinucleated cells per 400 cells 24 h after fusion.
*** Percent of cells plated forming macroscopic colonies 9–11 days after fusion.

TABLE 3

Yield of hybrid clones (G6PD heteropolymer) from synkaryon experiments between lines of human diploid skin fibroblasts.

Type of fusion	Clones prescreened*	Clones isolated** and tested	G6PD heteropolymer present
G6PD A male × G6PD B male	3,203	102	6
G6PD A/B × self	6,944	127	3
G6PD A male × Werner G6PD B	8,160	151	0

* Visual screening under inverted microscope for tentative ploidy diagnosis.
** G6PD electrophoresis and cytogenetics.

fusion experiments we have carried out for this purpose so far. They fall into three categories: 1) fusions between two different male strains, one G6PD A and one G6PD B; 2) self-fusions involving a genetically mosaic G6PD A/B heterozygotic female fibroblast line and 3) fusions between fibroblasts from a G6PD A male and a G6PD B female with Werner's syndrome. The latter represents a cross between a strain with comparatively high growth potential and one with exceedingly low growth potential [5]. From the frequencies of hybrid clones isolated in the first two types of fusions, averaging around one per thousand (table 3), one would have expected approximately eight hybrid clones among the more than 8,000 clones screened in the fusion experiment between the G6PD male strain with the Werner's syndrome strains. We found none, an observation indicative of a dominance of the Werner's syndrome senescent phenotype. This is also the conclusion from heterokaryon experiments which we shall shortly review. The in vitro life histories of the hybrid tetraploids and hybrid mixoploid isolates from the first two types of fusion experiments were investigated (table 4). Although we are still in the process of accumulating data for statistical analysis, our tentative conclusion is that their cumulative population cell doublings are in the same range as those of control diploids of these and similar strains in which there is a comparable degree of selection for clones big enough to be transferred and passaged. Certainly, none of the hybrid tetraploid and mixoploid clones surpassed the growth of the several hundred WI-38 clones studied by Smith and Hayflick [29]. We believe that this counts as evidence against somatic cell mutational theories of clonal senescence.

A few years ago [30] our laboratory presented evidence for diploidization within spontaneously occurring tetraploid clones isolated from fibroblast cultures

TABLE 4

In vitro life histories of 9 hybrid isolates

Clone	Cumulative population doublings	Tetraploid cells percentage	
		Initial	Final
1	35.3	100	100
2	cont.*	100	–
3	43.6	41	10
4	46.6	100	100
5	33.9	95	100
6	31.2	42	15
7	46.5	16	0
8	<25.0	100	100
9	senes**	–	–

* Loss due to contamination at third passage after isolation.
** No growth after third passage.

heterozygous for certain chromosomal markers. A few apparently recombinant homozygous diploid cells were observed, consistent with some sort of somatic segregation. Such a process would constitute an extremely valuable tool, given a sufficiency of markers, for the genetic analysis of clonal senescence. In the present material, however, the mixoploid clones almost certainly resulted from inter-colonial contamination. We have as yet no evidence of either diploidization or somatic segregation. Clearly, there were at least three tetraploid clones which remained chromosomally stable throughout their entire life-spans (table 4).

Interspecific synkaryons between normal diploid human somatic cells (fibroblasts or peripheral blood leukocytes) and neoplastoid animal cell lines have been carried out for years by investigators interested in gene mapping, as exemplified by Frank Ruddle's communication in this volume [31]. Depending upon the selective environment, one or more human chromosomes may be maintained in many such replicating hybrids for indefinite periods of time. Less is known about the maintenance of cytoplasmic organelles, although, in the case of mitochondrial DNA, its loss appears to parallel the loss of chromosomes of the hyperplastoid parent [32–34]. Recently, Danes [35] has claimed that proliferating hybrids between human progeric fibroblasts and mouse Ehrlich Ascites cells show decreased growth in comparison with control hybrids resulting from crosses between mouse Ehrlich Ascites cells and *normal* fibroblasts.

I would now like to review heterokaryon experiments carried out in our laboratory in order to test the hypothesis that senescent human fibroblasts can be regarded as terminally differentiated cells. Since Professor Henry Harris and his colleagues were able to employ heterokaryosis for the re-initiation of DNA synthesis in bonafide terminally differentiated cells [36,37], we reasoned that we should be able to do the same thing with senescent fibroblasts. In our first series

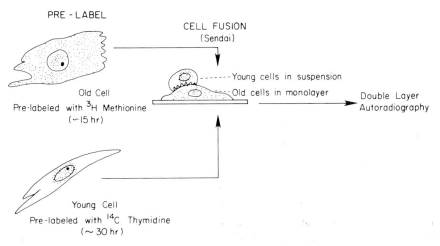

Fig. 3. Diagram of heterokaryon experiments. Methods are given in [38 and 40].

of experiments [37], we made heterokaryons between early and late passage fibroblasts using a double overlay autoradiographic technique for the diagnosis of heterokaryosis and for the detection of DNA synthesis after fusion (fig. 3). The actively replicating young partner was given a light nuclear prelabel with [^{14}C]thymidine, while the senescent partner received a light cytoplasmic and nuclear prelabel with [^{3}H]methionine. After Sendai virus-mediated fusion, three serial 24 hour test pulses of [^{3}H]thymidine were administered in order to evaluate nuclear DNA synthesis. Thymidine labeling indices were determined for the experimental crosses and for various controls. In the case of both homologous and isologous crosses, the results (Fig. 4) clearly indicated a *dominance* of the senescent phenotype. Not only had the young cells failed to resurrect DNA synthesis in the senescent nuclei, but DNA synthesis in the young nuclei had been markedly inhibited as a consequence of fusion with the old cells. This appeared to be evidence against our hypothesis, until we realized that, with one possible exception [39], all of Harris' experiments were carried out with *neoplastoid* cells, such as HeLa and Ehrlich ascites cells. We therefore repeated the experiments using HeLa and SV-40 transformed human fibroblasts and got entirely different results (figs. 5, 6) [40]. In both cases, there was a striking re-initiation of DNA synthesis by the senescent nuclei, although, in the case of HeLa, the extent of rescue fell during the second and third pulse periods. We believe that these results argue against error theories, unless one supposes that HeLa and SV-40 transformed cells possess very efficient error correction mechanisms. We believe that the various sets of data are consistent with the notion that postmitotic fibroblasts may synthesize specific repressor or inhibitor substances which block initiation of DNA synthesis. One possibility is that they synthesize a 'masking protein' which blocks plasma cell membrane receptors for serum mitogens. Such proteins might be highly specific to the given cell lineage. For example,

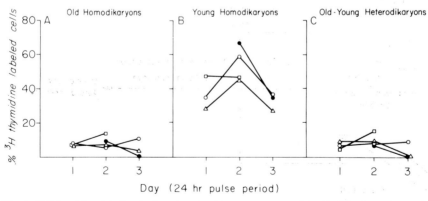

Fig. 4. [^{3}H]thymidine labeling indices in (A) old homodikaryons, (B) young homodikaryons and (C) old-young heterodikaryons. After Norwood, et al. [38]; reproduced with permission of the publishers.

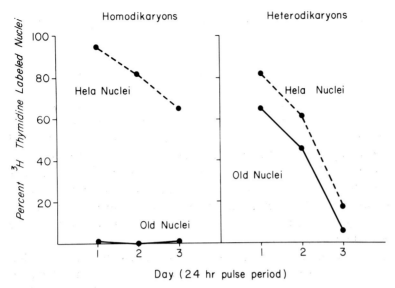

Fig. 5. [³H]thymidine labeling indices of nuclei within homodikaryons and heterodikaryons obtained by Sendai virus mediated fusion of senescent human skin fibroblasts and HeLa cells. After Norwood et al. [40]; reproduced by permission of the publisher.

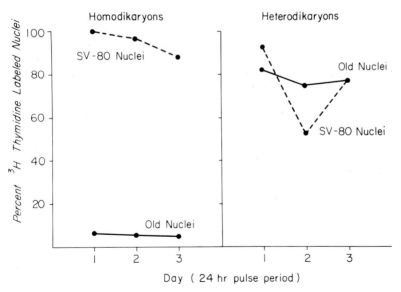

Fig. 6. [³H]thymidine labeling indices of nuclei within homodikaryons and heterokaryons obtained by Sendai virus mediated fusion of senescent human skin fibroblasts and a line of human skin fibroblasts which had been transformed by SV40 virus. After Norwood et al. [40]; reproduced by permission of the publisher.

terminally differentiated fibroblast progeny may not shut down DNA synthesis of myoblast nuclei in heterokaryons and myocytes might not shut down DNA synthesis in heterokaryons with fibroblasts.

It would of course be of great interest to biochemically characterize the substance in neoplastoid cell lines which appears to bypass our putative inhibitor or repressor. As a first step, one would want to know something about its relative concentrations within the major cellular compartments. We have therefore prepared karyoplasts and cytoplasts from HeLa cells, using a modification of cytochalasin-centrifugation enucleation techniques [20]. The preliminary results (table 5) reveal that HeLa cytoplasts are capable of reinitiating DNA synthesis in senescent fibroblasts, but the extent of such rescue is only about 20% of what is observable with intact HeLa cells. Control experiments (table 6) indicate that this loss in efficiency of rescue is not due to the pre-treatments of the HeLa cells with [^{3}H]methionine and cytochalasin.

There is unfortunately insufficient space here to review the interesting but rather complicated cell reconstruction experiments still being carried out in Len Hayflick's laboratory between old and young WI-38 cells and cytoplasts [21]. Suffice it to say that they interpret their results as evidence in favor of a nuclear control of clonal senescence consistent with programmed genetic events.

The putative genetic program does not control any simple biological clock whereby the numbers of cell divisions are counted, DNA synthesis ceasing after

TABLE 5

Percent [^{3}H]thymidine labeled nuclei in mononucleate, senescent human diploid fibroblasts fused with cytoplasms from enucleated HeLa cells. The senescent cells displaying cytoplasmic label (i.e., silver grains over the cytoplasm) were assumed to have fused with the HeLa cytoplasm, which had been heavily pre-labeled with [^{3}H]methionine.

Expt. No.	[^{3}H]thymidine pulse 0–24 h after fusion		[^{3}H]thymidine pulse 24–48 h after fusion	
	Cells without cytoplasmic label*	Cells with cytoplasmic label**	Cells without cytoplasmic label*	Cells with cytoplasmic label**
1	5.1	6.0	8.3	20.3
2	5.1	9.0	8.0	15.7
3	3.6	12.3	8.0	14.4
4	0.78	8.0	6.9	14.6
Mean	3.6	8.8	7.8	16.3
Control***	4.2		5.7	

* 250–500 cells counted.
** 25–100 cells counted.
*** Senescent human fibroblasts exposed to Sendai virus but not co-cultivated with HeLa cytoplasts.

TABLE 6

Percent [³H]thymidine labeled senescent fibroblast nuclei in: 1) monokaryons, 2) homodikaryons and 3) heterodikaryons resulting from the fusion of senescent cells to HeLa cells which had been exposed to [³H]methionine and to cytochalasin B as in the experiments summarized in table 5.

	monokaryons	homodikaryons	heterodikaryons
[³H]thymidine pulse 0–24 h after fusion	3.7	0	33
[³H]thymidine pulse 24–48 h after fusion	8.2	3.4	45

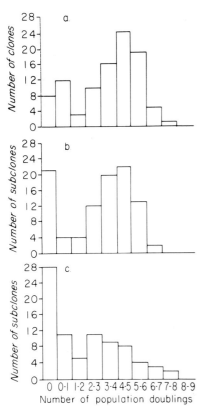

Fig. 7. The distribution of population doublings achieved by clones derived from a culture of human embryo skin fibroblasts. A. Primary clones, 5 days after isolation. B. Secondary clones of best growing primary clone, 5 days after secondary cloning. C. Quaternary clones of best growing tertiary sub-clone, 17 days after quaternary cloning, with feeding on day 9. After Martin et al. [17]; reproduced by permission of the publishers.

the limit is reached. This was shown in our laboratory by serial sub-cloning experiments (fig. 7), indicative of substantial epigenetic heterogeneity of growth potential.

As I indicated in my introduction, I will conclude this communication by mentioning two lines of more direct inquiry concerning the validity of the two contrasting theories of clonal senescence. Orgel's error theory predicts that we should discover abnormal forms of essentially *all* proteins within senescent cytoplasm. Indeed, reports of accumulations of abnormal proteins have been published [23,41]. Working in collaboration with Bill Pendergrass and Paul Bornstein, we have failed to detect increased levels of heat labile or immunologically unreactive glucose-6-phosphate dehydrogenase in senescent human skin fibroblasts, although our methods could detect such aberrant molecules in model systems [42]. This can be considered as strong evidence against the error catastrophe theory as an explanation for in vitro clonal senescence, although it is still worthwhile testing as an explanation of cell senescence in vivo, especially of postreplicative cells such as central nervous system neurones.

Finally, we have to mention evidence against the terminal differentiation hypothesis. That theory predicts qualitative and/or quantitative differences in gene expression in terminally differentiated cells. In other words, it predicts a different pattern of *normal* proteins rather than abnormal proteins [18]. Using high-resolution, thin-slab SDS polyacrylamide gel electrophoresis, we can detect at least 80 protein bands in extracts from mass cultures of late and early passage skin fibroblasts. To date, we have been unable to discern any consistent differences between the old and young cells, at least in mass cultures.

In conclusion, it is obvious that the problem of in vitro clonal senescence is far from resolved. We hope we have shown how a variety of somatic cell genetic methods can contribute to a solution. In the last analysis, however, we will require the more molecular methodologies to which Jim Bonner and his students have contributed so substantially over the years. We are fortunate in being able to hear an example of such an application in the following presentation by David Sabatini [43].

Acknowledgments

This work was supported by Institutional Cancer Grant IN-26 from the American Cancer Society, and NIH research grants AG 00257, AM 04826, GM 13543, and GM 15253.

References

[1] L. Hayflick and P.S. Moorhead, The serial cultivation of human diploid cell strains. Exp. Cell Res. 25, 585 (1961).
[2] G.M. Martin and C.A. Sprague, Symposium on in vitro studies related to atherogenesis: Life histories of hyperplastoid cell lines from aorta and skin. Exp. Mol. Pathol. 18, 125 (1973).
[3] L. Hayflick, The limited in vitro lifetime of human diploid cell strains. Exp. Cell Res. 37, 614 (1965).
[4] S. Goldstein, J.W. Littlefield and J.S. Soeldner, Diabetes mellitus and aging. Diminished plating efficiency of cultured human fibroblasts. Proc. Nat. Acad. Sci. 64, 155 (1969).
[5] G.M. Martin, C.A. Sprague and C.J. Epstein, Replicative life-span of cultivated human cells. Effects of donor's age, tissue and genotype. Lab. Invest. 23, 86 (1970).
[6] Y. LeGuilly, M. Simon, P. Lenoir and M. Gourel, Long-term culture of human adult liver cells: morphological changes related to in vitro senescence and effect of donor's age on growth potential. Gerontologia 19, 303 (1973).
[7] G. Martin, C. Ogburn and C. Sprague, Senescence and vascular disease. Adv. Exp. Med. Biol. 61, 163 (1975).
[8] G.M. Martin and C.A. Sprague, Clonal senescence and atherosclerosis. Lancet 2, 1370 (1972).
[9] S. Goldstein, Life-span of cultured cells in progeria. Lancet 1, 424 (1969).
[10] L. Hayflick, Cell biology of aging. Bioscience 25, 629 (1975).
[11] L.E. Orgel, The maintenance of the accuracy of protein synthesis and its relevance to aging. Proc. Nat. Acad. Sci. 49, 517 (1963).
[12] A.L. Goldberg, Degradation of abnormal proteins in *Escherichia coli*. Proc. Nat. Acad. Sci. 69, 422 (1972).
[13] M.R. Capecchi, N.E. Capecchi, S.H. Hughes and G.M. Wahl, Selective degradation of abnormal proteins in mammalian tissue culture cells. Proc. Nat. Acad. Sci. 71, 4732 (1974).
[14] F.M. Burnet, Intrinsic mutagenesis: a genetic basis of aging. Pathology 6, 1 (1974).
[15] F.M. Burnet, Intrinsic Mutagenesis. A Genetic Approach to Aging (John Wiley & Sons, New York) 1974.
[16] V.J. Cristofalo, Animal cell cultures as a model system for the study of aging. Adv. Gerontol. Res. 4, 45 (1972).
[17] G.M. Martin, C.A. Sprague, T.H. Norwood and W.R. Pendergrass, Clonal selection, attenuation and differentiation in an in vitro model of hyperplasia. Am. J. Pathol. 74, 137 (1974).
[18] G.M. Martin, C.A. Sprague, T.H. Norwood, W.R. Pendergrass, P. Bornstein, H. Hoehn and W.P. Arend, Do hyperplastoid cell lines 'differentiate themselves to death'? Adv. Exp. Med. Biol. 53, 67 (1975).
[19] J.A. Smith and L. Martin, Do cells cycle? Proc. Nat. Acad. Sci. 70, 1263 (1973).
[20] C.M. Croce and H. Koprowski, Enucleation of cells made simple and rescue of SV40 by enucleated cells made even simpler. Virology 51, 227 (1973).
[21] W.E. Wright and L. Hayflick, The regulation of cellular aging by nuclear events in cultured normal human fibroblasts (WI-38). Adv. Exp. Med. Biol. 61, 39 (1975).
[22] C.L. Bunn, D.C. Wallace and J.M. Eisenstadt, Cytoplasmic inheritance of chloramphenicol resistance in mouse tissue culture cells. Proc. Nat. Acad. Sci. 71, 1681 (1974).
[23] R. Holliday and G.M. Tarrant, Altered enzymes in aging human fibroblasts. Nature 238, 26 (1972).
[24] L.E. Orgel, Aging of clones of mammalian cells. Nature 243, 441 (1973).
[25] S.J. Fulder and R. Holliday, A rapid rise in cell variants during the senescence of populations of human fibroblasts. Cell 6, 67 (1975).
[26] J.W. Littlefield, Attempted hybridizations with senescent human fibroblasts. J. Cell. Physiol. 82, 129 (1973).

[27] H. Hoehn, E.M. Bryant, P. Johnston, T.H. Norwood and G.M. Martin, Non-selective isolation, stability and longevity of hybrids between normal human somatic cells. Nature 258, 608 (1975).
[28] C.A. Sprague, H. Hoehn and G.M. Martin, Ploidy of living clones of human somatic cells determined by mensuration at metaphase. J. Cell Biol. 60, 781 (1974).
[29] J.R. Smith and L. Hayflick, Variation in the life-span of clones derived from human diploid cell strains. J. Cell Biol. 62, 48 (1974).
[30] G.M. Martin and C.A. Sprague, Parasexual cycle in cultivated human somatic cells. Science 166, 761 (1969).
[31] F. Ruddle, Gene mapping in man and mouse by somatic cell genetics, in: Molecular Biology of the Mammalian Genetic Apparatus. Vol. B. Part III (Ed. Paul O.P. Ts'o), Elsevier-Exerpta Medica, North-Holland, 1977.
[32] D.A. Clayton, R.L. Teplitz, M. Nabholz, H. Dovey and W. Bodmer, Mitochondrial DNA of human-mouse cell hybrids. Nature 234, 560 (1971).
[33] B. Attardi and G. Attardi, Fate of mitochondrial DNA in human-mouse somatic cell hybrids. Proc. Nat. Acad. Sci. 69, 129 (1972).
[34] H.G. Coon, I. Horak and I.B. Dawid, Propagation of both parental mitochondrial DNAs in rat-human and mouse-human hybrid cells. J. Mol. Biol. 81, 285 (1973).
[35] B.S. Danes, Progeria: reduced growth of human progeric-mouse hybrids. Exp. Geront. 9, 169 (1974).
[36] H. Harris, J.F. Watkins, C.E. Ford and G.I. Schoefl, Artificial heterokaryons of animal cells from different species. J. Cell Sci. 1, 1 (1966).
[37] H. Harris, The reactivation of the red cell nucleus. J. Cell Sci. 2, 23 (1967).
[38] T.H. Norwood, W.R. Pendergrass, C.A. Sprague and G.M. Martin, Dominance of the senescent phenotype in heterokaryons between replicative and post-replicative human fibroblast-like cells. Proc. Nat. Acad. Sci. 71, 2231 (1974).
[39] H. Harris, E. Sidebottom, D.M. Grace and M.E. Bramwell, The expression of genetic information: a study with hybrid animal cells. J. Cell Sci. 4, 499 (1969).
[40] T.H. Norwood, W.R. Pendergrass and G.M. Martin, Reinitiation of DNA synthesis of senescent human fibroblasts upon fusion with cells of unlimited growth potential. J. Cell Biol. 64, 551 (1975).
[41] C.M. Lewis and G.M. Tarrant, Error theory and aging in human diploid fibroblasts. Nature 239, 316 (1972).
[42] W.R. Pendergrass, G.M. Martin and P. Bornstein, Evidence contrary to the protein error hypothesis for in vitro senescence. J. Cell Physiol. 87, 3 (1976).
[43] D. Sabatini, Studies on RNA metabolism and function in human fibroblasts, in: Molecular Biology of the Mammalian Genetic Apparatus. Vol. B, Part III (Ed. Paul O.P. Ts'o), Elsevier-Exerpta Medica, North-Holland, 1977.

Chapter 24

ENDOCRINE AND NEURAL FACTORS OF REPRODUCTIVE SENESCENCE IN RODENTS

CALEB E. FINCH

The Andrus Gerontology Center, and the Department of Biological Sciences, University of Southern California, Los Angeles, California 90007, U.S.A.

Endocrine and neural factors of reproductive senescence

Recent studies on the control mechanisms influencing gene function during the process of aging have focussed on the possibility that a number of cellular and genomic alterations which take place during the aging process in mammals are the consequence of hormonal changes [1]. Described in this brief review are studies of tissues whose response to hormones has been characterized in great detail and which provide particularly useful opportunities to study age-related alterations in the regulation of hormones and the response of target tissues to hormones.

Cell functions in the liver

Not all genomic functions show changes during aging [1]. For example, the liver enzyme, tyrosine aminotransferase, is equally inducible in old and young rodents by injection of glucocorticoids and insulin [2,3]. The action of glucocorticoids in inducing tyrosine aminotransferase is particularly significant, since this enzyme is induced by a mechanism involving translocation of a cytoplasmic receptor to the nucleus [4] and a subsequent selective increase in enzyme synthesis [5]. Consistent with the maintenance of some glucocorticoid mediated enzyme inductions is our recent finding that nuclear receptors in the mouse liver for glucocorticoids do not change with age [6]. Another persistent hepatic genomic function during aging is the capacity of healthy old rodents for complete hepatic regeneration [7,8].

The above studies indicate the maintained responsiveness of the liver to hormones given a sufficient direct stimulus. A number of examples show that age-related changes of a physiological nature may still occur in some enzyme systems. For example, the genomically-dependent induction of glucokinase after glucose in-

The Molecular Biology of the Mammalian Genetic Apparatus:
edited by P. Ts'o © 1977, Elsevier/North-Holland Biomedical Press

tubation [3] is impaired in aged rodents [3]. Because the basic mechanism for enzyme induction is not impaired as a consequence of aging, age-related changes in extra-hepatic (hormonal) controls may be inferred. At present, no intrinsic age-related change has been identified in hepatic cells [1].

Cell functions of the female reproductive system

The female reproductive system, which shows a well defined decline during mid-life in all mammals, is particularly amenable to analysis of the hormonal influences on gene activity during aging. The onset of reproductive senescence is marked by a well defined loss of fertility before mid-life. Then, after a varying

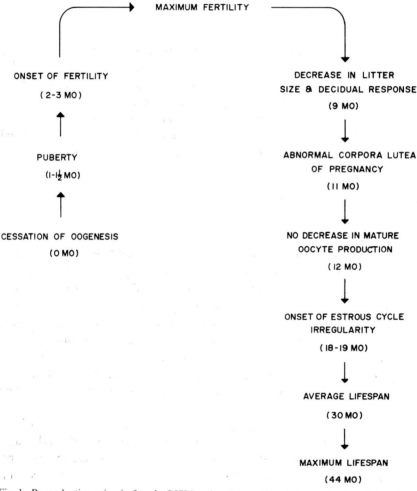

Fig. 1. Reproductive aging in female C57BL mice. Supporting data may be found in references 1, 11, 20, 22, 26.

interval, ovarian or estrous cycles become increasingly irregular and may stop altogether, as is the case after menopause in humans [9]. These changes are perhaps among the most general biological phenomena of aging described in mammals, and occur in all species during the mid-life period [1,10]. The chronology of events for a common strain of mice (C57Bl/6J) is shown in fig. 1.

An experimentally induced manifestation of the loss of fertility is the changing decidual cell response in the uterus. If young mice are hormonally primed with a sequence of estrogen and progesterone injections to simulate the hormonal levels in the early milieu of the pregnancy, then certain irritating stimuli such as injection of oil into the uterus induce a cellular reaction (deciduation) which is very similar to that which occurs during implantation of the blastocyst [11]. It has been (repeatedly) observed in rodents whose reproductive capacity has diminished e.g. in retired breeders, that there is a pronounced loss in the decidual cell response by the mid-life period [12,13]. As shown in table 1, the same change occurs in virgin mice (10 mo.) and involves not only a reduced increase in the weight of the uterine tissue but also in its glycogen content. The demonstration of this effect of age on virgin mice rules out its origin from a possible exhausting effect of pregnancy per se.

TABLE 1

Effect of age on the decidual cell response (C.F. Holinka and C.E. Finch, in preparation). Virgin C57BL/6J female mice were ovariectomized (day 1) and injected subcutaneously with estradiol-17β (E_2) and progesterone (P) in arachis oil in the following sequence employed by Shapiro and Talbert [15]. Day 3,4,5: 0.1 μg E_2; day 6: no injection; day 7,8,9: 1.0 mg P; day 10: 1.0 mg P and 0.025 μg E_2; day 11–13: 1.0 mg P. On day 10, mice were also 'decidualized' by injecting each uterine horn with 0.04 ml arachis oil. All mice were killed on day 14. The hormone replacements are designed to stimulate the hormonal sequence of pseudopregnancy [13]. Glycogen was assayed by the procedure of Demers et al. [44].

	4 mo.	10 mo.	Significance of age-difference**
A. Uterine weight (mean ± S.E.M.)			
1. ovariectomized controls	(17)* 13.7 ± .3	(9)* 17.1 ± .8	$P < .001$
2. hormone treated controls	(13) 29.3 ± .9	(11) 39.3 ± 1.2	$P < .001$
3. decidualized	(23) 78.9 ± 9.9	(14) 52.6 ± 2.4	$P < .001$
B. Glycogen (μg/100 mg uterus, mean ± S.E.M.)			
1. ovariectomized controls	(17) 28.0 ± 3.6	(9) 27.2 ± 4.0	not sig.
2. hormone treated controls	(13) 30.0 ± 4.3	(11) 22.8 ± 2.7	$P < .02$
3. decidualized	(23) 98.7 ± 12.2	(14) 26.2 ± 3.8	$P < .001$

* Number of mice.
** By 2 tailed t-test.

Underlying the decidual cell response is an interaction of ovarian hormones with the uterus [11]. The decidual cell response requires several days of priming with estrogen, which appears to be an inducer of uterine progesterone receptors [14]. Although the mechanisms of this sequential induction are not yet known, both estrogen and progesterone appear to be critical in initiating the selective proliferation of the epithelial and stromal layers of the uterus [11]. Because cytoplasmic steroid receptors appear to be major instruments in the uterine response to steroids, we therefore examined how aging alters uterine estrogen receptors. As shown in table 2, there is a progressive loss of cytoplasmic high affinity estrogen receptors in mice and rats during aging. Thus, it is plausible to consider that the reduction of estradiol receptors which occurs during the onset of fertility decline is also a component in the loss of the induced decidual response in the same age groups.

Other preliminary studies may provide clues concerning the loss of responsiveness of the aging uterus to hormonal stimuli. Also shown in table 2 are very old rats which were injected with estradiol (E_2) for a period of 7 to 11 days. The receptor levels appear to be substantially increased. This increase seems to

TABLE 2

Aging and uterine cytosol E_2 receptors (J.F. Nelson, C.F. Holinka, C.E. Finch, in preparation). 48–72 h after ovariectomy, uteri were homogenized and assayed for capacity to bind E_2 according to the procedure of Katzenellenbogen et al. [45]. This assay involves incubation for 1 h with ^3H-E_2 at 30 nM ±100-fold of unlabelled E_2, followed by removal of unbound steroid with charcoal. Specifically bound E_2 is calculated by subtracting the ^3H bound in the presence of 100 fold unlabelled E_2 and is tentatively considered to represent E_2 cytosol receptors. Holtzman rats in the treated group were injected after ovariectomy for 7–11 days with 12 μg E_2/100 g body wt to induce E_2 receptors. Binding was determined 72 h after the last injection. Samples analyzed in triplicate. C57BL/6J mice (3, 8, and 18 mo.). Samples analyzed in triplicate 48 h after ovariectomy. All mice were parous.

	^3H-E_2 specifically bound P moles E_2/uterus: means ± S.E.M.	
	Control	E_2 treated to induce receptor
A. Rats		
7 mo	1.56 ± 0.02 (n = 4)*	1.78 ± 0.28 (n = 6)*
27 mo	0.17 ± 0.02 (n = 4)*	1.49 ± 0.07 (n = 5)*
% decrease	−88% ($P < 0.05$)	−17% (not significant)
B. Mice		
3 mo	1.02 ± 0.09 (n = 8)	
8 mo	0.90 ± 0.10 (n = 10)	
18 mo	0.54 ± 0.08 (n = 13)	
$P < 0.001$, 1-way analysis of variance		

* n = number of animals, except in control group in 'A', where 3 of 4 uteri were pooled and analyzed in triplicate.

be a consequence of a basic control mechanism for estradiol receptors in the uterus, namely, that they are influenced by the level of ovarian steroids [15]. Thus, it seems reasonable to hypothesize that, as the ovarian production of steroids decreases, the estradiol receptor levels decrease as well. These results agree with studies on hamsters by Blaha in which the implantation frequency in middle-aged hamsters was greatly increased by transplantation of ovaries [16]. In humans, estradiol production is reduced before menopause [17] and completely ceases afterwards [18]. No such values are available in rodents at present.

The ovary is one of the few organs in mammals that undergoes a universal change as a consequence aging. A major feature of the ovary is that its store of egg cells or oocytes is fixed at birth. In a human there are more than 700,000 at birth [19] whereas a neonatal mouse [20] or a rat [21] has 10,000 or less. Because the numbers of primordial oocytes and active follicles decreases strikingly during puberty and at a slower rate thereafter [19,20,21], it is logical to consider to what extent the loss of fertility in middle-aged rodents is a consequence of the loss of oocytes. However, numerous studies show that, at the time of major fertility reduction in mice [22], rabbits [23], and hamsters [24], the number of mature, fertilizable (tubal) ova does not decrease. Even later, after the complete loss of fertility in mice [22], rats [21] and women [25], there is a residual store of histologically normal oocytes. Talbert and Krohn [26] have demonstrated that oocytes from infertile, but still cycling rodents can be fertilized and transplanted to young hosts with the same degree of success as young-to-young transplants. It may be that the declining numbers of hormone producing ovarian follicles [21,27] yield a suboptimal steroid output which then causes a marginal reduction of uterine estradiol receptors and in turn a failure of the induced decidual cell response in middle-aged mice (table 1). Very recent data from humans show that in the 4 or 5 years before menopause, while cycles are continuing in a normal fashion, estrogen levels drop significantly [17].

Age-related changes of catecholamines

At a later age, the ovarian cycles in rodents stop altogether [10]. Rats have been most extensively studied and it appears that their cycles may become arrested either in a state of constant estrous or pseudopregnancy [28–31]. It is a striking finding that ovarian cycles in old rats are reactivated after injection of either progesterone [28] or sympathomimetic drugs such as L-DOPA or iproniazide [28–32]. These findings have drawn attention to the neural mechanism which controls the ovarian cycle. In young mammals, ovarian cycles are critically dependent on levels of catecholamines in the hypothalamus [33,34]. Drugs, such as α-methyl-*p*-tyrosine, which reduce brain catecholamine levels also interrupt the ovarian cycle; hypothalamic administration of L-DOPA, which is converted to catecholamines, can overcome the blockade [34]. Since L-DOPA and iproniazide reactivate cycles in old rats as well as increase brain catecholamines [35], an age-related deficiency in catecholamines may underlie the loss of ovarian cycles in rodents. The effects of progesterone may also be related to its influence on catecholamine metabolism [33,36]. Whatever the mechanism involved in the reactivation of ovarian cycles, these studies also demonstrate that residual function in the ovary is sufficient to respond to the cycle of gonadotropin release by the pituitary. The maintained capacity of the pituitary itself in aging rodents is also suggested by these studies as well as by data from this laboratory (Finch, Jonec, Wisner, DeVellis, Sinha, and Swerdloff, in preparation).

We have recently obtained direct evidence that aging alters hypothalmic catecholamine synthesis. We have initially employed male mice, whose reproductive functions are intact, because interruption of ovulatory cycles in females will perturb catecholamine metabolism independently of age [33]. In male C57BL/6J mice, testicular function is not necessarily impaired with age. If animals are carefully selected for good health status (mice with tumors, kidney disease, respiratory disease, and other lesions are eliminated) plasma testosterone and testicular weight are stable throughout most of adult life [37]. In contrast, old mice with various diseases or with body weight loss show reduced plasma testosterone and testicular weight [37]. This finding demonstrates the importance of analyzing health factors in aging studies and suggests that some of the decrease in testicular function observed with increasing frequency in aging man [38] may be a consequence of age-related diseases. Using apparently healthy mice, we then studied the effect of aging on catecholamine metabolism. There were marked changes in the conversion of the catecholamine precursors, tyrosine, and L-DOPA in the healthy old mice (fig. 2). The metabolic differences cannot be attributed

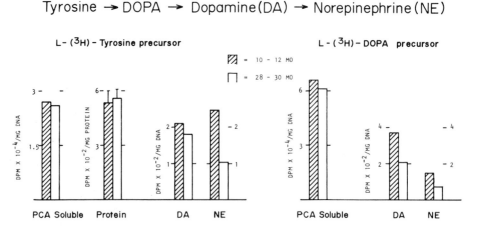

Fig. 2. The effect of age on the conversion of ^{3}H-L-tyrosine and ^{3}H-L-DOPA to catecholamines in the hypothalamus of C57BL/6J male mice, aged 10 and 30 mo. mice were injected i.p. and killed 45 min later. Complete description in ref. 39.

to changes in peripheral metabolism of L-DOPA, since the maximum levels and rate of decline of radioactivity in circulation were identical in both age groups after intraperitoneal injection of labelled L-DOPA [39]. Moreover, studies on L-tyrosine and L-DOPA transport by slices of hypothalamus showed no age difference [40]. The age-related reduction in precursor accumulation by the hypothalamus was about 10%, considerably less than the 30–50% difference in conversion to catecholamines observed in the same animals (fig. 2). The effect of age on catecholamine turnover was studied by following the specific activity of dopamine and norepinephrine after the peak plasma level of labeling. We observed a marked decrease in the turnover of hypothalamic norepinephrine in senescent mice (fig. 3). These measurements of catecholamine turnover reflect primarily the major catecholamine storage compartment. Of the several pools of catecholamines, one has high metabolic lability, with a turnover time of a few minutes or less, whereas the putative main storage compartment has a turnover time of several hours [41], and resembles the data in fig. 3. It is unclear which of these catecholamine compartments is pertinent to reproductive function, since not all catecholamine pools in the brain are of direct significance to the maintenance of the physiological processes [42]. We cannot infer from these studies which groups of nerve cells and which pools of catecholamines are replenished by L-DOPA in the aging female rats, which regain cycles after the administration of these drugs. Nonetheless, our studies on male mice support the rationale that a deficit of hypothalamic catecholamines during aging underlies the loss of ovarian cycles. Other reports of altered catecholamine metabolism in the brain during aging have been recently reviewed [1]. In the future, it should be possible to establish quantitative relationships between age-related changes of brain catecholamines and the endocrine glands which are controlled directly by the catecholamines.

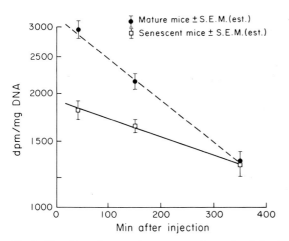

Fig. 3. The effect of age on the turnover of norepinephrine in the hypothalamus. Mice, as described in legend of fig. 2, were injected with ^{3}H-L-DOPA. Complete description in ref. 39.

Conclusion

The major issue in studying the regulation of cell function during aging is to determine the inter-relationships between the change of endocrine gland function (reviewed in [1]) and the effects of these changes on target cell function. Because the regulatory hormones of these systems are in many cases sensitive to other hormones as well as to those they regulate, an interlocking network of changes may develop during adult life. The reduced secretion of growth hormone after menopause [44] illustrates this phenomenon. Such inter-related changes may be described as a neuro-endocrine 'cascade' and could be pacemakers for many cellular and physiological changes during aging.

Acknowledgements

This research was supported by grants to Caleb E. Finch from NSF (GB-35236) and NIH (HD-07539 and AG-00446) by NIH training grant HD-157 and by a grant to Dr. Christian Holinka from the Population Council.

References

[1] C.E. Finch, The The regulation of physiological changes during mammalian aging. Q. Rev. Biol. 51, 49–83 (1976).
[2] C.E. Finch, J.R. Foster and A.E. Mirsky, Aging and the regulation of cellular activities during exposure of cold. J. Gen. Physiol. 54, 690–712 (1969).
[3] R.C. Adelman, Age-dependent effects in enzyme induction; a biochemical expression of aging. Exp. Geront. 6, 75–87 (1971).
[4] H.H. Samuels and G.M. Tomkins, Relation of steroid structure to enzyme induction in hepatoma tissue culture cells. J. Mol. Biol. 52, 57–74 (1970).
[5] F.T. Kenney, Induction of tyrosine-α-ketoglutarate transaminase in rat liver, IV. Evidence for an increase in rate of enzyme synthesis. J. Biol. Chem. 237, 3495–3498 (1962).
[6] K. Latham and C.E. Finch, Hepatic Glucocorticoid binders in mature and senescent C57BL/6J male mice. Endocrinol. 98, 1434–1443 (1976).
[7] N.L.R. Bucher and A.D. Glinos, The effect of age on regeneration of rat liver. Canc. Res. 10, 324–332 (1950).
[8] F. Bourlière and R. Molimard, L'action de l'age sur la regénération du foie chez le rat. C. R. Soc. Biol. 151, 1345–1348 (1957).
[9] A.E. Treloar, R.E. Bounton, B.G. Bohn and B.W. Brown, Variation of the human menstrual cycle through reproductive life. Int. J. Fertil. 12, 77–126 (1967).
[10] G.B. Talbert, Effect of maternal age in reproductive capacity. Am. J. Obstet. Gyn. 102, 451–477 (1968).
[11] C.A. Finn, The biology of decidual cells. Adv. Reprod. Physiol. 5, 1–26. A. McLaren (ed.) Logos Press: London (1971).
[12] G.C. Blaha, Effects of age, treatment, and method of induction on deciduomata in the Golden Hamster. Fertil. Steril. 18, 477–485 (1967).

[13] M. Shapiro and G.B. Talbert, The effect of maternal age on decidualization in the mouse. J. Geront. 29, 145–148 (1974).
[14] P.D. Feil, S.R. Glasser, D.O. Toft and B.W. O'Malley, Progesterone binding in the mouse and rat uterus. Endocrinol. 91, 738–746 (1972).
[15] P. Feherty, D.M. Robertson, H.B. Wayneforth and A.E. Kellie, Changes in the concentration of high affinity estradiol receptors in rat uterine supernate preparations during the estrous cycle, pseudopregnancy, pregnancy, maturation, and after ovariectomy. Biochem. J. 120, 837–844 (1970).
[16] G.C. Blaha, The influence of ovarian grafts from young donors on the development of transferred ova in the aged golden hamster. Fertil. Steril. 21, 268–273 (1970).
[17] B.M. Sherman and S.G. Korenman, Hormonal characteristics of the human menstrual cycle throughout reproductive life. J. Clin. Invest. 55, 699–706 (1975).
[18] H.G. Kaplan and M.M. Hreshchyshen, Gas-liquid chromatographic quantitation of urinary estrogens in non-pregnant women, postmenopausal women, and men. Am. J. Obstet. Gyn. 111, 386–390 (1971).
[19] E. Block, Quantitative morphological investigations of the follicular system in women. Variations at different ages. Acta Anat. 14, 108–123 (1952).
[20] E.C. Jones and P.L. Krohn, The relationships between age, numbers of oocytes, and fertility in virgin and multiparous mice. J. Endocrinol. 21, 469–496 (1962).
[21] A.M. Mandle and M. Shelton, A quantitative study of oocytes in young and old nulliparous laboratory rats. J. End. 18, 444–450 (1959).
[22] M.S. Harman and G.B. Talbert, The effect of maternal age on ovulation, corpora lutea of pregnancy, and implantation failure in mice. J. Reprod. Fert. 23, 33–39 (1970).
[23] L.L. Larson, C.H. Spilman, H.D. Dunn and R.H. Foote, Reproductive efficiency in aged female rabbits given supplemental progesterone and estradiol. J. Reprod. Fertil. 33, 31–38 (1973).
[24] I.H. Thornycroft and A.L. Soderwall, The nature of litter size loss in senescent hamsters. Anat. Rec. 165, 343–348 (1969).
[25] A. Costoff and V.B. Mahesh, Primordial follicles with normal oocytes in the ovaries of postmenopausal women. J. Am. Ger. Soc. 23, 193–196 (1975).
[26] G.B. Talbert and P.L. Krohn, Effect of maternal age on viability of ova and uterine support of pregnancy in mice. J. Reprod. Fert. 11, 399–406 (1966).
[27] R.D. Peppler, Effects of unilateral ovariectomy on follicular development and ovulation in cycling, aged rats. Am. J. Anat. 132, 423–428 (1972).
[28] J.W. Everett, The restoration of ovulatory cycles and corpus luteum formation in persistent estrous rats by progesterone. Endocrinol. 27, 681–686 (1940).
[29] J.A. Clemens, Y. Amenomori, T. Jenkins and J. Meites, Effects of hypothalamic stimulation, hormones, and drugs on ovarian function in old female rats. Proc. Soc. Exp. Biol. Med. 132, 561–563 (1969).
[30] S.K. Quadri, G.S. Kledzik and J. Meites, Reinitiation of estrous cycles in old constant-estrous rats by centralacting drugs. Neuroendocrinol. 11, 248–255 (1973).
[31] H.H. Huang and J. Meites, Reproductive capacity of aging female rats. Neuroendocrinol. 17, 289–295 (1975).
[32] M. Linnoila and R.L. Cooper, Reinstatement of vaginal cycles in aged non-cycling female rats (Abstr.). Gerontologist 15 (part II) 30 (1975).
[33] C.A. Wilson, Hypothalamic amines and the release of gonadotrophins and other anterior pituitary hormones. Adv. Drug Res. 8, 119–204 (1974).
[34] S.P. Kalra and S.M. McCann, Effect of drugs modifying catecholamine synthesis on LH release induced by preoptic stimulation in the rat. Endocrinol. 93, 356–362 (1973).
[35] S. Spector, D. Prockop, P.A. Shore and B.B. Brodie, The effect of iproniazide on brain levels of norepinephrine and serotonin. Science 127, 704 (1957).

[36] C.W. Beattie, C.H. Rodgers and L. Soyka, Influence of ovariectomy and ovarian steroids on hypothalamic tyrosine hydroxylase activity in the rat. Endocrinol. 91, 276–279 (1972).

[37] J.F. Nelson, K.R. Latham and C.E. Finch, Plasma testosterone levels in C57BL/6J male mice: Effects of age and disease. Acta Endocrinol. 80, 744–752 (1975).

[38] A. Vermeulen, R. Rubens and L. Verdonck, Testosterone secretion and metabolism in male senescence. J. Clin. Endocrinol. Metab. 34, 730–735 (1972).

[39] C.E. Finch, Catecholamine metabolism in the brains of aging male mice. Brain Res. 52, 261–276 (1973).

[40] C.E. Finch, V. Jones, G. Hody, J.P. Walker, W. Morton-Smith, A. Alper and G.J. Dougher, Aging and the passage of L-tyrosine and insulin into mouse brain slices in vitro. J. Geront. 30, 33–40 (1975).

[41] J. Glowinski, Some characteristics of the 'functional' and 'main storage' compartments in central catecholaminergic neurones. Brain Res. 62, 489–493 (1973).

[42] M.T. Hyyppä, Metabolic rates and multiple functional pool of brain biogenic amines: their significance in neuroendocrine regulation. Medical Biology 52, 170–175 (1974).

[43] A.G. Frantz and M.T. Rabkin, Effects of estrogen and sex difference on secretion of human growth hormone. J. Clin. Endocrinol. 25, 1470–1480 (1965).

[44] L.M. Demers, K. Yoshinaga and R.O. Greep, Uterine glycogen metabolism of the rat in early pregnancy. Biol. Reprod. 7, 297–304 (1972).

[45] J.A. Katzenellenbogen, H.J. Johnson, Jr. and K.E. Carlson, Studies on the uterine, cytoplasmic estrogen binding protein. Thermal stability and ligand dissociation rate. An assay of empty and filled sites by exchange. Biochemistry 12, 4092–4099 (1973).

Chapter 25

INTRODUCTION OF PROFESSOR JAMES BONNER, CALIFORNIA INSTITUTE OF TECHNOLOGY, PASADENA, CALIFORNIA – A SYMPOSIUM HONORING THE OCCASION OF HIS 65th BIRTHYEAR

JAMES L. LIVERMAN (PhD., 1952, Cal. Tech.)

Assistant Administrator for Environment and Safety, ERDA

There are many things that I could say about James Bonner, the Professor and the man. You have followed his distinguished scientific career for many years and can tell me much more about that aspect of his life than I. I believe that Gerald Fasman in his comments epitomized what most of us feel about James when he said that 'we are all certain that James Bonner will continue to provide that bountiful scientific intellectual feast for the next 40 years as well as he has for the past 40'. Few here question the easy flow of ideas from James and most of the time they end up hitting the jackpot.

It is about some other aspects of James Bonner that I want to spend my few introductory comments, however, for in my own case and certainly I feel in the cases of many others of those here it is James' general philogopy of the place of science in man's total endeavors that are as much a mark of his greatness as is his science.

I can best illustrate what I mean by taking a few instances from my own association with James in the early days of my scientific career for which I hope you will forgive me.

My first encounter with James personally was after I had entered Cal Tech. in the fall of 1949 and was run through that which I fondly recall as the 'inquisition' – a session with a group of senior professors to determine the extent of a new graduate student's knowledge base. There were a number of giants – Sturtevant, Borsook, Beadle, Mitchell, Haagen-Smit, Emerson, Horowitz, Nieman and one or two others – who gave me a four hour verbal quizzing. If the purpose was to make me feel humble – it did. Note that James Bonner was not there, he was on a world trek to some unknown place.

When he returned, of course, I rushed into his office to tell him what an ordeal I had been through but he disposed of it in a very straightforward manner: 'Jim, you NOW know what you don't know, so get busy with it and

The Molecular Biology of the Mammalian Genetic Apparatus:
edited by P. Ts'o © 1977, Elsevier/North-Holland Biomedical Press

learn it if you hope to become a first rate scientist. Don't loaf, don't lick your wounds, pick it up and move'.

Another of James' characteristics that has come to be of manifold usefulness relates to his simplistic approach to complex matters. A name by which you will recognize this characteristic is what has come to be called by those who know him well and EVEN by some of his 'detractors' as 'bonnerisms'.

Bonnerisms are those little phrases that paint a clear picture of the matter under discussion in a few direct words. One learns soon around James to eliminate a lot of words that only contribute fog to the discussion even if in the eyes of some there is injury done to the 'sacred dogmas' expressed with 'sacred jargons'. In discussing these matters with James upon occasion he made it very clear: 'Jim, the intent is to make the matter understandable to the people with whom you talk. If you can't tell the story in an uncomplex manner, there may even be a question as to whether you have a clear concept yourself of what you are talking about. Tell it straight, tell it simple, tell it with feeling and tell it understandably. If there is a black-board near, use that, too, if it will help tell the story'. My most recent successful negotiation with the Japanese government was totally due to the benefits I was able to point out by using simple direct words and the black-board to draw pictures.

Because of this characteristic of James', to simplify instead of make complex, he has been able to carry the complexities of science into practical usefulness, to make science a workhorse of man while at the same time continuing to explore the most basic concepts – being equally understood by the layman, the scientist, the housewife and by the Congressman who must vote to appropriate the funds you basic researchers need.

There is another characteristic about James that in my own view makes him stand out as a beacon in a world of darkness on a cold stormy night. That simple characteristic is – get new ideas often, give them away, get other people to work on them, and both you and the world will benefit. You will rest more comfortably at night, your friends will secretly thank you.

My own experience in learning that lesson comes from a time that I gave a seminar and discovered, as happens quite often, I'm afraid, that about 6 weeks later that idea had been picked up by a colleague at another university and a few experiments later a paper was published. Naturally, I was unhappy.

Do you know what James said to me, 'Jim, there are two lessons to learn from this event – when you have the data to publish, do it, and secondly, Jim, if that is the only idea you ever have, then you are no damn good as a scientist anyway and might as well know it now. If you have more ideas, then use them. Give this one away and the world will be better and you will not have lost but gained'.

Let me discuss only one more characteristic which has to do with James' own life's work and it goes back in some respects to his upbringing in Utah where I'm sure his father and mother instilled in him the simple thought that whenever possible, do good for mankind.

James seems to me to follow that precept. No matter how fundamental his research may be, he seeks out ways to convert that knowledge to useful application, thus he is putting science to work for the betterment of mankind at the same time that he continues to labor in the vineyard of basic research himself. Finally, in times of real need and sorrow James gives you the backbone and freshness to pick yourself up to begin again in spite of odds against recovery.

In summarizing my comments let me characterize James as a man of tremendous intellect, with a dedication to good science, with a commitment to telling the story of science to the lay public, with a determination to apply fundamental findings to man's use, and with the capacity to be a great simplifier of complex concepts.

Above all, when all is said and done, James Bonner is one of the finest human beings I have ever known.

It is with the greatest of pleasure then that I have been afforded the honor of introducing my mentor – James Bonner – statesman, scientist and outstanding fellow citizen to address you on a topic of his own particular choosing.

Chapter 26

MY LIFE AS A CHROMOSOMOLOGIST

JAMES BONNER

California Institute of Technology, Division of Biology, Pasadena, California 91125, U.S.A.

I learned from Paul O.P. Ts'o, in the early 1950's, that it is more interesting to study fundamental biological problems than to study plant hormones. My interest turned first to protein synthesis and then to ribosomes and ribosomal structure. In 1956, Robert Holley in my laboratory discovered what turned out to be transfer RNA. Due to all this interest in RNA an important question was pondered: Where do ribosomal RNA, transfer RNA and all the RNAs come from? As a result my colleague, Ru-Chih Huang and I discovered in early 1960 that homogenates of pea seedling axes possess the capability of turning nucleoside triphosphates into acid insoluble, RNase digestible material. Our first paper on this subject was published in Biochemical and Biophysical Research Communications in 1960 [1]. Dr. Huang and I then proceeded immediately to try to purify both the enzyme responsible for the polymerization of RNA and the basic entity which promotes such synthesis. We were able to purify the enzyme, RNA polymerase, to a considerable extent from pea seedling axes. The interphase chromosomes, and chromatin of the cells of pea seedling axes comprised the preparation from which we derived the enzyme, and the preparation also served as a template for the enzyme.

By 1962 we discovered how to dissociate proteins from DNA, a substance which we found necessary for RNA synthesis in our system, and, thus, were able to purify the DNA and separate if from the chromosomal proteins by cesium chloride density gradient centrifugation. Such purified DNA proved to be a much better template for RNA synthesis than interphase chromatin itself. We also found that upon reassociation of chromosomal proteins with DNA, the DNA became a considerably poorer template. Thus, we had visions of the control of gene expression [2]. We felt, as I think everyone does today, that comparing the function of chromatin as a template for RNA transcription with the DNA from deproteinized chromatin would demonstrate that in higher creatures only a portion of the genome is expressed or transcribed in any individual kind of cell.

The Molecular Biology of the Mammalian Genetic Apparatus:
edited by P. Ts'o © 1977, Elsevier/North-Holland Biomedical Press

We then learned how to purify chromatin by centrifugation through 1.8 M sucrose [3]. The proteins of such purified interphase chromosomes are mainly histones. At that time, however, very little was known of histone chemistry.

The exciting results of the previous two and one-half years convinced us that it was time to convene a worldwide symposium to consider histone chemistry and chromosome biology. As a result, Paul Ts'o and I organized a symposium held at the Rancho Sante Fe in May 1963 entitled 'The First World Conference on Histone Biology and Chemistry'. This resulted in a published volume, 'The Nucleohistones', edited by Bonner and Ts'o and published by Holden-Day in 1964 [4] in which each of the speakers at the symposium had an opportunity to have his talk printed. In essence, it turned out that little was known about histone chemistry, and that views about their function differed radically, ranging from histones are glue that sticks DNA strands together, to histones are specific repressors of individual genes. In general, it was clear that there was no unanimity about any aspect of the control of gene expression in eukaryotic chromatin.

Conferences with Professor Ts'o and Professor Vinograd convinced Ru-Chih Huang and myself that the most sensible approach was not to attempt to reform the current generation of histone chemists, but to educate a new generation of histone chemists who would separate histones and study them to determine their number, variety and structure. To this end we persuaded Douglas Fambrough to go to Stanford for a few weeks early in 1964 to study the separation of histones from one another by Amberlight IRC 50 chromatography (now called BioRex P70) with Kenneth Murray, then a postdoctoral fellow with Professor J. Murray Luck in the Department of Chemistry at Stanford (K. Murray is now Professor of Molecular Biology at the University of Edinburgh.) Douglas Fambrough returned full of enthusiasm. During the next several years he showed that there are five basic species of histones now called H1, H2A, H2B, H3 and H4 [5,6,7], and that the multiplicity of histones envisaged by earlier workers was due, in large part, to the fact that histone 3 contains cysteine, not previously believed to be a component in any histone [8]. As a result of the oxidation of cysteine residues, H3 could polymerize to form multimers of various sizes and kinds. Polyacrylamide gel electrophoresis was first developed and applied to histones by Douglas Fambrough and further elaborated using gels with greater lengths by my former colleague, Roger Chalkley. I recruited Roger from Oxford University. He was a physical organic chemist by education but a chromosomologist by inclination. He did many good experiments. In any case, with the electrophoretic tool and tryptic fingerprints Fambrough showed further that the histones of pea seedling axes and of calf thymus are exceedingly similar to one another.

By 1966 the histone chemistry detour had been pretty well accomplished. We now knew that the same basic histones are present in the chromatins of the different organs of each creature, that the histones of different creatures are very similar and that it is possible to purify individual histones into pure proteins. At the National Academy meetings of April 1966 I spoke with Emil Smith, Professor of Biological Chemistry, UCLA, about the interesting qualities of the histones: the fact that we could make pure histones, that they seemed to be conserved between different creatures, and that it would be of value to sequence them. Samples supplied to Emil Smith and turned over by him to Professor Robert DeLange proved to be pure by their criteria as well as by ours. For the sequencing for the first histone, namely, histone 4 (smallest and most readily purifiable) Bob DeLange and Emil Smith laid down the condition that we prepare 2 grams of each from pea seedlings and calf thymus. In the summer of 1967 Douglas Fambrough, with the help of three undergraduate assistants, succeeded in preparing two grams of pure histone 4 from pea seedlings. This required the germination of 25 tons of dried pea seeds, the mechanical separation of the seedling axes from the germinated pea seedlings, and the preparation of chromatin from the pea seedling axes thus produced, as well as production of two grams of pure histone 4 from calf thymus acquired from a local slaughterhouse. Upon obtaining two grams of each histone, Bob DeLange sequenced them and made the now famous discovery that histone 4 of peas and cows differ by only two conservative amino acid substitutions – a lysine for an arginine and a valine for an isoleucine. Thus, histone 4 is the most conserved protein found up to that time, having been conserved during the entire span from perhaps 600 million to 1 billion years ago to the present. This all appeared in PNAS in October 1968 [9] and in more detail in the JBC articles of 1968–69 [10,11,12]. As an acknowledgement, a small editorial in NATURE noted that 'This is the first evidence that histones are something other than glue' [13].

During all this time our group had been active on other fronts. We had been inquiring whether chromatin, as prepared by us, expresses the same sequences in vitro as are expressed in vivo. To this end we prepared chromatin from two different organs of the pea plant – the pea seedling axes, on the one hand, and the developing pea cotyledon on the other. The developing pea cotyledon makes pea seed globulin, which is about 9% of the total protein synthesized in the pea seedling cotyledon. The pea axes make no pea seed globulin. We arranged to detect pea seed globulin by an immunochemical method devised by Ray Gilden. We set up a transcription-translation system in which pea chromatin was transcribed by *E. coli* RNA polymerase, and the RNA thus transcribed was translated by a messenger RNA-free *E. coli* translational system. (We termed this system pea coli.) It was found that chromatin from pea seedling axes does not support the generation of messenger RNA for pea seed globulin, while the chromatin from pea cotyledon does support the synthesis of pea seed globulin [14]. This was the first demonstration that isolated chromatin possesses some fidelity with respect to which genes are turned on and which genes are turned off.

As everyone knows, serious biological experiments have to be done using either *E. coli* or mammals. We therefore undertook to show that what we had found for pea seedling chromatin and its transcription and translation is also true for mammals. We chose the rat. In 1963, a postdoctoral fellow, Keiji Marushige, arrived and set out to isolate chromatin from rat liver and study its properties. The first rat he tried to disliver bit him; but in spite of this disheartening start, he proceeded to find out how to isolate rat liver interphase chromosomes (rat liver chromatin), how to transcribe it, to learn that transcription by *E. coli* RNA polymerase is asymmetric, and to show that 20% or less of the DNA in rat liver chromatin is available for transcription. The method of Marushige, published in the Journal of Molecular Biology in 1966 (15), is an important milestone in the study of chromatin. A review article concerning the whole subject of chromatin (plants, animals and regulation of gene expression in higher organisms) was prepared by our group in 1967 and appeared in Science in January 1968 [16]. This review article also served as a sort of focus for our thoughts at the time and had some influence on others regarding chromatin, as some 4,000 reprints were requested and distributed. We had also prepared an article which appeared in Methods in Enzymology (Vol. 12B: pp. 3–60, 1968) on the isolation of chromatin and the study of individual components. This article has, I think, also had a considerable impact on chromatin research.

One of the important questions in the study of chromatin impressed upon us by our early results in 1962 was how DNA sequences, initially unavailable for transcription, become available for transcription. There are many ways to approach this and we have tried several of them. The very first way employed was the activation of liver chromatin by dihydrocortisone in rats previously deprived of their adrenal glands, resulting in decreased levels of cortisone. Michael Dahmus showed in 1965 that the administration of dihydrocortisone to such adrenalectomized rats resulted in an increase by about 35% in the amount of rat liver chromatin DNA available for transcription by *E. coli* polymerase [18]. Similar results have been found in subsequent years by many investigators studying a vast variety of hormone systems. This method is well-founded. The inner workings of the cell from the time the hormone enters the cell, binds to its receptor protein, which in turn binds to the chromosome and, at this point, turns on DNA sequences to transcriptional form are as yet unknown, despite our continuing attempts to unravel the mystery. A further similar attempt was made in the 1970's by William T. Garrard. We tried to find out more about how the massive turning on of the genes in regenerating liver occurs. When two-thirds of the liver of the rat is removed, the remaining one-third starts to synthesize DNA after 18 h and undergoes mitosis after 24 h, then again approximately every 24 h until all the liver is regenerated. Bill Garrard showed that after 6 h the template activity (transcribability) of rat liver regenerating chromatin increases about 40% over that of normal adult liver chromatin; that is, a vast number of new DNA sequences have been made available for transcription. He showed very clearly that the histone-to-DNA ratio of the chromatin decreases when some histones are removed by proteolysis. We have subsequently shown that there is a chromosomal protease which rapidly reduces histones to little peptides [20]. The mode of activation of this enzyme is unknown, as is the mechanism involved in the recognition of particular sequences of DNA which are to be depleted of their histones by the protease. In general, up to the present time and in spite of a great deal of work on the mechanism of hormone activation of gene expression and many others, we still do not understand how a small molecule such as an estrogen or other hormone can bring about the activation of a DNA sequence previously repressed. But we will find out.

It is now fashionable to suppose that since the histone molecules bind to DNA without sequence specificity and cannot discriminate between the DNA sequences they bind to, the non-histone chromosomal proteins must, therefore, play the major role in determining which genes are expressed and which are not expressed. A vast spate of papers have been published which attempt, in one fashion or another, to support this point of view. Interestingly enough, it has turned out that about half of the total mass and all of the major non-histone chromosomal proteins of chromatin are structural proteins – actin, myosin, tropomyosin, alpha and beta tubulin, the RNA packaging proteins and so on [21,22]. Nonetheless, the experiments of Hnilica, Stein, Paul and others do persuasively argue that the transplantation of one population of non-histone chromosomal proteins to another kind of chromatin causes genes to be turned on which were previously turned off. This can be tested by the method known as chromatin reconstitution. Such reconstitution was first suggested by Huang and Huang [26] and by Bekhor, Kung and Bonner [27]. The idea was to disassociate all of the chromosomal proteins from DNA by 2 M NaCl 5 M urea, which disassociates essentially all protein and other components from DNA. One can then gradient-dialyze the salt away and finally the urea also. This method causes chromatin to be reformed by reassociation of all components to about the same level of transcriptional activity as that of native chromatin. Crude measures of sequence specificity, that is, the study of the middle repetitive sequences expressed, indicated at that time some sequence fidelity of the reconstitution. This method has been used subsequently by all those who have attempted to determine whether or not non-histone chromosomal proteins influence sequence specific chromosomal reconstitution. The whole subject is presently in a state of flux. My guess is that the method works, but imperfectly. Our own findings are that the physical structure of the nonexpressed portion of the genome is not perfectly reformed. We will return to this later in the discussion. Whether the expressed portions are perfectly reformed is a matter requiring more discussion and a matter in which discrepancies arise.

Chromatin contains not only DNA, histones, and non-histone chromosomal proteins, but also RNA. A portion of this RNA is clearly nascent RNA, that is,

RNA that is being transcribed at the moment that the investigator harvests the chromatin. A further portion, about 4 or 5% of the genome in the case of pea seedling buds and rat ascites tumor cells, consists of a small, slowly labeled species of RNA, which we have termed chromosomal RNA. Chromosomal RNA has the following characteristics: (a) small, less than 100 nucleotides in length, (b) hybridizes to about 4% of the genome in the two cases mentioned above, (c) hybridizes almost exclusively to repetitive sequences, and (d) appears to have a vast sequence diversity, that is, to hybridize to many different families of repetitive genes [28,29,30,31]. Chromosomal RNA (as we have termed it) may also be a candidate for a role in the control of gene expression, although no experiments have rigorously shown that this is so. It has been argued that chromosomal RNA is an artifact and is produced by the breakdown of ribosomal or transfer RNA. However, this is not true. Chromosomal RNA is a real entity and we must elucidate its role. The small RNAs of the nucleus are coming into fashion again, as witnessed by the small RNAs sequenced by Harris Busch and his group and now are known to be associated with the packaging of heterogenous nuclear RNA.

Let us turn from the biology of chromatin to structure. We have spent a considerable amount of energy and effort to separate the transcribed portion of the genome from the non-transcribed portion. These efforts were first reported by Marushige and Bonner [32], although the experiments extended over a period of several years before the publication of our results. The method consists of shearing or breaking chromatin with DNAse II, which preferentially makes double stranded breaks. We then separated the DNA which is attacked readily by nuclease from that DNA which is much more resistant to nuclease, by precipitation of the latter in 0.15 M NaCl or later [33] by 2 mM magnesium chloride. Certainly, the physical properties of these two fractions with respect to precipitation are vastly different from one another. The precipitable fraction has a histone-to-DNA ratio of about 1 and is deficient in non-histones. The non-precipitated fractions, which form about 10% of the total in the case of liver chromatin, is lacking in histones and is very much enriched in non-histone chromosomal proteins. We have dwelt upon this matter and have shown (Gottesfeld et al. [34,35]) that the 10% of the genome which is not precipitated by magnesium chloride and which is deficient in histones consists of about 10% of the single copy sequence DNA of the whole genome and also of a subset of all the repetitive sequence families of the rat genome. About 60% is hybridizable to whole cell RNA, whereas the DNA of the magnesium precipitable fraction is not. We have shown also that the two fractions, which I will term the transcriptionally active and transcriptionally inactive portions of the chromatin, differ in physical properties. Thus, the CD spectra of transcriptionally active DNA is that of B-form DNA. The DNA of the transcriptionally inactive fraction has CD spectra of C-form DNA. The DNA of the transcriptionally active fraction is not stabilized against melting to any considerable degree, whereas the DNA of transcriptionally inactive chromatin is stabilized greatly against melting by the presence of histones. It seems to be generally agreed that fractionation of chromatin into template active and inactive fractions by DNAse II is a good method. It should aid us in understanding the control of gene expression.

By the same token that transcriptionally active sequences of chromatin can be broken and physically separated from the remainder by mild nuclease attack, so too can the expressed sequences be attacked and destroyed by more vigorous DNAse treatment of chromatin. This has been shown particularly clearly by Axel [36] and by Weintraub [37].

Beginning in 1974 Don and Ada Olins [38], Ken Van Holde [39] and others showed that interphase chromatin is largely organized in a 'beads on a string' structure, a finding which previously had been semi-evident in the electron micrographs of earlier investigators but not correctly interpreted. The beads are about 110 Å in diameter and are separated by about 0–100 Å (depending on the ionic strength) from the next bead by a spacer. The work of Jack Griffith had shown earlier that histone 1, which differs in many chemical and biological properties from other histones (it makes up only about 10% of the total histone molecules and its C-terminal half is basic, rather than the N-terminal half as in the other histones), is deposited at regular intervals about 150 base pairs apart, histone 1 to histone 1, along the DNA chain. This demonstration was accomplished by the high resolution electron microscopy of Jack Griffith and utilized DNA from chromatin in which histone 1 had been bound covalently to DNA by formaldehyde treatment by the method of Brutlag, Schlehuber and Bonner [40]. Histone 1 reacts about one order of magnitude more rapidly than do the other histones. H1 becomes covalently bound to DNA and the other histones can be eluted from the complex by density gradient centrifugation in cesium chloride. It would now appear that the other four histone species amalgamate into an octamer containing two molecules of each [41,42] and that the histone 1 occupies the spacer DNA between the DNA complexed to each octamer. The packing ratio, that is the extended length of DNA compared to the length of DNA in the structure of chromatin organized in this 'beads on a string' configuration, is about 7 to 1 [43,44].

Some 90% or more of the DNA of chromatin is organized in the 'beads on a string', or, as we now know it, nubody or nucleosome configuration.

The structure of the remaining 10% or so, which is the transcribable portion of the genome, is in a more extended configuration. It is now under investigation.

There are many more facets of chromatin structure and function which are of interest and importance, some of which have been reviewed in these volumes. An important aspect is the study of the arrangement of DNA sequences in the genome and the detailed mapping of gene arrangement that has been done so elegantly for ribosomal and for histone genes as by my former colleague,

Max Birnstiel. In conclusion, chromatin, the eucaryotic genome, still provides us with a small infinity of interesting questions requiring resolution.

One of the most satisfying aspects of my life as a chromosomologist has been the marvelous colleagues with whom I have had the opportunity to work. Among these, I note particularly those who have gone out into the world and established their own centers of chromosomology, starting with Paul Ts'o, Ru-Chih Huang, Keiji Marushige, Roger Chalkley, Douglas Fambrough (now defected to neurobiology), Max Birnstiel, Michael and Grace Dahmus, Douglas Brutlag, Sally Elgin, Ric Firtel, William Garrard, Sandy Sevall, Tom Shih, David Holmes, Isaac Bekhor, Gary Flamm, Jack Griffith, John Hearst, John Mayfield, David McConnell, Jack Widholm, Lynda Uphouse and Yuri Sivolap, to name only a few. I am proud of them and I still have many more colleagues coming along who will sooner or later join the above list, namely: Jung-Rung Wu, William Pearson, Tony Bakke, Michael Savage, Tom Sargent, Angie Douvas and Bruce Wallace. There is, however, one colleague who will not leave to go out into the world and establish a competing center and that is my colleague, companion and wife, Ingelore Bonner, without whose help all of the above could not have been accomplished.

References

[1] R.-Chih Huang, Nirmala Maheshwari and James Bonner, Enzymatic synthesis of RNA. Biochem. Biophys. Res. Comm. 3, 689–694 (1960).
[2] Ru-Chih Huang and James Bonner, Histone, a suppressor of chromosomal RNA synthesis. Proc. Nat. Acad. Sci. 48, 1216–1222 (1962).
[3] Ru-Chih Huang and James Bonner, Properties of chromosomal nucleohistone. J. Mol. Biol. 6, 169–174 (1963).
[4] James Bonner and Paul O.P. Ts'o (Eds.), The Nucleohistones. Holden-Day, Inc. San Francisco, California (1964).
[5] Douglas M. Fambrough and James Bonner, On the similarity of plant and animal histones. Biochemistry 5, 2563–2570 (1966).
[6] Douglas M. Fambrough, Frank Fujimura and James Bonner, Quantitative distribution of histone components in the pea plant. Biochemistry 7, 575–585 (1968).
[7] Douglas M. Fambrough and James Bonner, Limited molecular heterogeneity of plant histones. Biochim. Biophys. Acta 175, 113–122 (1969).
[8] Douglas M. Fambrough and James Bonner, Sequence homology and role of cysteine in plant and animal arginine-rich histones. J. Biol. Chem. 243, 4434–4439 (1968).
[9] Robert J. DeLange, Emil L. Smith, Douglas M. Fambrough and James Bonner, Amino acid sequence of histone IV: presence of εN-acetyllysine. Proc. Nat. Acad. Sci. 61, 7–8 (1968) (Abstract).
[10] Robert J. DeLange, Douglas M. Fambrough, Emil L. Smith and James Bonner, Calf and pea histone IV. I. Amino acid compositions and the identical COOH-terminal 19-residue sequence. J. Biol. Chem. 243, 5906–5913 (1968).
[11] Robert J. DeLange, Douglas M. Fambrough, Emil L. Smith and James Bonner, Calf and pea

histone IV. II. The complete amino acid sequence of calf thymus histone IV; presence of ε-N-acetyllysine. J. Biol. Chem. 244, 319–334 (1969).
[12] Robert J. DeLange, M. Douglas Fambrough, Emil L. Smith and James Bonner, Calf and pea histone IV. III. Complete amino acid sequence of pea seedling histone IV; comparison with the homologous calf thymus histone. J. Biol. Chem. 244, 5669–5679 (1969).
[13] Our Cell Biology Correspondent. Nature 223, 892 (1969).
[14] James Bonner, Ru-Chih C. Huang and Raymond V. Gilden, Chromosomally directed protein synthesis. Proc. Nat. Acad. Sci. 50, 893–900 (1963).
[15] Keiji Marushige and James Bonner, Template properties of liver chromatin. J. Mol. Biol. 15, 160–174 (1966).
[16] James Bonner, Michael E. Dahmus, Douglas Fambrough, Ru-Chih Huang, Keiji Marushige and Dorothy Y.H. Tuan, The biology of isolated chromatin. Science 159, 47–56 (1968).
[17] James Bonner, Roger Chalkley, Michael Dahmus, Douglas Fambrough, Frank Fujimura, Ru-Chih C. Huang, Joel Huberman, Ronald Jensen, Keiji Marushige, Heiko Ohlenbusch, Baldomero Olivera and Jack Widholm, Isolation and characterization of chromosomal nucleoproteins. Methods Enzymol. 12, 3–65 (1968).
[18] Michael E. Dahmus and James Bonner, Increased template activity of liver chromatin as a result of hydrocortisone administration. Proc. Nat. Acad. Sci. 54, 1370–1375 (1965).
[19] William T. Garrard and James Bonner, Changes in chromatin during liver regeneration. J. Biol. Chem. 249, 5570–5579 (1974).
[20] Ming Ta Chong, William T. Garrard and James Bonner, Purification and properties of a neutral protease from rat liver chromatin. Biochemistry 13, 5128–5134 (1974).
[21] Sarah C.R. Elgin and James Bonner, Limited heterogeneity of the major non-histone chromosomal proteins. Biochemistry 9, 4440–4447 (1970).
[22] Angeline S. Douvas, Christina A. Harrington and James Bonner, Major non-histone proteins of rat liver chromatin: Preliminary identification of myosin, actin, tubulin, and tropomyosin. Proc. Nat. Acad. Sci. 72, 3902–3906 (1975).
[23] T. Spelsberg and L. Hnilica, Proteins of chromatin in template restriction I. RNA synthesis in vitro. Biochim. Biophys. Acta, 228 202–211 (1971).
[24] G. Stein, J. Stein, C. Thrall and W. Park, Regulation of histone gene transcription during the cell cycle by non-histone chromosomal proteins. In: Chromosomal Proteins and Their Role in the Regulation of Gene Expression, G. Stein & L. Kleinsmith (Eds.) Academic Press, N.Y. pp. 1–17 (1975).
[25] S. Gilmour and J. Paul, The in vitro transcription of the globin gene in chromatin. Idem. pp. 19–33.
[26] Ru-Chih Huang and P.C. Huang, Effect of protein-bound RNA associated with chick embryo chromatin on template specificity of the chromatin. J. Mol. Biol. 39, 365–378 (1969).
[27] Isaac Bekhor, Grace M. Kung and James Bonner, Sequence-specific interaction of DNA and chromosomal protein. J. Mol. Biol. 39, 351–364 (1969).
[28] Ru-Chih Huang and James Bonner, Histone-bound RNA, a component of native nucleohistone. Proc. Nat. Acad. Sci. 54, 960–967 (1965).
[29] James Bonner and Jack Widholm, Molecular complementarity between nuclear DNA and organ specific chromosomal RNA. Proc. Nat. Acad. Sci. 57, 1379–1385 (1967).
[30] Y. Sivolap and James Bonner, Association of c-RNA with repetitive DNA. Biochemistry 10, 1461–1470 (1970).
[31] David S. Holmes, John E. Mayfield, Gernot Sander and James Bonner, Chromosomal RNA: Its properties. Science 177, 72–74 (1972).
[32] Keiji Marushige and James Bonner, Fractionation of liver chromatin. Proc. Nat. Acad. Sci. 68, 2941–2944 (1971).

[33] Ronald James Billing and James Bonner, The structure of chromatin as revealed by deoxyribonuclease digestion studies. Biochim. Biophys. Acta 281, 453–462 (1972).

[34] J.M. Gottesfeld, W.T. Garrard, G. Bagi, R.F. Wilson and J. Bonner, Partial purification of the template-active fraction of chromatin. A preliminary report. Proc. Nat. Acad. Sci. 71, 2193–2197 (1974).

[35] Joel M. Gottesfeld, György Bagi, Beckie Berg and James Bonner, Sequence composition of the template-active fraction of rat liver chromatin. Biochemistry 15, 2472–2483 (1976).

[36] E. Lacy and R. Axel, Analysis of DNA of isolated chromatin subunits. Proc. Nat. Acad. Sci. 72, 3978–3982 (1975).

[37] H. Weintraub and M. Groudline, Chromosomal subunits in active genes have altered conformation. Science 193, 848–856.

[38] A. Olins and D. Olins, Spheroid chromatin units (v-bodies). Science 183, 330–332 (1974).

[39] C.G. Sahasrabuddhe and K. Van Holde, The effect of trypsin on nuclease resistant chromatin fragments. J. Biol. Chem. 249, 152–161 (1974).

[40] Douglas Brutlag, Cameron Schlehuber and James Bonner, Properties of formaldehyde-treated nucleohistone. Biochemistry 8, 3214–3218 (1969).

[41] J. DiAnna and I. Isenberg, Interactions of histone LAK with histones KAS and GAK. Biochemistry 13, 2098–2104 (1974).

[42] R. Kornberg and J. Thomas, Chromatin structure: Oligomers of the histones. Science 184, 865–868 (1975).

[43] M. Noll, Subunit structure of chromatin. Nature 251, 249–251 (1974).

[44] Jack Griffith, Chromatin structure deduced from a minichromosome. Science 187, 1202–1203 (1975).

SUBJECT INDEX

Aberrations, 208, 209
Absorption cytophotometry, mammalian chromosome, 191, 192
Acetabularia, tDNA 79
Adenovirus, 5, 280
   effects on chromatin, 217–220
Aging
   clonal senescence, 289–300
   endocrine control, 303–310
   error catastrophy theory, 290, 292, 300
Aneuploidy, 205, 230, 234
Antigens, cell surface, 171–176
Antisera
   against human cells, 174, 175
   fluorescent antibody, 176
Aorta, aging in, 290, 291
Autoradiography, 215
Ara-cytidine transformation, 230, 231
Avian leukosis viruses, 280
Avian sarcoma virus, 277–285
   $cDNA_{sarc}$, 278–282
   $cDNA_{gp}$, 278–283
   d strains, 278
   gene, 277–285
   RNA, 280
   td strains, 278
8-Azagaunine resistance, 245, 248

Benzo(a)pyrene, 206, 208, 212, 213, 244
   effects on human leukocytes, 212
Biotin-tRNA conjugate, 2, 3
Bromodeoxyuridine, 244, 259

Cancer, various types, 206, 207, 229–237
Carcinogens, effect, 207, 208
Catecholamine turnover, aging effect, 309, 310
$cDNA_{gp}$, 278, 279, 283
$cDNA_k$, 125
   hybridization with nuclear Hg-RNA, 130
   reverse transcribed copy, 125
$cDNA_{sarc}$, 278–282
Cell fusion experiments on aging, 291, 295
   complementation, 292, 293
   hybridization, 292
   karyoplasts, 298
   mixploid clones, 294, 295
   population doubling, 290
   recombinant homozygous diploid cells, 295
   synkaryon experiments 292, 295
      interspecific synkaryons, 295
   tetraploid hybrids, 292, 294
   thymidine labeling index, 299
Centromere, 43
Chinese hamster
   cells, 166–168
      chromosomal DNA distribution, 198, 202
   chromosomes, 195, 199, 200
   ovary cell, (CHO), 173–177
Chromatin
   centromeric heterochromatin, 43
   digestion by deoxyribonuclease II, 149, 155
   erythroid cells, 141–145
      fractionation, 143–145
   interphase nucleus, 262
   nuclease digestion, 149–156
   preparation, 150

reassociation kinetics, 157, 158
reconstitution, 149–160
  nuclease digestion, 156
  reassociation kinetics, 157, 158
Chromosome
  absorption cytophotometry, 192
  Chinese hamster, 195, 198–200
  DNA content, 193–203
  flow fluorometry, 191, 196, 201, 202
  fractionation, 196–203
  gene transfer, 163–168
  human, 193, 202
    loss of in cell hybrid, 174
  image analysis, 191–196
  Indian muntjac, 201
  markers (transformed cell)
    human, 233
    mouse, 232
    Syrian hamster, 219, 220, 231
  in neoplastic transformation, 205–224, 229–237
    adenovirus effect, 217, 219, 220
    ara-cytidine transformation, 230
    carcinogen effect, 207, 208
    guinea pig, 220, 221
    human leucocytes, 212
    irradiation effect, 208, 213, 214
    marker chromosome, 219–223, 231
    rat cells, 215
    suppressor gene, 236
    Syrian hamster karyotypes, 210, 213, 230
  Philadelphia chromosome, 194
  sorting, 196–203
Colcemid, 177
Col El, 104
C_0t curve, see Reassociation kinetics
Cyclic AMP, 171, 176, 177
  DBcAMP (dibutylcyclic AMP), 177, 178
  effect on dendrites, 178
  effect on microtubular systems, 179
  effect on transformation, 177
  prostaglandin, 177
Cytochalasin, 298
  cytochalasin B, 177
Cytoplasts, 298

Dendrites, 178
Deoxyribonuclease II, chromatin digestion, 149, 155
Dictyostelium discoideum DNA, restriction study and R loop formation, 26

7,12-dimethyl-benz(a)anthracene, 206, 215, 221
  effect on rat cell, 214, 215
Dimethylsulphoxide (DMSO), 138–141
Diploidization, 294, 295
DNA
  5S DNA
    injection to fertilized eggs, 113
    in vitro transcription, 126
    Xenopus, 111
  aggregates, plant, 70–72
  angle of unwinding, by ethidium, 106
  cDNA, globin, 137, 143
  $cDNA_{gp}$, 278, 279, 283
  $cDNA_k$, 125
    hybridization with nuclear Hg-RNA, 130
  $cDNA_{sarc}$, 278–282
  closed circular, 101
  content in chromosomes, 193–196, 199, 202, 203
  cyclization reaction, 269, 270
  Dictyostelium discoideum, 26
  distribution, chromosome, 195, 198, 199
  DNA:DNA duplex, nucleic acid hybridization, 4
  duplex rotation angle, 106
  flow-fluorometric analysis, 201, 202
  genome size, see DNA sequence organization
  globin cDNA, 137, 143
    in situ hybridization, 142
  globin gene, 141, 142, 145
  immunoglobulin kappa chain gene, 132, 133
  injection into fertilized eggs, 113–122
    intracellular location, 115
    persistence, 115, 116, 117
    toxicity, 114
    transcription, 120, 121
  ligase, 99–109
  organization, *Xenopus* 29, see *Xenopus*
  plant, see Plant DNA
  polymerase I, deficient cells for cloning studies, 19, 20
  reassociation kinetics,
    Apis mellifera (honeybee), 31–34
    Musca domestica (housefly), 30, 33–35
    rat 52, 57, 59
  repetitive, fraction in short elements, 30, 31
    see DNA sequence organization
  reverse transcribed copy ($cDNA_k$), 125
DNA:RNA hybridization, nucleic acid hybridization, 1–10

Subject index

DNA:rRNA hybridization, nucleic acid hybridization, 3, 5
rotation angle, temperature dependence, 106
sequence loss in transformation, 254–256
sequence organization, 29–37, 51–62
  Drosophila melanogaster, highly repeated sequences, 43–46
  genome size, 30, 31
  highly repeated sequences, 43–49
  in neoplastic transformation, 254–261
  plant DNA, *see* Plant DNA
  rat, 51–62
    DNA reassociation kinetics, 52
    foldback, 51–53, 58, 61
      electron microscopy, 54–56
      most rapidly reannealing sequences, 51, 53, 55
      repetitive DNA sequence, 51–53, 55, 57–61
        distribution, 57
  reassociation kinetics, 34, 35
  repetition frequencies, 32, 33
  repetitive DNA sequences
    fraction in short elements, 30, 31
    frequency of families, 30, 31
    functions of short interspersed, 36
    highly repeated, 43
    intermediately repetitive DNA sequences, 55, 57, 65
    length, 29, 31–36
  sea urchin histones genes 87–97, *see* Histone gene organization
single copy DNA sequence
  complexity, 32
  fraction interspersed, 30–34
  length, 29, 32
supercoils, 99, 101–104, 109
superhelix free energy, 104
Syrian hamster, 258
  mDNA, 258
  nmDNA, 258
  repetitive sequence transcript, 258
topological isomers, 101
topological winding number, 100
transcriptional, system
  in vitro, 126, 127
transfecting heteroduplex, 271, 272
turns
  duplex, 101
  superhelical, 101, 108

Xenopus pattern, 29
DOPA, 308
Dopamine, 308
Drosophila
  DNA sequences and organization, 29, 30, 43
  hybrid plasmids, 47–49
  restriction enzyme site, 47–49
  satellite DNA, 43–46, 49
    sedimentation equilibrium, 44, 45

E. coli, RNA polymerase, 141, 142
Eco RI restriction sites, 15
Electron micrographs
  ϕx-174 DNA, 8
  plant DNA, 71
  rat
    foldback DNA, 54, 56
    repetitive DNA, 54
  rDNA *Tetrahymena*, 82
  sea urchin histone DNA, 94
Electron microscopy
  ferritin labeling method, 1, 2
    biotin-avidin reaction, 2, 3
    ferritin-avidin reaction, 3
  formamide-cytochrome c spreading method, 1
  gene cloning, R loop, 22–25
  gene mapping, 1–12
  rat DNA sequence organization, 54–56
  foldbacks, 51–53, 58, 61
Endocrine control, in aging, 303–310
Error catastrophe theory on aging, 290, 292, 300
Erythroid cells, 135, 136, 138, 145
  euchromatin, 143–145
  erythrocytes, 137
  heterochromatin, 143–145
  hybrids, 138–140
  lymphoma, 140
  maturation, 137
  mutants, 138
  non-histone proteins, 143, 144
  variants, 138, 139
Estrogen, 305, 306
  uterine receptor, 306
Ethidium, effect on DNA conformation, 106
Euchromatin, erythroid cells, 143–145
Evolution, plant DNA changes, 73–76

Female reproductive system
  aging, 304
  decidual cell response in the uterus, 305, 306

epithelial and stromal layers of the uterus, 306
neural mechanism control of the ovarian cycle, 307
ovarian follicles, 307
ovarian production of steroids, 307
uterine estrogen receptors, 306
Fibrinolytic activity, 246, 250, 254
Fibroblast growth, in aging experiments, 289, 295–300
Flow fluorometry, 191, 196, 197, 202
analyses, DNA loss in transformation 256 chromosome, 198
Friend erythroleukaemia cells, 136–138, 143

Galactokinase, 165–167
Gene, oncogenic virus
avian sarcoma virus *env*, 277, 283
avian sarcoma virus *onc*, 277, 285
avian sarcoma virus *gag*, 277
avian sarcoma virus *sarc*, 281–285
Gene 32 method for hybrid identification, 7
Gene cloning, 15–27
complementary RNA, 22
DNA polymerase I deficient cell plating, 19, 20
lysogenic selection, 21
nucleic acid hybridization, 23, 24
phage lambda, 15–21
physical screening and selection, 21, 22, 24
R-loop, 22–25
yeast DNA hybrid, 16–18
Gene mapping, 1
electron microscopy, 1–12
gene 32 method for identifying nucleic acid hybridization, 7
Hela mitochondrial 4S RNA genes, 3, 6 mapping, 4–6
Hela mitochondrial tRNA genes, 3
histone genes, 12
histone mRNA: chimeric plasmids, hybrid, 7, 9, 10
mRNAs, 9
R loop method, 7, 10, 11
rRNA genes, 3, 6
spacer, 6
Gene transfer, 163–168
chromosome, 163–168
human cells, 165–168
thymidine kinase, 165–167

Glucocorticoids, 303
Glucose-6-phosphate dehydrogenase, 292, 293, 300
Glycophorin, 176

Haem, 140
Haemoglobin, 136, 137
Hela
cell, 289, 296–298
mitochondrial 4S RNA genes, 3, 6
tRNA gene, 3
gene mapping, 4–6
Herpes simplex I, 280
Heterochromatin, 43, 44, 47
erythroid cells, 143, 145
Heterokaryosis and heterokaryons, 292, 294–297
Heteroplasmosis, 292
HGPRT locus, 245, 248, 260
Hg-RNA, 130, 131
hybridization with cDNA$_k$, 130
Hg-UTP, 129
Histone, 1
relationship to nick-closing enzyme, 107, 108
Histone DNA
denaturation map, 95
electron microscope, 94
restriction map, 91–93
sea urchin, 89, 91, 94
Histone genes
mapping by electron microscopy, 12
sea urchin, 87–97
Histone mRNA
composition, 96
frequency, 96
sea urchin, 87
transcription direction, 97
Histone production, sea urchin, 88, 91, 96, 97
Human chromosomes, 193, 202
Human diploid
cells, 232–234
fibroblasts, 289, 292–294, 298
Hutchinson–Gilford syndrome, 290
Hybrid plasmids, *Drosophila* satellite DNA, 47–49
Hydroxyurea, 259
Hyperplastoid cell lines, 289
Hypothalmic catecholamine synthesis, aging effect, 308
Hypoxanthine-guanine phosphoribosyltransferase (HGPRT), 164, 165

Subject index

Image analysis, chromosomes, 191–196
Immunoglobulin, mouse mRNA, 125, 128–132
Insect, DNA sequence organization, 30–33, 36
    Apis mellifera (honeybee), 33–35
    Drosophila melanogaster (fruitfly), 29, 30
    Musca domestica (housefly), 33, 34
Insulin, 303
Interspecific synkaryons, 295
Irradiation effect, 208, 213, 214

Karyoplasts, 298
Karyotyping
    C-banding, 216
    G-banding, 218, 221
    trypsin banding, 216
K^+/Na^+ membrane ATPase locus, 245, 248, 260

Liver, aging, 303, 304
Lysogenic selection, gene cloning, 21

Macronucleus, *Tetrahymena*, 79, 80, 83
Menopause, 305
Messenger RNA (mRNA), 37–40, 257, 258
    complexity, 37–40
    gene mapping, 9
    globin mRNA, 136–145
    Hg-mRNA, 129–131
    kappa chain, 128
    mouse immunoglobulin, 125
    sea urchin embryos, 37–40
    total single copy sequence, 40
    transcription specificity, 37–39
Messenger RNA_k ($mRNA_k$), hybridization, 131, 132
Microtubular system, 177–179
Minicol DNA, 104
Morphological transformation, 249, 254, 261
Myeloma, mouse, transcriptional unit 125–133

Neuro-endocrine cascade, in aging, 310
Nicking-closing enzyme, 99–109
    enzyme purification, 106, 107
    reaction products, 105–108
    relationship to Hl, 107–109
4-nitroquinoline-l-oxide, 233
N-methyl-N'-nitro-nitrosoguanidine, 244
N-methyl-N'-nitro-N-nitrosoguanidine, 208, 209, 220, 233

Non-histone proteins, 141, 142, 144
    erythroid cells, 143–145
Norepinephrine, 308, 309
Nu body, rat liver chromatin, 154
Nuclear RNA
    in vitro transcription, 127
    hybridization with $cDNA_k$, 130, 131
Nuclei labeling indices, 297, 298
Nucleic acid hybridization
    DNA:DNA duplex, 4
    DNA:RNA hybrid, 1, 3, 10
    DNA:rRNA hybrid, 5
    gene cloning, 23, 24
    identification
        gene 32 method, 7
Nucleoli, *Tetrahymena*, 79, 83
Nucleosomes 145

Oncogene hypothesis, 285
Ouabain resistance, 248
Ovarian follicles, 307

Phage
    ϕX 174 DNA, 7, 8
    lambda, gene cloning, 15–21
Philadelphia chromosome, 194
Physarum, rDNA, 79
Plant DNA
    aggregates, 70–72
    DNA sequence organization
        repeated DNA sequences, 63–65, 75, 76
            intermediately repetitive, 63–65
        satellite, 63–77
            kinetic complexity, 70, 74
    electron micrographs, 71
    evolutionary history, 73–76
    homogeneous families of sequences, 65–69, 74–77
    networks, 70
    reassociation kinetics, 65–76
    satellite DNA, 63, 64, 70, 73, 74, 76, 77
    simple-sequence DNA, 63
    T_m, 69
    translocated DNA sequences, 63, 75
Plasmid
    hybrid with *Drosophila* DNA, 47–49
    gene mapping, 7, 9, 10
Plasminogen activator, 233
PM2-DNA, 102–104

Polyploids, 214
Progeria, 290, 295
Progesterone, 305, 306

R loop method, 7, 10, 11
  gene cloning, 22, 24, 25
Rat
  chromatin
    digestion, 149–156
    nu bodies, 154
    preparation, 150
    reassociation kinetics, 157–158
    reconstitution, 150, 151, 157, 158
  DNA reassociation kinetics, 52, 59
  DNA sequence organization, 51–62
Reassociation kinetics, 65–74
  of DNA – insects, 34, 35
  rat liver chromatin, 157, 158
  Syrian hamster DNA, 256, 257
Recombinant DNA, see Gene cloning, 15–27
  plasmid with *Drosophila* satellite DNA, 47–49
  sea urchin histone DNA
    hybrid with lambda phage, 94
    hybrid with plasmid, 93
  Restriction studies
    Drosophila DNA site, 47, 49
    Hela mitochondrial DNA, HpaII duplex fragment, 3, 4
    histone gene DNA, 89, 91
    ribosomal DNA, 84
    sea urchin DNA, 91–93
    yeast DNA, 17, 18
Ribosomal DNA (rDNA)
  Tetrahymena, 79–84
  amplification, 83
  electron microscopy, 82
  palindromic nature, 80, 81
  repeated sequences, 82
  restriction studies, 84
  Xenopus, 111
    injection to fertilized eggs, 113
RNA, complementary, gene cloning, 22
RNA polymerase
  E. coli, 131, 132, 141, 142
  myeloma, 131, 132
ribosomal RNA
  gene mapping, 3, 6
  synthesis in injected embryos, 113
  Tetrahymena, 79, 83
Rous associated virus (RAV-O), 280–283

Satellite DNA, 46, 49
  Drosophila, 43, 45, 46, 49
    hybrid with plasmids, 47–49
    nucleotide sequences, 44–47
    sedimentation equilibrium 44, 45
Sea urchin
  denaturation map, 94, 95
  embryo messenger RNA, 37–40
  histone DNA, 89, 91, 95
  histone gene organization, 87–97
  restriction map, 91–93
Sedimentation equilibrium
  satellite DNA, *Drosophila melanogaster*, 44, 45
  sodium iothalamate gradient, 3
Senescence, clonal, 289–300
  biological clock 298
  error catastrophe theory, 290, 292, 300
  growth potential in culture, 290
  nuclear control, 298
  somatic mutation theory, 290, 292
  specific DNA synthesis inhibitor, 296
  terminal differentiation, 300
Simian virus 40, 269–275, 280
  aging experiments, 296
  hetero and homoduplexes, 270–274
  marker rescue experiments, 271
  mismatched bases, 271
  mutants, recombination, 270
  mutants, deletion, 270
  open circular Form II DNA, 272
  transfection, 270, 271
  tsA/tsA heteroduplexes, 272–274
Single copy sequences
  fraction, 29–34
  mRNA total amount, 40
  rat, 51, 53, 55, 58, 59
Somatic cell genetics, 163–168, 173
  aging, 289–300
  auxotrophic mutants, 173
  hybrid cells, 174
    human-CHO hybrid, 175
Somatic cell hybrid, 164
Somatic mutation, 245–261
  8-azaguanine resistance, 245, 248
  HGPRT locus, 245, 248
  K^+/Na^+ membrane ATPase locus, 245, 248
  ouabain resistance, 248
  6-thioguanine resistence, 245, 248
Sorting, chromosomes, 196–203
Spacer, gene mapping, 6

Staphylococcal nuclease, 149, 151, 159
  chromatin digestion, 152, 153
Surface antigens, cell, 171–176
Synkaryosis and synkaryons, 291, 292, 295
Syrian hamster cell, 207–211, 216, 218, 231
  DNA reassociation kinetics, 256, 257
  karyotype
    ara-C transformation, 230
    marker chromosomes, 219, 220
    trypsin banding, 216
  mRNA, 257
  neoplastic transformation, 243–260

Tetrahymena
  macronucleus, 79, 80, 83
  nucleoli, 79, 83
  ribosomal RNA, 79–83
Tetraploids, 292, 294
6-Thioguanine resistance, 245, 248
Thymidine kinase, 181–187
  cytosol enzyme, 183
  electrophoretic forms, 182, 183
  gene transfer, 165–167
  herpes virus induction, 185
  mitochondrial enzyme, 183
  thymidine uptake, 184
T_m
  cDNA$_{sarc}$ hybridization, 282
  plant DNA, 69
Transcription, mRNA specificity, 37–39
Transcriptional control, 135–145
Transfection of SV40 DNA, 269–271
Transformation, neoplastic,
  avian sarcoma virus, 277–285
  chromosome relationship, 205–224, 229–237
    adenovirus effect, 217, 219, 220
    ara C transformation, 230
    carcinogen effect, 207, 208
    guinea pig, 220, 221
    human leucocytes, 212
    in vitro transformation, 230
    irradiation effect, 208, 213
    marker chromosome, 219–223, 231
    rat cells, 215
    Syrian hamster karyotypes, 210, 213, 214, 230
    suppressor gene, 236
  DNA sequence organization, 254–261
  DNA sequence loss, 256
  fibrinolytic activity, 246, 250, 254
  growth in soft agar, 246, 250, 254
  mDNA, 258
  mRNA, 257
  morphological, 246, 254, 261
  oncogene hypothesis, 285
  progression, 246, 253
  reassociation kinetics, 256–257
  transcription, 257, 258
  tumorigenicity, 251
Transforming gene avian sarcoma virus, 280
tRNA-biotin conjugate, 2, 3
Tyrosine, 308
Tyrosine aminotransferase, in aging studies, 303

Uterus and uterine, effect of aging and hormones 305, 306

Virus
  avian leukosis, 280
  avian sarcoma virus, 277–285
  endogenous virus, 281
  oncogenic, 235
  protovirus, 282
  RAV-O, 280–283
  Simian virus, 40, 269–275

Werner's syndrome, 290, 294

Xenopus
  DNA organization, 29
  rDNA, 79, 83
Xenopus eggs, 111, 114, 115
  rDNA, 111
  5S DNA, 111
Xenopus laevis, histone mRNA, 87

Yeast DNA, restriction spectra, 17, 18